# INTRODUCTION TO MOLECULAR BIOLOGY

# Introduction to Molecular Biology

Peter Paolella

Boston, Massachusetts Burr Ridge, Illinios Dubuque, Iowa
Madison, Wisconsin New York, New York San Francisco, California St. Louis, Missouri

*WCB/McGraw-Hill*

*A Division of The* ***McGraw-Hill*** *Companies*

INTRODUCTION TO MOLECULAR BIOLOGY FIRST EDITION



1 2 3 4 5 6 7 8 9 0 QPD/QPD 0 9 8 7

Library of Congress Catalog Number: 96-83781

Recycled/acid-free paper

 This book is printed on recycled, acid-free paper containing 10% postconsumer waste.

ISBN 0–697–20939–3

Editorial director: Kevin T. Kane
Publisher: James M. Smith
Sr. Sponsoring editor: Ronald E. Worthington, Ph.D.
Sr. Developmental editor: Elizabeth M. Sievers
Developmental editor: Terrance Stanton
Marketing manager: Thomas C. Lyon
Project manager: Marilyn M. Sulzer
Production supervisor: Sandy Hahn
Designer: K. Wayne Harms
Photo research coordinator: Lori Hancock
Art editor: Joyce Watters
Compositor: Shepherd, Inc.
Typeface: Goudy 10/12
Printer: Quebecor, Inc.



http://www.mhcollege.com

To my wife, Cees, and our children,
Christopher, Andreana, and Eileen
for their love, patience,
and understanding.

# A Note from the Editor

It is with great pleasure that I am able to write this note celebrating the publication of *Introduction to Molecular Biology*. Peter would be so proud. We lost our dear friend and colleague, Peter Paolella, in the spring of 1996, only a month after the manuscript for *Introduction to Molecular Biology* was put into production. His love of teaching and passion for writing (he was a novelist at heart) are the enduring qualities found in this text. The devotion and hard work of Peter's friends at Stonehill College carried the book through production and to fruition. It is with my deepest appreciation that I thank Robert Peabody, Craig Almeida, Sandra McAlister, and Maura Geens Tyrrell, all of Stonehill College, for carrying on Peter's dream by taking the manuscript through production—it was a labor of love. I am also greatly indebted to Cecile Paolella, Peter's wife, for putting me in contact with his colleagues at Stonehill College. Without her help, I would never have found such a caring group of individuals who would preserve Peter's heart and soul in this book. Thank you all, and finally, thank you, Peter.

Elizabeth M. Sievers
Editor

# CONTENTS

# 10

## Eukaryotic Genes: Basic Structure and Function 145



# 11

## Gene Manipulation: Genetic Engineering 175



# 12

## The Promises and the Perils of the New Genetics 201

# Preface

This book is aimed at undergraduate biology or science majors, sophomores to seniors, who are not genetics majors. It is intended for a one-semester introductory course for students requiring a knowledge of molecular biology as a basis for the study of other areas of biology. A beginning biology course and a basic organic chemistry course are the only prerequisites.

In the text, molecular genetics is placed within the general framework of metabolism. In this way, the processes and mechanisms of gene function and control are seen as phenomena connected to other cell activities.

The text begins with the structure of genetic material, moves to the function of that material (e.g., expression and regulation of genes), and concludes with the application of the knowledge gained in the study of the first two sections in the relatively new field of genetic engineering, or gene manipulation. In its entirety, then, the text is concerned with the structure, expression, regulation, and manipulation of genetic material, in particular, DNA.

Theoretical considerations and experimental data, which are better left to courses for genetics majors, are kept to a minimum. Instead, the idea is to give the student a frame of reference to either continue the study of molecular biology or genetics or to use the information acquired to further an understanding of other courses. When, however, the experimental evidence is crucial to an understanding of how advances in the field occurred, the experiments are described in some detail (e.g., the Meselson-Stahl experiments proving the semiconservative model of DNA replication).

No book is the product of a single person's labors. I have had the assistance of numerous reviewers. Their comments and recommendations have been carefully considered, though not always followed. I am, nonetheless, grateful to them.

Helen H. Benford
Tuskegee University

Richard Crawford
Trinity College

Beth DeStasio
Lawrence University

Karen Kurvink
Moravian College

Roger Sloboda
Dartmouth College

Douglas J. Burks
Wilmington College

Christopher Cullis
Case Western Reserve University

Carolyn Jones
Vincennes University

Joel Piperberg
Millersville University

Steven Woeste
Scholl College

Last, I thank Meg Johnson, who first showed an interest in this project and ultimately got it approved by Wm. C. Brown, and thanks to Robin Steffek, my most capable and enthusiastic editor, who guided me to the completion of the book with as little pain as possible.

Peter Paolella
Johnston, Rhode Island

CHAPTER ONE

# THE NATURE OF DNA

## OUTLINE



## CHAPTER OBJECTIVES

*This chapter will discuss:*

- The contributions of various scientists to our understanding of the nature of DNA
- The chemical and physical characteristics of DNA
- The biological functions of genetic material
- How to calculate the number of nucleotides in a DNA molecule
- The nature of the gene

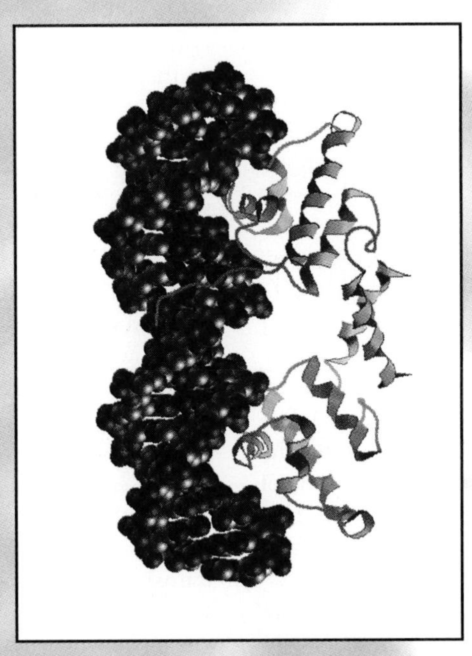

# INTRODUCTION

On April 25, 1953, in the British journal *Nature*, a paper, two columns in length, appeared. It was entitled "Molecular Structure of Nucleic Acids: A Structure for Deoxyribose Nucleic Acid" and was authored by the American James D. Watson and the Englishman Francis H. C. Crick (see Figure 1.1). The structure they proposed has, they say in the first paragraph, "novel features which are of considerable biological interest." And at the end of the paper is the statement, "It has not escaped our notice that the specific pairing we have postulated immediately suggests a possible copying mechanism for the genetic material."

This paper was the culmination of work that stretched back 85 years to Friedrich Miescher, the German scientist who had reported his discovery of a nucleic acid. He called it nuclein because it was isolated from nuclei of pus cells and salmon sperm. Miescher reported his findings in 1871. In 1866, Gregor Mendel had published his work that led to the principles of independent segregation and assortment of genes.

The late 1800s are considered the time of the birth of genetics. And at its birth, the new science was already started in two directions. Mendel's work would lay the foundation of what has been called classical genetics, and Miescher's had begun what is now called molecular genetics. The two scientists apparently worked without knowledge of the other's discoveries.

The classical geneticists have focused on how genes are transferred from one generation to the next **(inheritance),** gene location within **chromosomes,** chromosomal rearrangements, and the concept of dominance. The molecular geneticists, on the other hand, have focused on the structure of genes and on how genes work and are regulated.

For nearly a century, the work of trying to elucidate the structure of Miescher's nuclein went on. Nuclein's role in inheritance and in the metabolism of the cell was not universally accepted. In fact, once the two components of chromosomes, nucleic acid and **protein,** were discovered, it was argued that, of the two, the nucleic acid was much too simple a component to explain the differences between chromosomes and, thus, the differences between individuals whether of the same or of different species. Rather, the protein component, itself composed of a larger variety of subunits, the **amino acids,** was thought to more easily account for the differences between chromosomes and individuals.

Watson and Crick's paper ended one search while simultaneously beginning another. The search for the molecular structure of inheritance had ended; the search for the molecular functions of the nucleic acid had begun. One of these functions, **replication,** by which information is passed on to the next generation was proposed in the paragraph quoted earlier: "It has not escaped our notice . . . " The other functions—how genes work and how they are regulated as an integral part of the cell's metabolism—are still under intense study.

After Miescher's discovery of nuclein (DNA), another 60 years were to elapse before the basic components of DNA were identified and their relationships to one another determined.

At the turn of the century, A. Kossel had demonstrated that a nucleic acid was composed of a nitrogenous base **(adenine, guanine, cytosine, thymine,** or **uracil)** (Figure 1.2) a sugar, and a phosphate group (Figure 1.3).

Then, by the early 1930s, largely as the result of the work of P. A. T. Levene, the arrangement of the bases, sugar, and phosphate was discovered. A single base is linked to the sugar, which in turn is linked to the phosphate. The resulting structure is a **nucleotide** (Figure 1.4), the fundamental unit of nucleic acids.

Levene, along with other workers, also discovered that the sugar of nuclein is **deoxyribose.** And he discovered that there are in fact two nucleic acids: **ribonucleic acid,** or **RNA** (actually discovered by Kossel), and **deoxyribonucleic acid,** or **DNA** (Miescher's nuclein).

The discovery of the components of nucleic acids, in particular DNA, then led to the first models of the structure of DNA. Takahashi, in 1930, proposed the "tetranucleotide" structure for DNA. In this model, the nucleotides of adenine, guanine, cytosine, and thymine repeat in a regular pattern. Thus, the idea that DNA is composed of simple parts arranged in a simple way was born.

The year 1953 proved to be the year of the structure of DNA. In addition to James Watson and Francis Crick, L. Linus Pauling and R. Corey, and Fraser also proposed models. Both these attempts, however, consisted of three intertwining chains. The Pauling and Corey model also suggested that the phosphates were located along the axis of the molecule and the bases were placed on the inside. Although Fraser had positioned the phosphates and bases correctly, Watson and Crick considered the model to be "rather ill-defined."

# THE PATH TO THE WATSON AND CRICK MODEL

As we have seen, by the 1930s, much evidence had accumulated regarding the components of DNA, but what was it that was driving scientists in the mid-twentieth century to search for the structure of DNA? The answer is: there was a preponderance of evidence pointing to DNA as the genetic material.

As early as the 1880s, there had been speculation that the chromosomes were involved in inheritance. Wilhelm Roux suggested that it was unlikely that the mechanisms of **mitosis** and **meiosis** had evolved without some good reason. He proposed the chromosomes as the genetic material.

# MOLECULAR STRUCTURE OF NUCLEIC ACIDS

## A Structure for Deoxyribose Nucleic Acid

WE wish to suggest a structure for the salt of deoxyribose nucleic acid (D.N.A.). This structure has novel features which are of considerable biological interest.

A structure for nucleic acid has already been proposed by Pauling and Corey[1]. They kindly made their manuscript available to us in advance of publication. Their model consists of three intertwined chains, with the phosphates near the fibre axis, and the bases on the outside. In our opinion, this structure is unsatisfactory for two reasons: (1) We believe that the material which gives the X-ray diagrams is the salt, not the free acid. Without the acidic hydrogen atoms it is not clear what forces would hold the structure together, especially as the negatively charged phosphates near the axis will repel each other. (2) Some of the van der Waals distances appear to be too small.

Another three-chain structure has also been suggested by Fraser (in the press). In his model the phosphates are on the outside and the bases on the inside, linked together by hydrogen bonds. This structure as described is rather ill-defined, and for this reason we shall not comment on it.

We wish to put forward a radically different structure for the salt of deoxyribose nucleic acid. This structure has two helical chains each coiled round the same axis (see diagram). We have made the usual chemical assumptions, namely, that each chain consists of phosphate di-ester groups joining β-D-deoxyribofuranose residues with 3′,5′ linkages. The two chains (but not their bases) are related by a dyad perpendicular to the fibre axis. Both chains follow right-handed helices, but owing to the dyad the sequences of the atoms in the two chains run in opposite directions. Each chain loosely resembles Furberg's[2] model No. 1; that is, the bases are on the inside of the helix and the phosphates on the outside. The configuration of the sugar and the atoms near it is close to Furberg's 'standard configuration', the sugar being roughly perpendicular to the attached base. There is a residue on each chain every 3-4 A, in the z-direction. We have assumed an angle of 36° between adjacent residues in the same chain, so that the structure repeats after 10 residues on each chain, that is, after 34 A. The distance of a phosphorus atom from the fibre axis is 10 A. As the phosphates are on the outside, cations have easy access to them.

The structure is an open one, and its water content is rather high. At lower water contents we would expect the bases to tilt, so that the structure could become more compact.

The novel feature of the structure is the manner in which the two chains are held together by the purine and pyrimidine bases. The planes of the bases are perpendicular to the fibre axis. They are joined together in pairs, a single base from one chain being hydrogen-bonded to a single base from the other chain, so that the two lie side by side with identical z-co-ordinates. One of the pair must be a purine and the other a pyrimidine for bonding to occur. The hydrogen bonds are made as follows: purine position 1 to pyrimidine position 1; purine position 6 to pyrimidine position 6.

If it is assumed that the bases only occur in the structure in the most plausible tautomeric forms (that is, with the keto rather than the enol configurations) it is found that only specific pairs of bases can bond together. These pairs are: adenine (purine) with thymine (pyrimidine), and guanine (purine) with cytosine (pyrimidine).

In other words, if an adenine forms one member of a pair, on either chain, then on these assumptions the other member must be thymine; similarly for guanine and cytosine. The sequence of bases on a single chain does not appear to be restricted in any way. However, if only specific pairs of bases can be formed, it follows that if the sequence of bases on one chain is given, then the sequence on the other chain is automatically determined.

This figure is purely diagrammatic. The two ribbons symbolize the two phosphate—sugar chains, and the horizontal rods the pairs of bases holding the chains together. The vertical line marks the fibre axis.

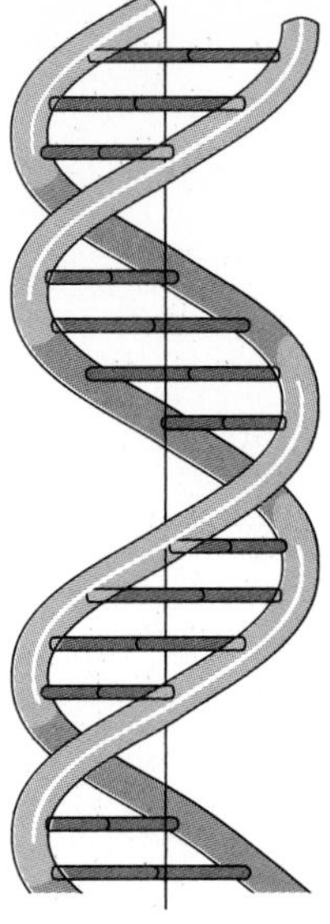

It has been found experimentally[3,4] that the ratio of the amounts of adenine to thymine, and the ratio of guanine to cytosine, are always very close to unity for deoxyribose nucleic acid.

It is probably impossible to build this structure with a ribose sugar in place of the deoxyribose, as the extra oxygen atom would make too close a van der Waals contact.

The previously published X-ray data[5,6] on deoxyribose nucleic acid are insufficient for a rigorous test of our structure. So far as we can tell, it is roughly compatible with the experimental data, but it must be regarded as unproved until it has been checked against more exact results. Some of these are given in the following communications. We were not aware of the details of the results presented there when we devised our structure, which rests mainly though not entirely on published experimental data and stereo-chemical arguments.

It has not escaped our notice that the specific pairing we have postulated immediately suggests a possible copying mechanism for the genetic material.

Full details of the structure, including the conditions assumed in building it, together with a set of co-ordinates for the atoms, will be published elsewhere.

We are much indebted to Dr. Jerry Donohue for constant advice and criticism, especially on inter-atomic distances. We have also been stimulated by a knowledge of the general nature of the unpublished experimental results and ideas of Dr. M. H. F. Wilkins, Dr. R. E. Franklin and their co-workers at King's College, London. One of us (J.D.W.) has been aided by a fellowship from the National Foundation for Infantile Paralysis.

J. D. WATSON
F. H. C. CRICK

Medical Research Council Unit for the Study of the Molecular Structure of Biological Systems,
Cavendish Laboratory, Cambridge.
April 2.

[1]Pauling, L., and Corey, R. B. *Nature*. 171. 346 (1953); *Proc. U.S. Nat. Anad. Sci.*, 38, 84 (1953).
[2]Furberg, S., *Acta Chem. Scand.*, 6, 634 (1952).
[3]Chargaff, E., for references see Zamenhof, S., Brawernian, G., and Chargaff, E., *Biochim. et Biophys. Acta*, 9, 402 (1952).
[4]Wyatt, G. R., *J. Gen. Physici.*, 36, 201 (1952).
[5]Astbury, W. T., Symp. Soc. Exp. Biol. 1, Nucleic Acid, 66 (Camb. Univ. Press, 1947).
[6]Wilkins, M. H. F., and Randall, J. T., *Biochim. et Biophys. Acta*, 10, 182 (1953).

FIGURE 1.1
Watson and Crick's 1953 article on the structure of DNA.

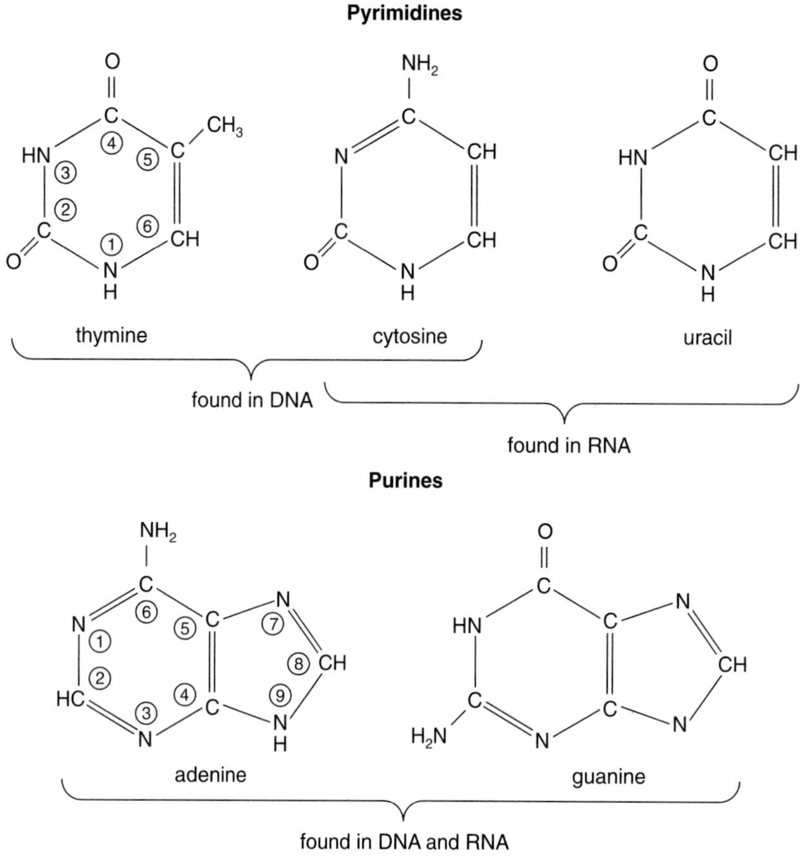

Pyrimidines

thymine cytosine uracil

found in DNA

found in RNA

Purines

adenine guanine

found in DNA and RNA

FIGURE 1.2
The nitrogenous bases of nucleic acids. Pyrimidines contain one carbon-nitrogen ring; purines contain two rings.

And, apparently with no knowledge of Mendel's work, he proposed a "linear arrangement of hereditary units along the chromosomal threads."

But not until Frederick Griffith's 1928 discovery of transformation in pneumococci was there experimental evidence pointing to DNA as the genetic material. Griffith was an English physician working in the British Public Health Service. His work laid the foundation for the investigations of other scientists including O. T. Avery and coworkers, E. Chargaff, A. D. Hershey and M. Chase, and Franklin, about whom we will hear later.

Let's now take a brief look at the experimental evidence produced by these workers to see how it led to the Watson and Crick model of the structure of DNA and to the hypothesis of a "possible copying mechanism."

Griffith worked with the pneumococcal bacteria. These organisms come in a variety of "types" that are distinguishable by the nature of the polysaccharide that surrounds the cell. The types are designated by roman numerals. This enveloping polysaccharide is called the capsule. Whether a particular bacterium is encapsulated determines whether it will be able to produce a pneumonia that can kill its host. Cells surrounded by a capsule are virulent (able to produce pneumonia); those without the capsule are not.

The capsular material enables an investigator to quickly determine whether a colony growing on agar media is virulent or avirulent (unable to produce pneumonia). The virulent encapsulated cells produce colonies that have a smooth, mucoid appearance; these are designated *S* colonies. Nonencapsulated, avirulent cells, on the other hand, produce colonies that have a rough, dry appearance; these are designated *R* colonies.

Griffith's experiments required injecting mice with the different cell types and, after a day or so, sampling the heart blood of dead mice to determine what organisms were present. The experiments can be summarized as follows:

1. Injection with type$_{\text{III}}$ S cells produced dead mice + Type$_{\text{III}}$ S cells.
2. Injection with type$_{\text{II}}$ R cells produced live mice.
3. Injection with heat-killed type$_{\text{III}}$ S cells produced live mice.
4. Injection with heat-killed type$_{\text{III}}$ S cells + type$_{\text{II}}$ R cells produced dead mice + type$_{\text{III}}$ S cells.

Griffith concluded that the presence of the heat-killed S cells must have caused the "transformation" of the living R cells. The living $R_{\text{II}}$ cells were transformed into $S_{\text{III}}$ cells. **Mutation** did not explain the results, because when cells

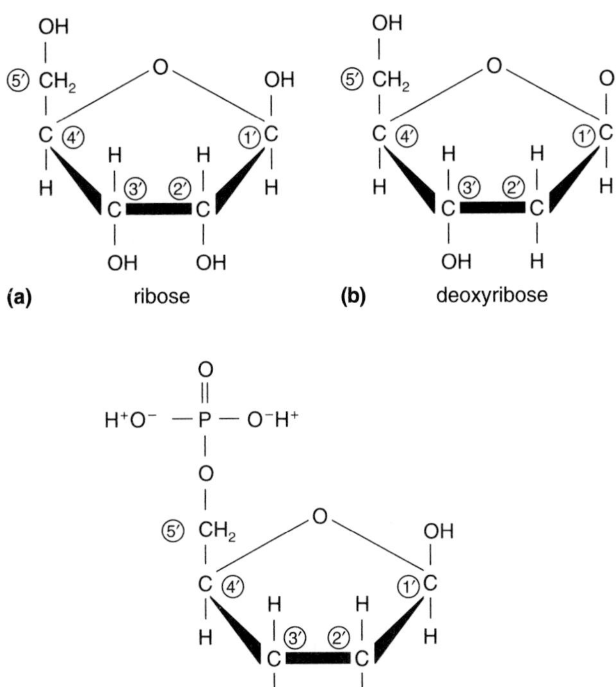


FIGURE 1.3
(a) The sugar ribose has an —OH linked at the 2´ carbon. (b) the sugar deoxyribose has a single hydrogen at the 2´ carbon. (c) Deoxyribose with a phosphate group.

mutated from the R to the S form or from the S to the R form the capsular type was not lost. For example, a mutation would cause $R_{II}$ cells to become $S_{II}$ cells not $S_{III}$ cells.

Griffith's results were published in the British *Journal of Hygiene* in 1928. A few years later, J. L. Alloway reported that crude extracts of S cells from which all cells were removed by filtration was capable of transforming R cells.

Then, in 1944, O. T. Avery, C. M. MacLeod, and M. McCarty published the results of their experiments. The results were, in Watson's phrase, "Avery's bombshell." The conclusion "met with great surprise and disbelief, because at that time hardly anyone was prepared to accept such an informational role for DNA." Avery and his colleagues had demonstrated that Griffith's transforming principle was DNA.

Avery, Macleod, and McCarty used a variety of enzymatic reactions designed to destroy particular kinds of molecules. They reported that "only those preparations shown to contain an enzyme capable of depolymerizing authentic samples of desoxyribonucleic [*sic*] acid were found to inactivate the transforming principle" (Avery et al. 1944). By the use of chemical, enzymatic, and serological reactions, electrophoresis, ultracentrifugation, and UV (ultraviolet) spectroscopy, they could eliminate protein, unbound lipid, RNA, and polysaccharides as the transforming principle.

It might be assumed that this work led to universal acceptance of DNA as the genetic material. But there was still some criticism that Avery's preparations might have been

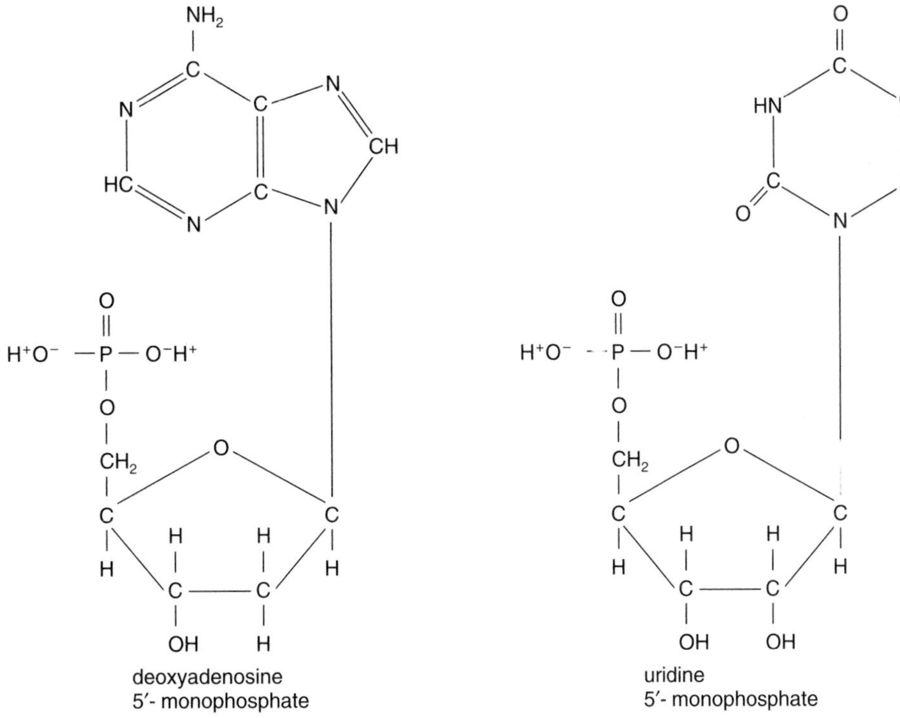

deoxyadenosine
5′- monophosphate

uridine
5′- monophosphate

FIGURE 1.4
Nucleotides.

contaminated by undetectable amounts of protein. In fact, Avery, MacLeod, and McCarty concluded their paper with the statement: "*If* the results of the present study on the chemical nature of the transforming principle are confirmed, then nucleic acids must be regarded as possessing biological specificity the chemical basis of which is as yet undetermined" [emphasis added]. Their work and data were, nonetheless, a stimulant for the biochemist Erwin Chargaff.

A technique called paper chromatography is used to determine the identity of unknown materials. Substances dissolved in various solvents move up or down paper strips; the distance they move relative to the starting point and to one another controls their identities.

Chargaff gathered DNA from several sources, subjected the samples to acid hydrolysis at high temperatures and high pressures to release the adenines, guanines, cytosines, and thymines. He identified the compounds by means of paper chromatography and determined the amounts of each in the 2–40 μg range using absorption spectra techniques.

In 1950, he published his results. Chargaff noted that DNA extracted from different species was "of a composition constant for different organs of the same species and characteristic of the species." Further, he said, "The results serve to disprove the tetranucleotide hypothesis." That idea required that the amounts of adenine, guanine, cytosine, and thymine be equal.

Instead, he wrote, "It is noteworthy that in all the desoxypentose [*sic*] nucleic acids examined thus far the molar ratios of total **purines** to total **pyrimidines,** and also of adenine to thymine and of guanine to cytosine were not far from 1." This observation has come to be known as **Chargaff's rule.**

This was the first published report of so important a structural feature of DNA since Levene's work on nucleotides in the 1930s. Still, the idea that the genetic material might be protein and not DNA would not die.

The research of Alfred Hershey and Martha Chase finally put that notion to rest. Their experiments culminated in a paper published in 1952 in *The Journal of General Physiology* entitled "Independent Functions of Viral Protein and Nucleic Acid in Growth of Bacteriophage." This may have been the paper that spurred Watson and Crick to redouble their efforts to propose a model of DNA in 1953.

A number of bacterial viruses, called bacteriophage, or simply phage, were known to be constructed of two macromolecules, protein and DNA. Like all viruses, bacteriophage are intracellular parasites; they can replicate only within a living cell. The viruses first adsorb to the cell surface and then penetrate the cell. Once inside, however, the viruses are noninfective. That is, shortly after infection of the cell, if the cells are ruptured, no viral particles capable of infecting other cells can be found. This noninfective, intracellular form was seen "as the connecting link between parental and progeny phage, and the elucidation of its structure and function became the central problem in the study of viral growth."

Hershey and Chase hit on the idea of using bacteriophage T2 grown in such a way that both the protein and DNA would be made radioactive but with different radioisotopes. They were able to do this because the viral proteins contain sulfur in the amino acids methionine and **cysteine,** and the viral DNA contains phosphorus. What is also critical is that the proteins of T2 contain no phosphorus and its DNA contains no sulfur.

The radioactive T2 phage was allowed to infect its host cell, the bacterium ***Escherichia coli*** (abbreviated ***E. coli***), and at a predetermined time the cells were put into a Waring blender and subjected to violent shearing forces in order to release the attached viruses.

The investigators then checked for the following:

1. The ability of the infected cells to yield viral progeny: it was unaffected, so the intracellular form was present and functioning.
2. The removal of sulfur: the shearing forces stripped away 70%–80% of the attached radioactive sulfur from infected cells.
3. The removal of phosphorus: only 20%–35% of the radioactive phosphorus was stripped from the bacteria. (This result can be explained by the fact that not all bacteriophage succeed in penetrating the host cell. So some radioactive phosphorus is expected to be left on the cell surface and subsequently released by the shearing forces.)

Hershey and Chase interpreted these results to mean that the viral DNA leaves the protein coat of the virus and enters the cell. To bolster this interpretation, they allowed viral particles to adsorb to isolated bacterial membranes or to heat-killed bacteria and showed that adsorption to both caused release of DNA. The nucleic acid was identified by means of its susceptibility to the enzyme DNase.

From these experiments, Hershey and Chase (1952) concluded "that one of the first steps in the growth of T2 is the release from its protein coat of the nucleic acid of the viral particle, after which the bulk of the sulfur-containing protein has no further function."

In other words, DNA, not protein, is the genetic material of T2. The argument was at last settled in a decisive manner in favor of DNA. From this time on, no one again seriously questioned the genetic role of DNA.

The last piece of evidence Watson and Crick would need to convince themselves that they were on the right track was the work of Rosalind Franklin. She used a technique first demonstrated in 1913 by W. H. Brag and

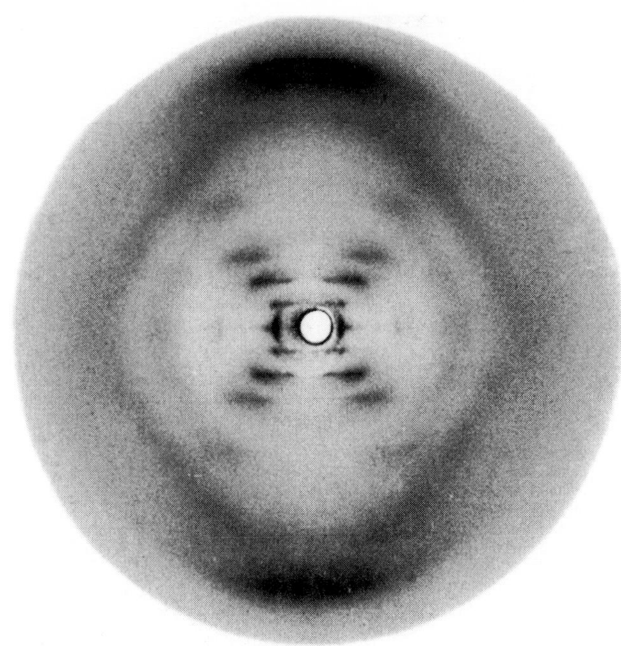

FIGURE 1.5
X-ray diffraction photograph of DNA, by Franklin.
Courtesy M. H. F. Wilkins, Kings College, London.

W. L. Brag, who showed that x-ray diffraction patterns can be used to reveal the three-dimensional structure of molecules. The technique is called x-ray crystallography. Crystalline forms of substances are subjected to x-ray beams, which are then deflected from the crystal onto a film surface. The film is, thus, exposed in a pattern characteristic for a particular structural arrangement of atoms.

In the winter of 1952-1953, Franklin produced the now-famous x-ray "picture" of DNA (Figure 1.5). Watson (1976) wrote that the picture was

> the key x-ray photograph involved in the elucidation of the DNA structure. It experimentally confirmed the then current guesses that DNA was helical. The helical form is indicated by the crossways pattern of x-ray reflections in the center of the photograph. The heavy black regions at the top and bottom tell that the 3.4 Å thick purine and pyrimidine bases are regularly stacked next to each other, perpendicular to the helical axis.

## THE COMPONENTS OF DNA

As we have already seen, there are three basic components of DNA: a phosphate, a sugar, and a nitrogenous base. The phosphate group gives DNA its acidic properties and a negative charge. In vivo, unless these charges are neutralized, it would not be possible to pack the huge DNA molecule into the cell's nucleus. Neutralization is brought about by reaction of basic proteins with the acidic DNA in both **eukaryotes** (organisms with cells that have a nucleus surrounded by a membrane), such as human beings, and **prokaryotes,** the bacteria. In eukaryotes, histones are the basic proteins involved in packaging DNA (see Chapter 10), and in prokaryotes, polyamines are the basic proteins.

In fact, the smallest of the human chromosomes, with a DNA molecule of approximately 1.4 cm, in its most condensed state during mitosis, is about 1/7000 its extended length. So the packing ratio, the length of the DNA molecule divided by the length of the unit that contains it (the chromosome in this case), is on the order of 7000.

The sugar of DNA is the five-carbon 2′-deoxyribose (Figure 1.3). The "prime" is used to distinguish the sugar carbons from the nitrogenous base atoms (numbered in Figure 1.2).

The commonly occurring nitrogenous bases in DNA (Figure 1.2), usually referred to simply as "bases," are adenine, guanine, cytosine, and thymine. There are other bases sometimes called "rare" bases, but we will not be discussing those.

From these components—phosphate, sugar, base—cells construct the precursors of DNA, the **deoxyribonucleotides,** most often referred to simply as nucleotides. Any one of the nitrogenous bases, plus the deoxyribose yields a **nucleoside** (see deoxyadenosine, Figure 1.6) and if to this a phosphate is added, a nucleotide is formed (see deoxyadenosine 5′-monophosphate, Figure 1.4). (Later, we will discuss nucleotides and nucleosides formed with ribose, also shown in Figures 1.4 and 1.6.)

**Deoxyribonucleic acid (DNA)** is a **polymer,** or chain, of these nucleotides (Figure 1.7). The sugar is attached via its 1′ carbon to a nitrogen atom of the base, and the phosphate group is attached to the 5′ carbon of the same sugar. In fact, all nucleotides are synthesized as the 5′-monophosphate nucleotides. So, for example, the nucleotide containing adenine as the base is really 2′-deoxyriboadenine-5′-monophosphate, and the nucleotide containing cytosine is 2′-deoxyribocytidine-5′-monophosphate.

In vivo, synthesis of the nucleotides takes place via two pathways, namely, the purine and pyrimidine pathways (Table 1.1). All the nucleotides are synthesized with the phosphate group attached to the 5′ carbon of the sugar. This fact is especially important in the structure and replication of DNA, as we shall see.

For convenience and by consensus, the nucleotides are designated by the capitalized first letter of the base they contain. Thus **A, G, C,** and **T** represent the nucleotides containing adenine, guanine, cytosine, and thymine, respectively. We shall use this notation system throughout the text.

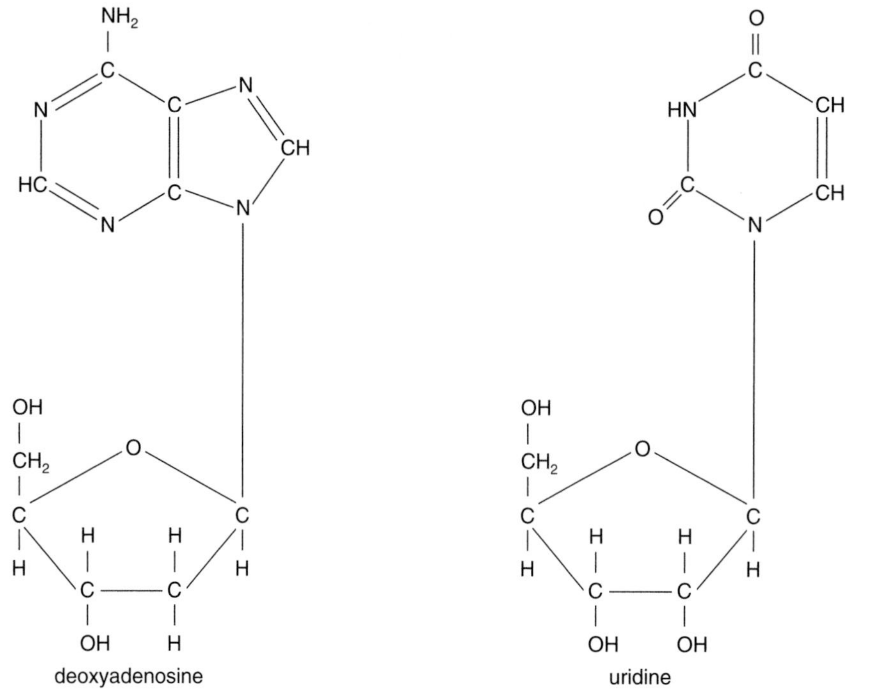

FIGURE 1.6
Nucleosides.

## THE STRUCTURE OF DNA

Double-stranded DNA is held together by two types of chemical bonds: phosphodiester bonds and hydrogen bonds. Nucleotides are joined by **phosphodiester bonds,** and the strands are held together by hydrogen bonds.

Phosphodiester bonds link adjacent carbon-containing groups via a phosphorus atom attached to an oxygen atom from each group (Figure 1.7). In this way a polynucleotide chain is formed. We have, then, a single strand of nucleotides. The phosphate of the 5′ carbon of one nucleotide is linked to the preceding nucleotide at its 3′ carbon, which does not have a phosphate group joined to it. Any one nucleotide in the chain, except the first and last, then, is joined to two other nucleotides at its 5′ and 3′ carbons via the phosphodiester bond. So each linear, open-ended chain would have a nucleotide at one end with a 5′ carbon not joined to another nucleotide and at the other end a 3′ carbon not joined to another nucleotide. These are called the **5′** and **3′ ends.**

In fact, DNA synthesis always moves from the 5′ end to the 3′ end of the molecule. Remember, nucleotides are synthesized as the 5′-monophosphates with a 3′ hydroxyl group, so the 5′ end of one nucleotide attaches to the 3′ end of another nucleotide. This idea must be understood because of its importance in DNA and protein synthesis. Terms such as right and left are meaningless in describing the cell, so, when molecule orientation is important, we designate direction or orientation by referring to the ends of molecules, such as the 5′ end and the 3′ end of nucleic acids.

As Figure 1.7 shows, the phosphate of the phosphodiester bond is connected to the carbon atoms indirectly, and oxygen separates the phosphorus from the carbons to make an ester. Because there are two such esters, one on either side of the phosphorus for each two nucleotides joined, the bond is called a phosphodiester bond. So the polymer, or polynucleotide, **DNA,** is a sequence of **nucleotides** joined by phosphodiester bonds.

Double-stranded DNA has two of these polymer chains, or strands. The strands are held together by hydrogen bonds between apposing nucleotides, the second of the two types of chemical bonds.

A molecule of double-stranded DNA, then, is held together by phosphodiester linkages forming the individual strands and hydrogen bonds across the strands forming the double-stranded molecule. The phosphodiester bond is much the stronger of the two because it is a covalent bond whereas the other is a hydrogen bond.

The hydrogen bonds between pairs of apposing nucleotides across the double-stranded DNA molecule hold together Chargaff's base pairs. These pairings are always adenine-thymine (A=T) and guanine-cytosine (G≡C) (Figure 1.8). The capital letters indicate the bases, and the number of lines between the letters indicate the number of hydrogen bonds between the bases. Between adenine and

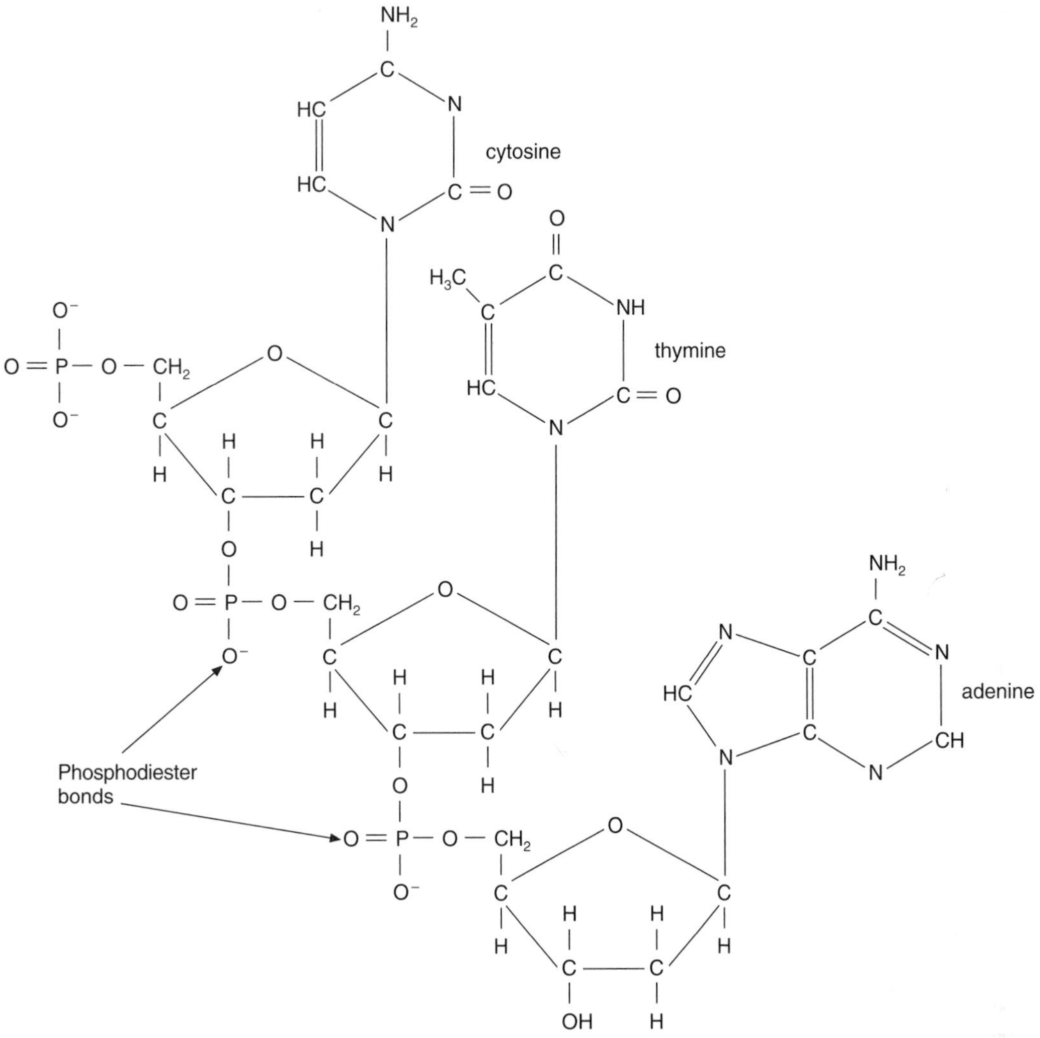

FIGURE 1.7
Phosphodiester bonds of a polynucleotide.

**table 1.1**

THE TWO PATHWAYS OF NUCLEOTIDE SYNTHESIS

| PATHWAY | BASE | NUCLEOSIDE | NUCLEOTIDE |
|---|---|---|---|
| Purine | Adenine (A) | Adenosine | Deoxyadenosine 5′-monophosphate |
| | Guanine (G) | Guanosine | Deoxyguanosine 5′-monophosphate |
| Pyrimidine | Thymine (T) | Thymidine | Deoxythymidine 5′-monophosphate* |
| | Cytosine (C) | Cytidine | Deoxycytidine 5′-monophosphate |
| | Uracil (U) | Uridine | Uridine 5′-monophosphate† |

* Found in DNA *not* in RNA.
† Found in RNA *not* in DNA.

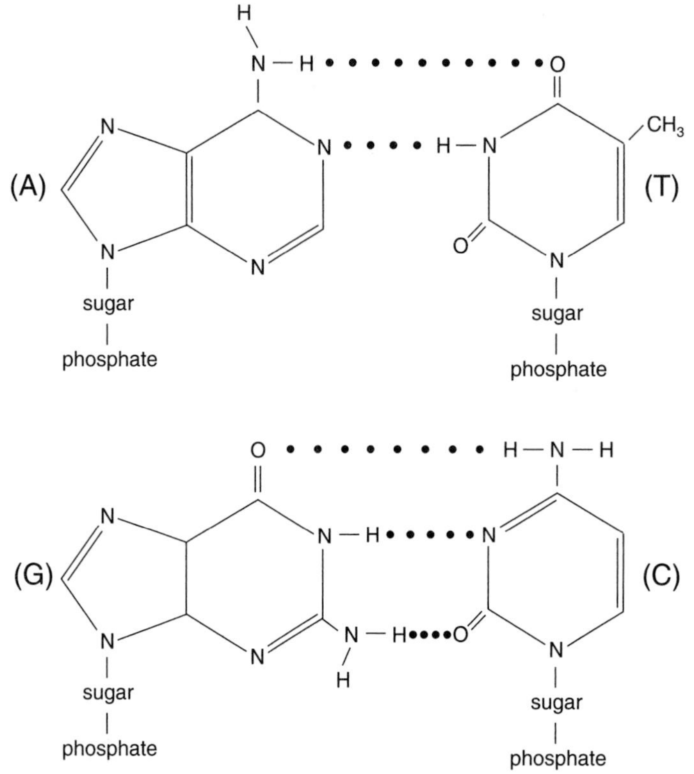

FIGURE 1.8
Hydrogen bonding between nucleotides.

thymine there are two hydrogen bonds, and between guanine and cytosine there are three such bonds. For a DNA molecule containing millions of nucleotide pairs, there would be 2–3 times that number of hydrogen bonds holding the strands together.

Regardless of which strand a purine or pyrimidine happens to be in, the complementary nucleotide of Chargaff's base pairs always appears opposite it in the other strand. So we could have, for example, A=T or T=A and G≡C or C≡G.

This **complementarity** of one strand with the other is sometimes also known as Chargaff's rule. Watson and Crick correctly interpreted complementarity to be the key to the way in which information stored in the DNA molecule, in the sequences of nucleotides the two strands, can be transmitted to the next generation. This is what they meant by "the specific pairing we have postulated immediately suggests a possible copying mechanism for the genetic material."

They proposed that each strand of the double-stranded DNA molecule acts as the **template** for a new strand; the nucleotides of each strand attract their complementary nucleotides. In this way, two copies of the original molecule are produced: Each is an exact replica of the parental molecule. And that, as they saw, is how inheritance is accomplished. This method of replication has become known as **semiconservative replication.** Each strand, or one-half of the parental molecule, is conserved in a daughter molecule.

Besides the phosphodiester and hydrogen bonds, the stacking of the base pairs one above the other, as seen in Franklin's picture (Figure 1.5), also helps to hold the double helix together. Both the hydrogen bonds between the bases and the hydrophobic stacking (meaning that the nonpolar nucleotides will arrange themselves so that they are not in contact with water molecules) help to release energy, thereby stabilizing the molecule.

In this arrangement, the bases, which are flat, lie on the inside of the double-stranded DNA. They lie perpendicular to the molecule's long axis, and the sugar-phosphate groups form the **backbone** in a repeating sugar-phosphate-sugar-phosphate . . . alternating order along both strands. In this way, double-stranded DNA is like a spiral staircase in which the base pairs represent the steps and the sugar-phosphate backbones represent the railings (Figure 1.9).

The spiral nature of DNA is what gave the molecule its other name of "the double helix." The two strands are woven or twisted around one another, producing a right-handed **double helix.**

This helical structure gives rise to two other features of the molecule: **major** and **minor grooves.** The first groove results from the twisting of the two strands and is large com-

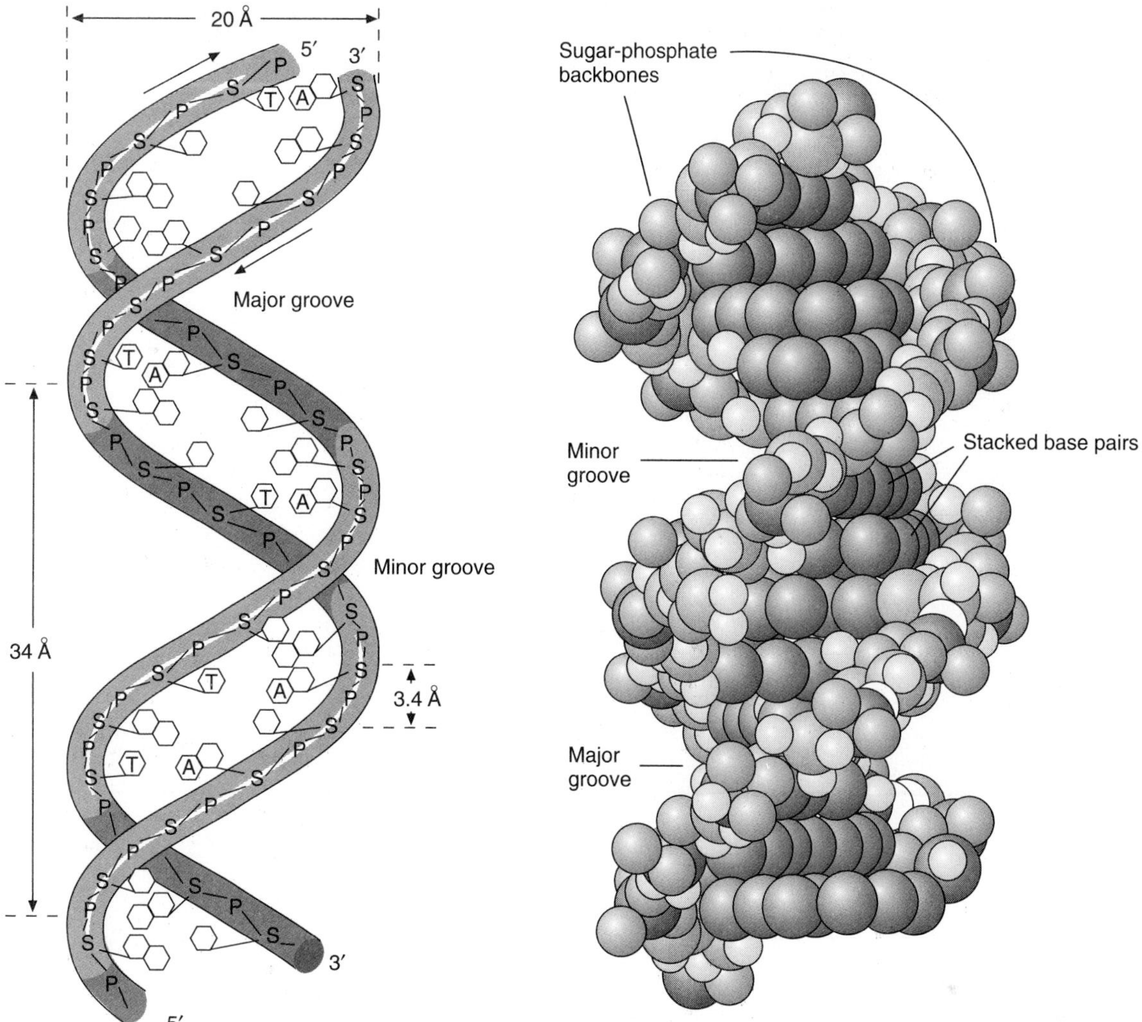


FIGURE 1.9
Diagram (left) and space-filling model (right) of the Watson-Crick double-helix model of the structure of DNA.

pared with the minor groove, which results from the space separating the two strands along the backbone. Studies in many types of organisms indicate that regulatory proteins, which control the activity of genes, have special segments of amino acids within their structures that are able to fit in the major groove and there form hydrogen bonds with bases of specific nucleotide sequences (Chapters 7 and 10). This interaction of proteins with DNA is the common method of regulating gene activity in both prokaryotes and eukaryotes. It is a theme to which we will continually refer in explaining the regulation of gene activity.

The minor groove may serve as the attachment site for a special class of proteins that provide a surface around which DNA is wrapped, thereby helping to package the eukaryotic DNA molecule (see Chapter 10). These proteins, the histones, contain large amounts of the basic amino acids arginine and lysine, which can interact with the negatively charged phosphate groups of the DNA molecule.

## SOME PHYSICAL CHARACTERISTICS OF DNA

The positioning of the base pairs within the molecule at right angles (perpendicular) to the long axis of the molecule and the fact that the bases are ring compounds account for DNA's absorption of UV light at 260 nm. The absorption characteristic was important in the Meselson-Stahl experiments (Chapter 2), which proved the Watson and Crick model of semiconservative DNA replication, in which each strand of the double-stranded DNA acts as a template.

When attempts were first being made to determine the structure of double-stranded DNA, it was the density of the molecule and the x-ray diffraction data that suggested that the molecule was built of two strands. As noted earlier, the strands are complementary. But to achieve this complementary two-strandedness, the

**table 1.2**

FEATURES OF A, B, AND Z FORMS OF DNA

| | Form of DNA | | |
|---|---|---|---|
| Feature | A | B | Z |
| Helical rotation | Right-handed | Right-handed | Left-handed |
| Conditions | 75% relative humidity | 92% relative humidity | High salt concentration and methylation of cytosines |
| Diameter | 26 Å | 20 Å | 18Å |
| Number of base pairs per helical turn | 11 | 10 | 12 |
| Helical twist per base pair | 33° | 36° | 60° |
| Rise per turn of helix | 28 Å | 34 Å | 45 Å |
| Helix rise per base pair | 2.6 Å | 3.4 Å | 3.7 Å |
| Base tilt normal to the helix axis | 20° | 6° | 7° |
| Major groove | Narrow and deep | Wide and deep | Flat |
| Minor groove | Wide and shallow | Narrow and deep | Narrow and deep |

nucleotides of one strand must be rotated 180 degrees with respect to the nucleotides of the other strand as the molecule is being synthesized. This rotation is necessary for the formation of hydrogen bonds. In the absence of rotation, the atoms needed for the bonds would not be in the same plane, and, therefore, no bonding would take place. So, if one strand is running in the 5′ to 3′ direction, say from top to bottom, the other strand's orientation will be from 3′ to 5′, also from top to bottom. At the top of the double-stranded molecule, then, a nucleotide with an open 5′ position would be base paired with a nucleotide having an open 3′ position on the other strand; at the bottom of the molecule the opposite would be the case. Such a molecule (Figure 1.9) is said to be polarized, or the strands are **antiparallel.**

Depending on the amount of water surrounding the molecule, referred to as the relative humidity, and the salt concentration of the DNA preparation the molecule can take on a number of conformations (Table 1.2).

The most common of these are the "A" form **(A-DNA),** produced when the relative humidity is about 75%, and the "B" form (**B-DNA;** the "ideal" Watson-Crick molecule), produced when the relative humidity is 92% and the counter-ion is the alkali metal $Na^+$.

If one were to look at the "ideal" B-DNA form using x-ray crystallography techniques, the following dimensions could be determined. The diameter of the double helix, from backbone to backbone, is 20 Å (angstroms). The regularity of the width all along the length of the double-stranded DNA molecule results from the bonding of a purine with a pyrimidine, Chargaff's base pairing. In one complete turn of the molecule, 360 degrees, there are 10 base pairs arranged so that each pair is out of phase with the pair above and below by 36 degrees. A complete turn is equivalent to a run along the length of the molecule of 34 Å. And, since there are 10 base pairs in one complete turn, the base pairs must be 3.4 Å apart.

Both A-DNA and B-DNA are right-handed helices. If, however, DNA is placed in an environment of high salt concentration, a most unusual conformational change occurs. A left-handed helix forms, and, because the pattern of the sugar-phosphate backbone is a zigzag, this conformation is called **Z-DNA.** The biological significance of such a structure is still unresolved. It is of some interest that Z-DNA can be produced in vitro by the methylation of cytosines. Methylation is a common biological modification of DNA. In eukaryotes, it may be a way of controlling gene activity; methylation may prevent the decoding of genes (more on methylation in Chapter 10).

Double-stranded DNA is a linear molecule. That is, it is unbranched. In bacteria, DNA is a circular molecule. And, though our knowledge of eukaryotic chromosomal DNA is still fragmentary, we do know that, like the bacterial DNA, human DNA is an unbranched molecule.

table 1.3

SOME CHARACTERISTICS OF HAPLOID CELL DNA

| ORGANISM | DNA CONTENT ($\times 10^{-12}$g) | NUMBER OF BASE-PAIRS ($\times 10^{3}$) | LENGTH* (μM) |
|---|---|---|---|
| VIRUSES | | | |
| SV40 | 0.0000051 | 5.1 | 1.8 |
| *Herpes simplex* | 0.00011 | 110.0 | 38.0 |
| Lambda | 0.000055 | 55.0 | 19.0 |
| $T_2$, $T_4$, $T_6$ | 0.0002 | 200.0 | 69.0 |
| BACTERIA | | | |
| *Escherichia coli* | 0.0047 | 4700 | 1620 |
| *Staphyloccus aureus* | 0.007 | 7000 | 2414 |
| *Streptococcus pneumoniae* | 0.002 | 2000 | 690 |
| YEAST | | | |
| *Saccharomyces cerevisiae* | 0.245 | 245,000 | 84,483 |
| PLANTS | | | |
| Corn | 7.5 | 7,500,000 | 2,586,000 |
| Tobacco | 1.2 | 1,200,000 | 413,790 |
| ANIMALS | | | |
| *Drosophila melanogaster* | 0.18 | 180,000 | 62,070 |
| Frog | 6.5 | 6,500,000 | 2,241,380 |
| Chicken | 1.3 | 1,300,000 | 448,275 |
| Mouse | 2.5 | 2,500,000 | 862,070 |
| Cattle | 3.0 | 3,000,000 | 1,034,483 |
| Human | 3.2 | 3,200,000 | 1,103,448 |
| Lungfish | 102.0 | 102,000,000 | 3,517,241 |

*Number of base pairs can be converted into physical lengths by the formula $2.9 \times 10^{6}$ base pairs of double-stranded DNA = 1 m.

There appear to be no theoretical restrictions on either the length of DNA or the sequence of nucleotides along a strand. Except, of course, that the sequence of nucleotides in one strand must be the complement of the sequence in the other strand according to Chargaff's rule. And, as a general rule, the more complex an organism the more information it must contain and, therefore, the more DNA necessary. There are some exceptions to this general rule. For example, the cells of lungfish have more DNA than do human cells (Table 1.3). But from the bacterium *Escherichia coli* to mammalian cells, the amount of DNA increases approximately 800-fold (see the C-paradox, Chapter 10).

The length of all the DNA found in the 46 human chromosomes of a single **diploid** cell, if stretched out end to end, is estimated to be about 2 m, or 6 ft. That works out to approximately 6 billion nucleotide base pairs per diploid cell, and there are trillions of cells in each individual human being!

The number of nucleotides in a cell can be estimated using the following information: 1 g nucleic acid = $2.0 \times 10^{21}$ nucleotides. The haploid complement of human chromosomes (23 chromosomes) yields $3.2 \times 10^{-12}$ DNA. So $(2.0 \times 10^{21}) \times (3.2 \times 10^{-12}) = 6.4 \times 10^{9}$ nucleotides. But, since our DNA is double-stranded, we have $6.4 \times 10^{9}/2 = 3.2 \times 10^{9}$ nucleotide pairs. This is the haploid number; the number of nucleotides found in the DNA of egg and sperm. Somatic cells would have twice that number because they are diploid cells and contain twice as many chromosomes. Somatic cells, therefore, would contain about 6 billion base pairs ($6.4 \times 10^{9}$ base pairs).

## SOME BIOLOGICAL QUESTIONS

How can we account for the differences between species, between individuals of the same species, and between cells of the same individual if all DNA is made of the same material (i.e., the four nucleotides adenine, guanine, cytosine, and thymine). How can we account for the differences between normal and cancer cells?

The percentages of each of the nucleotides in a DNA molecule do not account for the differences. Organisms with vastly different amounts of DNA may still have the

table 1.4

BASE COMPOSITION OF DNA EXPRESSED AS PERCENTAGES IN SELECTED SPECIES

| SPECIES | ADENINE % | GUANINE % | CYTOSINE % | THYMINE % |
|---|---|---|---|---|
| VIRUSES | | | | |
| Bacteriophage T2 | 32.6 | 18.1 | 16.6 | 32.6 |
| *Herpes simplex* | 13.8 | 37.7 | 35.6 | 12.8 |
| Lambda | 26.0 | 23.8 | 24.3 | 25.8 |
| *Pseudorables* | 13.2 | 37.0 | 36.3 | 13.5 |
| *Vaccinia* | 31.5 | 18.0 | 19.0 | 31.5 |
| BACTERIA | | | | |
| *Escherichia coli* | 26.0 | 24.9 | 25.2 | 23.9 |
| *Diplococcus pneumoniae* | 29.8 | 20.5 | 18.0 | 31.6 |
| *Micrococcus hysodeikticus* | 14.4 | 37.3 | 34.6 | 13.7 |
| *Ramibacterium ramosum* | 35.1 | 14.9 | 15.2 | 34.8 |
| FUNGI | | | | |
| *Neurospora crassa* | 23.0 | 27.1 | 26.6 | 23.3 |
| *Aspergillus niger* | 25.0 | 25.1 | 25.0 | 24.9 |
| *Saccharomyces cerevisiae* (baker's yeast) | 31.7 | 18.3 | 17.4 | 32.6 |
| HIGHER EUKARYOTES | | | | |
| *Arachis hypogaea* (peanut) | 32.1 | 17.6 | 18.0 | 32.2 |
| *Bombyx mori* (silkworm) | 30.7 | 18.9 | 19.4 | 31.1 |
| *Drosophila melanogaster* (fruit fly) | 30.7 | 19.6 | 20.2 | 29.4 |
| *Homo sapiens* (human) | | | | |
| liver | 30.3 | 19.5 | 19.9 | 30.3 |
| spermatozoa | 29.8 | 20.2 | 18.2 | 31.8 |
| thymus | 30.5 | 19.9 | 20.6 | 28.9 |
| *Nicotiana tabacum* (tobacco) | 29.3 | 23.5 | 16.5 | 30.7 |
| *Rana pipiens* (frog) | 26.3 | 23.5 | 23.8 | 26.4 |
| *Zea mays* (corn) | 25.6 | 24.5 | 24.6 | 25.3 |

same percentage of each of the nucleotides (Table 1.4). For example, one organism with 10,000 base pairs in its DNA may have a total of 1000 adenines, while another organism with 1,000,000 base pairs may have 100,000 adenines. But both organisms have DNA composed of 10% adenine. And such completely different organisms, in terms of complexity, as bacteriophage T2, ***Saccharomyces cerevisiae*** (baker's yeast), and human cells have nearly equal percentages of the bases (Table 1.4).

The differences between species can be accounted for by (1) the differences in base sequence and (2) the total amount of DNA present in the cells. The differences between members of the same species can be accounted for by (1) the differences in base sequences and (2) the nature of the "satellite" DNA, so named because upon centrifugation this DNA settles out in a band different from genomic DNA (see Chapter 10). It is uncommon for satellite DNA to represent more than about 5% of the total DNA of the cell, but its base composition can be different from that of the cell's genomic DNA. For all members of the same species, however, the total amount of DNA would be about the same.

The difference between cell types within the same multicellular individual is accounted for by differences in gene expression patterns; that is, different cell types have different genes and sets of genes in an active state.

Differences between normal and cancer cells can be seen in the almost universal change in the cancer cell's chromosomal complement. Cancer cells are not true diploids. These cells are said to be aneuploid, a term meaning that the chromosomal number has changed by loss or duplication of chromosomes or chromosomal segments (see Chapter 9 for more on cancer).

## THE BIOLOGICAL FUNCTIONS OF DNA

To function as genetic material, a molecule must have four characteristics or abilities:

1. The molecule must be the repository of all the information required by the cell (see Chapter 4).

2. The molecule must be capable of faithful replication so that the information it contains can be passed on to the next generation by reproduction of the organism, by cell division, or by both (see Chapter 2, Replication Fidelity).
3. The molecule must be capable of transferring its information to the cell on demand.
4. The molecule must be stable enough so that **mutations,** changes in the nucleotide sequence of the molecule, are rare (Chapter 9).

In the cells of eukaryotes, all the information required by the cell is divided unequally between nuclear, or genomic, DNA and the DNA found in the mitochondria (the organelles responsible for the synthesis of the high-energy compound adenosine triphosphate, ATP) or in the chloroplasts (the chlorophyll-containing, photosynthesizing organelles of plants). But in both cases, the vast majority of the information is found in nuclear DNA.

As we shall see later (Chapters 6, 7, 8), transference is done by means of two processes: **transcription** and **translation.** Transcription involves the transfer of information from the DNA molecule to a **messenger RNA (mRNA)** molecule. The nucleotide sequence of DNA is converted into a nucleotide sequence of mRNA. The second process then converts the mRNA nucleotide sequence into the amino acid sequence of a protein molecule. The result is information transfer, mediated by mRNA, from DNA to a protein. This is the critical connection between DNA and proteins; the one provides the information and the other the means by which cells can carry out their functions. This flow of information is sometimes referred to as the central dogma of molecular biology (Chapter 8).

The rarity of mutations provides the stability that makes life, as we know it, possible. Frequent mutations could render beneficial information useless as harmful mutations accumulated. Ironically, however, some instability must be possible, because mutations are needed for (1) the correction of mistakes (harmful mutations) in existing information, (2) the accumulation of more genetic information in the form of nucleotide sequences, and (3) the possible change in current nucleotide sequences resulting in the generation of new information (beneficial mutations). So the molecule must be stable enough to preserve necessary information and yet unstable enough to allow for correction and for an increase in information.

To function as genetic material, the molecule requires a favorable environment. That environment is provided by the cell, which gets its instructions to produce such an environment from the genetic material. The genetic material, the master molecule of the cell, must provide for its own replication and for the proper functioning of the cell in which it resides and on which it depends for survival and replication. The instructions for these activities are found, primarily, in nuclear DNA.

An analogy can be made to the computer. The disk of the computer is the DNA of the machine; it carries the computer's instructions. The machine itself is the cell, ready to execute those instructions. Computer disks can be corrected, information can be added to them, and current information can be altered to provide new information.

## THE GENE

We can now, finally, define a gene. By the late 1940s, it was known that the DNA component of the chromosomes is present in a constant amount in the different somatic cells of an organism and it is present in one-half that amount in the **germ cells** (i.e., egg and sperm). The conclusion drawn from those observations was that DNA was at least a part of whatever the gene would prove to be.

In their second paper on DNA, entitled "Genetical Implications of the Structure of Deoxyribonucleic Acid," also published in 1953 in *Nature*, Watson and Crick began by saying: "The importance of deoxyribonucleic acid (DNA) within living cells is undisputed. It is found in all dividing cells, largely if not entirely in the nucleus, where it is an essential constituent of the chromosome. Many lines of evidence indicate that it is the carrier of a part of (if not all) the genetic specificity of the chromosomes and thus of the gene itself."

Implicit throughout this chapter has been the notion that the information stored in nucleic acids is in the sequences of nucleotides of the molecule. And that is, indeed, the case. More precisely, however, the information is in the sequence of bases, the sugar-phosphate residues are not important in this regard. Nevertheless the phrase "sequence of nucleotides" is commonly used when speaking of the gene or the information contained therein.

So a gene may be defined as a sequence of nucleotides (bases) that codes for a protein. Later we will modify that definition somewhat.

What we have, then, is one sequence determining another. A nucleotide sequence (DNA) codes for a sequence of amino acids (protein). It is this sequence of amino acids, the primary structure of a protein, that determines the biological activity of the protein (Chapter 8).

The size or length of a gene depends on the amount of information it contains (Chapter 4). Large proteins contain more amino acids than small ones. Consequently, large proteins require more nucleotides in their DNA to store the information concerning the sequence of amino acids. Remember, what is crucial is the sequence of nucleotides of the gene, which, in turn, determines the sequence of amino acids of the protein.

Although DNA and, therefore, the gene, consists of only four components, the four nucleotides, as a molecule of DNA increases in length (i.e., as nucleotides are added), the number of possible combinations of these nucleotides or sequences increases dramatically. In fact, the increase is exponential. And, because different combinations of nucleotides would code for different combinations of amino acids, the number of different possible proteins also increases dramatically. But not all possible proteins are of use to the cell. Evolution has determined which of the many different proteins are useful and so which genes or sequences of nucleotides to keep as genetic information.

Some sequences of DNA do not code for amino acids but instead are necessary as signal sequences. Such sequences serve, for example, as the origins for DNA replication, as regulatory sequences, and as sites for chromosome folding. In addition, there is still a large amount of DNA for which no function is as yet ascribed.

Let us now look at what happens to the number of possible nucleotide combinations as the DNA molecule increases in length. The number of possible combinations of nucleotides is given by the term $4^n$, where 4 is the number of nucleotides (A, T, G, C) and $n$ is the number of nucleotides in the molecule (single strand); that is, $n$ is the length of the molecule. So, for a molecule made up of 3 nucleotides there are 64 possible combinations ($4^3 = 64$). A molecule of 5 nucleotides could have 1024 different combinations ($4^5 = 1024$). A chain just 10 nucleotides long could be sequenced to yield more than a million different molecules ($4^{10}$). Consider, then, the possible combinations from the billions of nucleotides found in human cells!

## Summary

The discovery of the genetic material, its chemical and physical properties, took nearly a century of work, beginning with Miescher's work and culminating with the Watson and Crick double-helix model.

Essentially two tracks were taken in the study of the DNA. One involved studies that led to an understanding of the molecule's chemical and physical features, and the other was intended to determine DNA's biological characteristics. The two tracks can be summarized as the "structure-function" concept familiar to all biologists. In the case of DNA, however, much of the work of the two tracks ran parallel.

The work of Levene, in the 1920s and 1930s, for example, showed that nucleotides were composed of bases (adenine, guanine, cytosine, and thymine), a sugar (deoxyribose), and a phosphate group. Meanwhile, Griffith's investigations were laying the foundation for subsequent discoveries that would prove DNA to be the genetic material.

Although much evidence was accumulating early on from studies of chromosome distribution in eukaryotes, the definitive experiments needed to prove that DNA is the genetic material were not done until the 1940s and 1950s. The experiments of Avery, MacLeod, and McCarty and of Hershey and Chase concerning the identity of the genetic material played a major role in spurring on the search for the structure of DNA.

Meanwhile, other scientists, most notably Chargaff and Franklin, in the early 1950s, made crucial discoveries about the structure of double-stranded DNA. Chargaff discovered the base-pairing linkages of nucleotides; Franklin discovered the arrangements of the nucleotides in DNA. The data supplied by these investigators were directly incorporated into the Watson and Crick model.

Watson and Crick, in 1953, proposed a double-stranded model for DNA. The structure showed a backbone of alternating sugar and phosphate and the bases of the nucleotides lying on the inside of the molecule at right angles to the long axis. Further, the proposed structure indicated that Chargaff's base pairs could explain how the molecule replicated. Each strand, Watson and Crick suggested, acted as the template for a new strand during semiconservative replication.

As it has turned out, DNA meets the four criteria that are required of genetic material: (1) it is the repository of information; (2) it is capable of faithful replication; (3) its information can be transferred to the cell; and (4) it is a stable molecule but subject to correction and change.

The conclusion that DNA is the genetic material of cellular organisms and the elucidation of its structure then made possible a definition of the gene. The gene could now be defined at the molecular level as a sequence of nucleotides coding for a sequence of amino acids, the primary structure of proteins.

## Study Questions

1. What is the composition of a nucleotide?
2. Name the purines and the pyrimidines.
3. What purines and pyrimidines appear in DNA?
4. Write out a purine-pyrimidine base pair and a pyrimidine-purine base pair.
5. What chemical bond joins nucleotides within a strand of DNA, and what chemical bond joins strands of DNA?
6. How was Griffith's experiment concerning the "transforming principle" done, and how did Avery improve upon it?
7. Explain Chargaff's major contribution to an understanding of the structure of DNA. How did this contribution help Watson and Crick construct their model?

8. What biological evidence did Hershey and Chase supply to the debate about the chemical nature of the genetic material?
9. What were Franklin's data that played such a central role in the construction of the Watson and Crick model?
10. Describe the chemical and physical properties of the Watson and Crick double helix.
11. Explain the term *semiconservative*.
12. What makes possible the differences between human beings? What makes possible the differences between bacteria and human beings?
13. To serve as genetic material, a substance must meet four requirements. What are they?
14. How many different DNAs are possible in a molecule of DNA composed of 8 base pairs?
15. Define a gene at the molecular level.

## Readings and References


Avery, O. T., C. M. MacLeod, and M. McCarty. 1944. Studies on the chemical nature of the substance inducing transformation of pneumococcal types. Induction of transformation by a deoxyribonucleic acid fraction isolated from *Pneumococcus* Type III. *J. Exp. Med.* 79:137–158.

Chargaff, E. 1950. Chemical specificity of nucleic acids and mechanism of their enzymatic degradation. *Experientia* 6:201–09.

*The chemical basis of life: An introduction to molecular and cell biology*. 1973. *Sci. Amer.* (entire issue).

Franklin, R. E., and R. Gosling. 1953. Molecular configuration in sodium thymonucleate. *Nature* 171:740–41.

Hershey, A. D., and M. Chase. 1952. Independent functions of viral protein and nucleic acid in growth of bacteriophage. *J. Genet. Physiol.* 36:39–56.

Lewin, B. 1997. *Genes VI*. New York: Oxford University Press.

*The molecules of life*. 1985. *Sci. Amer.* (October, entire issue).

Stent, G. S., and R. Calendar. 1978. *Molecular genetics: An introductory narrative*. 2d ed. San Francisco: W.H. Freeman.

Stern, C., and E. R. Sherwood, eds. *The Origins of Genetics: A Mendel Source Book*. San Francisco: W.H. Freeman.

Watson, J. D. 1976. *Molecular biology of the gene*. 3rd ed., Menlo Park, W.A. Benjamin.

Watson, J. D., and F. H. C. Crick. 1953. Genetical implications of the structure of deoxyribonucleic acid. *Nature* 171:964–67.

Watson, J. D., and F. H. C. Crick. 1953. Molecular structure of nucleic acids: A structure for deoxyribose nucleic acid. *Nature* 171:737.

Weaver, R. F., and P. W. Hedrick. 1997. *Genetics*. 3rd ed. Dubuque, Iowa: Wm. C. Brown.

Wilkins, M. H. F. 1956. Physical studies of the molecular structure of deoxyribose nucleic acid and nucleoprotein. *Cold Spring Harbor Symp. Quant. Biol.* 21:75–90.

CHAPTER TWO

# THE REPLICATION OF DOUBLE-STRANDED DNA

## OUTLINE



## CHAPTER OBJECTIVES

*This chapter will discuss:*

- The proof of the semiconservative model of DNA replication
- The components needed for the replication of double-stranded DNA
- The differences between leading strand synthesis and lagging strand synthesis
- How synthesis of prokaryotic DNA differs from synthesis of eukaryotic DNA
- How DNA duplication fidelity is achieved

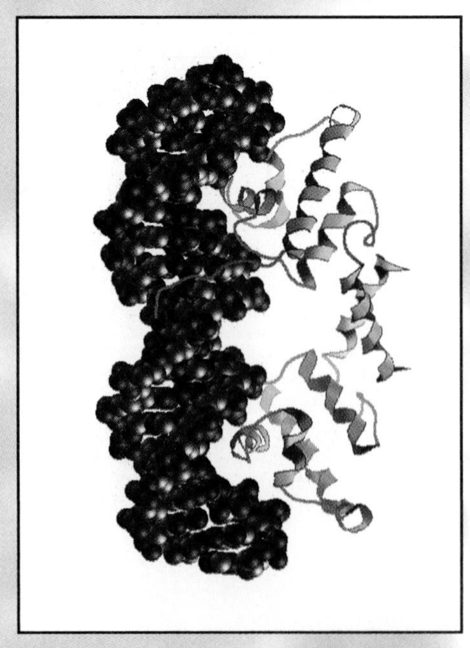

## INTRODUCTION

For any cell or cellular organism to secure its continuity requires that sooner or later it must reproduce. Reproduction ensures genetic continuity, which in turn requires that DNA replicate itself faithfully. That is, each kind of cell or organism will give rise to like progeny provided all the necessary genetic information in the parent (or parents) is transmitted to the next generation completely and unchanged, or nearly so. Offspring, then, inherit genetic information from their parents. The inheritance of this information begins when DNA is replicated in parental cells. DNA replication is the copying mechanism, suggested by Watson and Crick, in which most of the information to be passed on is stored in nuclear DNA and some in other organelles (Chapter 1). This chapter discusses the duplication of information via DNA replication. (Modification of nucleotides, primarily by methylation, may also serve an information function and is also inherited. That topic will be dealt with in greater detail in Chapter 10.)

As was alluded to in Chapter 1 and as will be discussed in greater detail in Chapter 9, some flexibility or change in the informational content of DNA is a prerequisite if new species or new biochemical processes are to come into being. The changes, however, must not be so numerous or so located within genes as to be lethal. A fine balance must be struck between changes that are beneficial and those that are deadly.

In this chapter, we deal with the problem of the accurate transmission of genetic information from one double-stranded DNA molecule to another—the process known as inheritance. The next chapter will look at the replication of other types of nucleic acids.

The paradigm of double-stranded DNA replication is that found in the bacterium *Escherichia coli*. But even in this relatively simple organism, which contains a single, circular, double-stranded molecule of DNA, the complexity of DNA replication is daunting. Table 2.1 gives some idea of the number of genes and gene products (proteins) required to replicate this comparatively small DNA molecule.

All cellular life forms studied so far contain as their genetic material double-stranded DNA. (Viruses are not considered to be cellular life forms.) It is the kind of DNA that Watson and Crick investigated and that they suggested replicates semiconservatively. That is, each strand acts as the template for the synthesis of a new strand so that finally each of the two daughter molecules contains one parent strand and one newly synthesized strand. One strand from the parent is conserved in each new double-stranded molecule, hence the term *semiconservative*.

But Watson and Crick's suggested method of replication required proof. The evidence for such a method was provided by an ingenious set of experiments done by Matthew Meselson and Frank W. Stahl. Their experiments opened the door to the detailed investigation of how DNA replicates and, thus, to the molecular mechanisms underlying inheritance.

The two new strands are synthesized in two different ways. One strand is produced in a continuous operation in which nucleotides are added one after another to the growing chain. The other strand is produced in a discontinuous manner; small pieces of DNA, called **Okazaki fragments,** after their discoverer, are first made and then joined together.

## THE WATSON AND CRICK MODEL

Watson and Crick were aware that the double-helix structure they were proposing gave a clue to one of the functions of DNA—the ability to replicate and to do so faithfully. In other words, they suspected that DNA's structure, its chemical and physical makeup, determined to a considerable extent its "copying mechanism." The structure of the molecule allowed it to function as the material of inheritance. And the way the double helix replicated was the way in which information was passed on from one generation to the next for all cellular organisms.

Watson and Crick had postulated a molecule composed of two complementary strands, and that was the clue to how replication occurred. Each strand acted as the template for the synthesis of its complementary strand. This was the "specific pairing" they referred to in their 1953 paper. In fact, this was the pairing first noted by Chargaff—base pairs. Watson and Crick published another paper a month after their first paper. In the second paper, entitled "Genetical Implications of the Structure of Deoxyribonucleic Acid," they more fully explained the template idea of replication and suggested that an explanation for errors in replication—spontaneous mutations—involves what are called **tautomeric shifts** of the bases (a rearrangement of hydrogen atoms that produces another isomer of the molecule; see Chapter 9).

This notion of specific base pairings gives us a method by which replication might occur but none of the details, none of the molecular mechanisms by which one molecule of DNA is capable of producing a faithful copy of itself. That elucidation of the mechanism of DNA replication has taken years and countless experiments. In this chapter, we look at the picture of DNA replication that has emerged from those efforts.

**table 2.1**

**REPLICATION GENES AND PROTEINS OF *ESCHERICHIA COLI***

| GENE | PROTEIN AND IN VITRO FUNCTION |
|---|---|
| *dnaA* | DnaA; initiation at the origin |
| *dnaB* | DnaB; helicase |
| *dnaC* | DnaC; complex with dnaB |
| *dnaE(polC)* | Alpha subunit of PolIII holoenzyme |
| *dnaG* | Primase |
| *dnaJ* | DnaJ; phage λ initiation; heat-shock response |
| *dnaK* | DnaK; phage λ initiation; heat-shock response |
| *dnaN* | Beta subunit of PolIII holoenzyme |
| *dnaQ* | Epsilon subunit of PolIII holoenzyme |
| *dnaT* | DnaT (protein i); primosome assembly |
| *dnaX* | Gamma and tau subunits of PolIII holoenzyme |
| *dnaY* | Arginine tRNA for rare codon |
| *dnaZ* | Gamma subunit of PolIII holoenzyme; renamed *dnaX* |
| *dut* | dUTPase |
| *grpE** | Initiation of phages lambda and P1; heat-shock response |
| *gyrA (nalA)* | DNA gyrase subunit alpha; nicking–closing |
| *gyrB (cou)* | DNA gyrase subunit beta; ATPase |
| *lig* | DNA ligase; covalently seals DNA nicks |
| *nrdA (dnaF)* | R1 subunit of ribonucleotide reductase |
| *nrdB* | R2 subunit of ribonucleotide reductase |
| *ori* | Origin of chromosomal replication |
| *polA* | PolI; gap filling, RNA excision |
| *polB* | DNA polymerase II |
| *priA*† | Protein PriA (n′); primosome |
| *priB*† | Protein PriB (n); primosome |
| *priC*† | Protein PriC (n″); primosome |
| *rep* | Helicase |
| *rnhA* | RNase H1; removal of RNA primers |
| *rpoA* | Alpha subunit of RNA polymerase |
| *rpoB* | Beta subunit of RNA polymerase |
| *rpoC* | Beta′ subunit of RNA polymerase |
| *rpoD* | Sigma subunit of RNA polymerase |
| *ssb* | SSB (single-strand binding protein) |
| *ter* | Terminus of chromosomal replication |
| *topA* | Topoisomerase I |
| *trxA* | Thioredoxin: coenzyme of ribonucleotide reductase, subunit of phage T7 DNA polymerase |
| *tus(tau)* | *ter*-binding protein; termination |

## MODELS OF DNA REPLICATION: CONSERVATIVE, SEMICONSERVATIVE, AND DISPERSIVE

Before scientists could attack the problem of the mechanism of DNA replication, a very basic question had to be answered. Was the Watson and Crick model of replication correct? In other words, even if the double helix was the true structure of the DNA molecule (and all the contemporaneous physical and chemical data indicated that it was), was Watson and Crick's suggested method of replication also true? Or were there other possibilities that would be consistent with the data on the structure of DNA?

After the Watson and Crick papers had been published, a number of attempts were made to answer those questions. The most successful of these attempts was a series of experiments conducted by Meselson and Stahl and published in 1958.

The problem centered around the distribution of the parental atoms in the progeny molecules. How the atoms were distributed would help to explain what role the parent DNA played in its own replication and by what method DNA replicated.

There were three possibilities concerning the distribution of parental atoms. They have been referred to as the conservative, semiconservative, and dispersive models. Each of these models predicts a different distribution of parental atoms in the progeny molecules (Figure 2.1). The distribution of atoms in the products of replication are referred to as parental if identical to the parent and progeny if novel. The distribution of atoms according to the three models of replication would be as follows:

**Semiconservative model:** All of the atoms of one strand of the parent molecule are transferred intact and without rearrangement to one strand of the progeny DNA molecule; the other strand is formed entirely of new atoms.

**Conservative model:** Atoms of the two parental DNA strands serve as a template for two new progeny strands. The two parental strands remain intact (without rearrangement) and remain together following replication, as do the two progeny strands.

**Dispersive model:** All of the atoms of each strand of the parent molecule appear in the progeny DNA, but they appear as large sections scattered throughout the length of both strands of the progeny DNA.

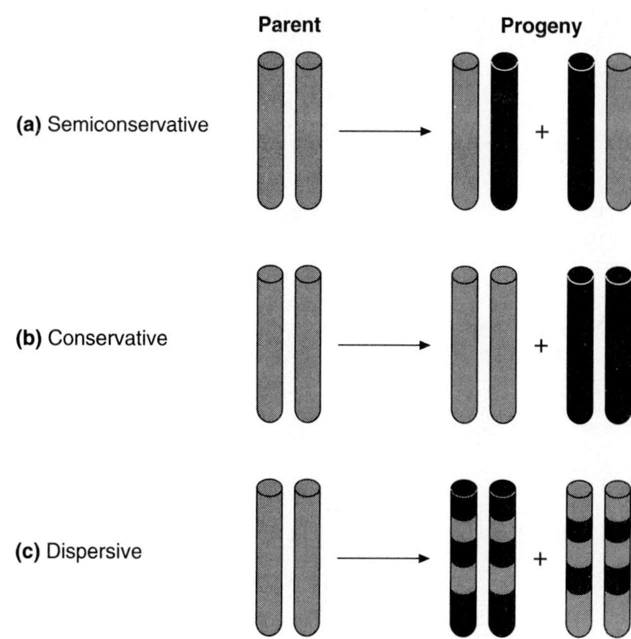


FIGURE 2.1
Three hypotheses for DNA replication. (a) Semiconservative replication; daughter molecules each contain one intact parental strand and one strand entirely of new atoms. (b) Conservative replication; one daughter molecule is identical to the original parental molecule, the other is composed entirely of new atoms. (c) Dispersive replication; each daughter molecule contains some sections derived from the parental molecule and some sections composed entirely of new atoms.

Meselson and Stahl attempted to distinguish among these possibilities by using a technique called density-gradient centrifugation. This technique involves the layering of molecule solutions on the surface of another solution in centrifuge tubes. The other solution is often one of cesium chloride (CsCl) whose concentration increases from top to bottom—this is the density gradient.

Then as the tubes are centrifuged at very high speeds (50,000 rpm) and for long periods of time (hours) the molecules are driven by centrifugal force into the gradient until a region is reached where the cesium chloride's density equals the molecule's buoyant density, at which point no further movement of the molecules takes place. As centrifugation is going on, ultraviolet absorption photographs of the DNA solution are taken so that the migration of the DNA band can be monitored. (Remember, DNA absorbs UV light at 260 nm; Chapter 1). However, this technique will work only if the molecular weights and configurations of different molecules are different. Molecules of different weights or configurations settle out at different positions within a CsCl gradient.

The problem, then, for Meselson and Stahl was to find a way by which parent and progeny DNA molecules of different molecular weights could be made (configuration of the DNAs would necessarily be the same—double helices). It could be done, they found, by incorporating into newly synthesized DNA one or the other isotope of nitrogen—either the heavy isotope ($^{15}$N) or the light isotope ($^{14}$N).

The Meselson and Stahl experiments were done as follows:

*Escherichia coli B* was grown in a glucose-salts medium containing the heavy nitrogen isotope ($^{15}$N), as ammonium chloride, for 14 generations. Under these conditions, as *E. coli* reproduced it would be forced to use the ammonium chloride as the source of nitrogen to synthesize its required purines and pyrimidines for DNA replication. At the end of 14 generations, virtually all the cells would have DNA containing the $^{15}$N. This cell population represented the parent DNA. A sample of the cells was taken, and the DNA was extracted and set aside.

Then, an abrupt change of medium was made. An addition of $^{14}$N, the light isotope, to the medium was made to give a 10-fold excess of light nitrogen ammonium chloride along with the nucleosides needed for DNA synthesis. These additions diluted out the $^{15}$N and provided a large source of both $^{14}$N and preformed DNA precursors. From this point on, newly synthesized DNA would contain the light isotope.

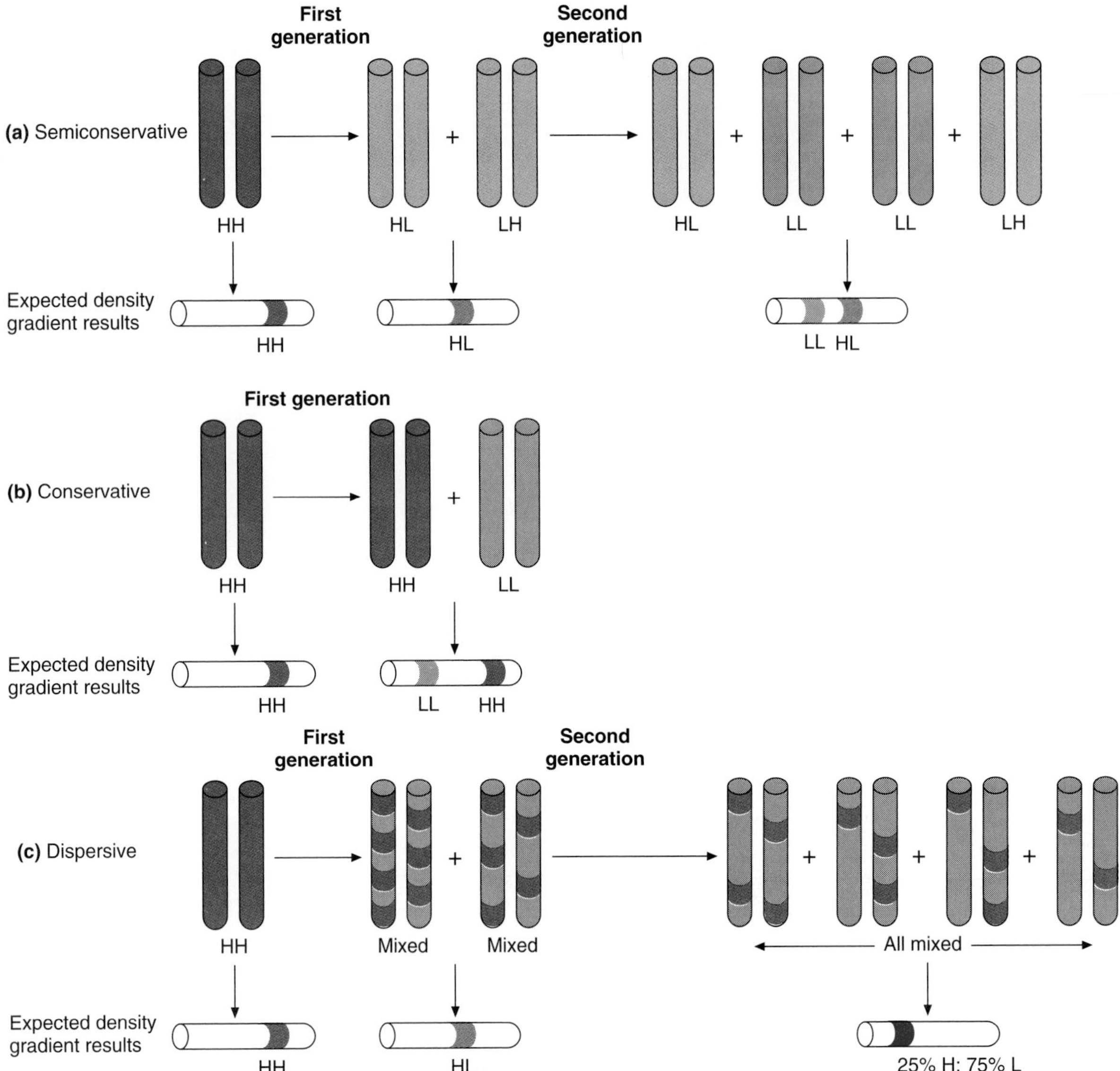


FIGURE 2.2
Three replication hypotheses predict different results for Meselson and Stahl's experiment. The conservative hypothesis (b) can be ruled out because it predicts equal amounts of two different DNAs after one generation (50% light/light + 50% heavy/heavy). In fact, all the DNAs after one generation are of one type (density intermediate between light/light and heavy/heavy). The dispersive hypothesis (c) can be ruled out after the second generation because it predicts that all DNAs will be of one type (density consistent with a molecule that is 25% heavy and 75% light). In fact, after the second generation two types of DNA are produced (50% light/light + 50% heavy/light). Only the semiconservative hypothesis (a) is consistent with Meselson and Stahl's results.

At intervals for several generations, cell samples were taken. All samples were then treated to extract the DNA. The DNA samples, including the parental DNA containing the heavy isotope, were then centrifuged in the CsCl density gradient and the sedimentation positions in each sample were noted using UV absorption.

The results were that the parent DNA, being composed of heavy nitrogen, settled out closest to the bottom of the density gradient. The first generation of cells grown in the presence of the light nitrogen had DNA that settled above the parent DNA. And as each successive sample of cells was taken—that is as the second, third, and so on generation arose—the DNA settled out in two bands. One band was identical to that of the first-generation DNA. But the second band was new; it appeared still higher up in the density gradient, above the first-generation DNA (Figure 2.2).

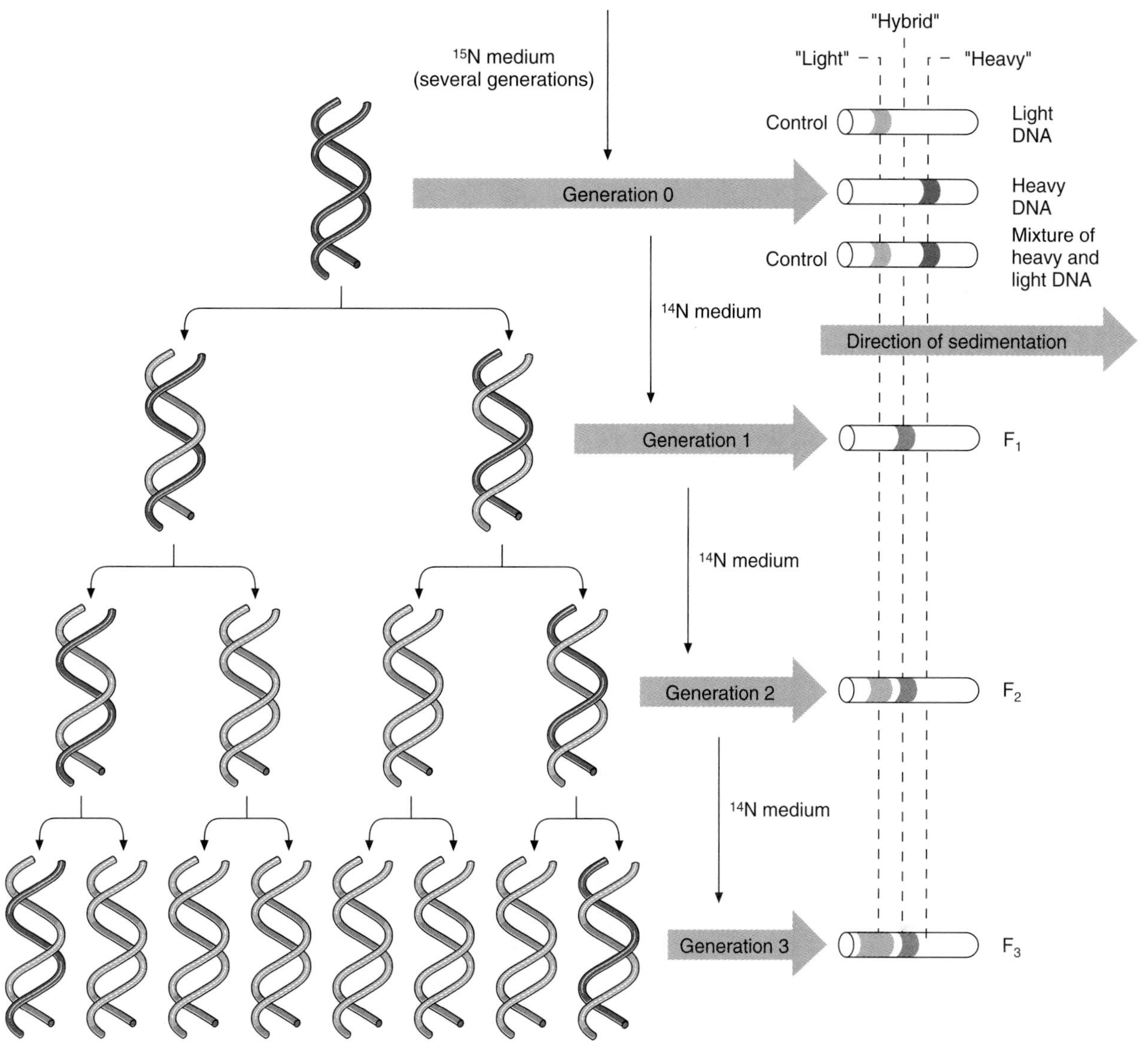


FIGURE 2.3
The results of Meselson and Stahl's experiment (right) are consistent with an interpretation based on the semiconservative hypothesis of DNA replication (left).

As the generations passed, the middle band (first-generation DNA) disappeared while the uppermost band (second-, third- generation DNAs, etc.) became more pronounced because it was accumulating more and more DNA. A schematic interpretation of the results is shown in Figure 2.3.

Note in Figure 2.3 that parent DNA is made of two heavy strands, thus, it settles toward the bottom of the density gradient. The first-generation molecules are hybrids; one strand, from the parent DNA, is heavy (made of $^{15}$N), and the other, its newly synthesized complement, is light (composed of $^{14}$N). This hybrid DNA molecule settles above the parental DNA. At the second generation, a new molecule appears. It is composed exclusively of light strands, because only light nitrogen is available for use in their synthesis. Of the total second-generation DNA, half the molecules are hybrids and the other half are light. This distribution occurs because each strand of the hybrid, first-generation DNAs acted as a template for a new DNA molecule. So each hybrid DNA gives rise to another hybrid and a light DNA. Then, at each successive generation, there is more and more of the light DNA as a proportion of the total DNA and less and less of the hybrid. (Remember, the only source of nitrogen is the light isotope, $^{14}$N). So, as the

DNA of later and later generations is centrifuged, the uppermost light band becomes more and more prominent while the hybrid, middle band fades.

Meselson and Stahl concluded that "the results of the present experiments are in exact accord with the Watson-Crick model for DNA duplication" (Meselson and Stahl 1958). That is, their data supported the semiconservative model of DNA replication.

What then of the conservative and dispersive models? These data eliminated the conservative model because that model predicts that the first-generation DNA molecules would consist of half heavy, parental DNA and half light DNA, the progeny DNA. If that distribution were correct, each parent DNA molecule would have given rise to a new molecule composed solely of light strands. In other words, two bands would have appeared at the first generation, a band of heavy DNA and one of light DNA, but there would have been no middle or hybrid band.

The dispersive model could also be eliminated. The first-generation DNA would be hybrid, just as in the semiconservative model, but so would the DNA from each succeeding generation. The uppermost light band would not appear in the second or following generations, because each parental strand would be parceled among the progeny DNA molecules, and thus all the molecules produced at each replication would be hybrid.

So the mode of replication for double-stranded DNA is semiconservative. Each new molecule is composed of one strand donated by the parent molecule and one newly synthesized strand. Put another way, during DNA replication, each strand acts as the template for the synthesis of a new strand. DNA replication, therefore, conserves one parent strand in each of the two new daughter molecules.

# The Components Required for Prokaryotic DNA Replication

Throughout the book, components—such as building blocks, enzymes, etc.—are listed and described before the processes involving them are discussed.

For the replication of double-stranded DNA, the following components are necessary:

1. **template DNA:** The parent sequence of nucleotides to be used as information in the synthesis of the complementary strand.
2. **origin:** A specific sequence of nucleotides on the parent DNA that is recognized as the initiation site of synthesis.
3. proteins:

   **DnaA, DnaB, and DnaC:** Needed to recognize the origin and to separate the strands of the parent DNA (Table 2.1).

   **Rep protein** and **single-strand binding proteins (SSB):** These molecules, working together, help to uncoil the tightly packaged parent DNA (Rep protein) and unwind the double helix (SSB) in preparation for replication. These activities require ATP as an energy source.
4. **nucleotides:** The triphosphate deoxyribonucleotides and ribose forms of the bases adenine, guanine, cytosine, thymine, and uracil are required as building blocks, energy sources, and information units.
5. enzymes: Many enzymes are needed for replication (Table 2.1), but we will deal only with the most important ones and in the order they are used:

   **gyrases:** Uncoil the DNA in preparation for the activity of the helicase(s).

   **helicases:** Unwind the double helix.

   **DNA-directed RNA polymerase** and **primase:** Needed for the synthesis of RNA primers, a requirement of DNA synthesis.

   **DNA-directed DNA polymerase III:** The actual replicating enzyme that synthesizes DNA.

   **DNA-directed DNA polymerase I:** Removes the RNA primers and replaces them with DNA. Is also the "proofreading" enzyme.

   **DNA ligase:** Joins the Okazaki fragments by way of phosphodiester bonds.

## *Template DNA*

Neither DNA polymerase I nor III is capable of synthesizing DNA de novo. The enzymes will not place nucleotides in a random order. There are, also, no known DNA polymerases that will initiate DNA synthesis. And, since the DNA enzymes cannot know a predetermined order of nucleotides, a template is necessary.

As the new DNA strand is being synthesized, each new nucleotide to be added is selected by first matching it to the nucleotide on the template strand—matching is the formation of a Chargaff base pair.

It is, then, the sequence of nucleotides of the parent DNA strand that determines the sequence of nucleotides of the new DNA strand. So the information of one strand, the nucleotide sequence of the parent DNA, is used to create the information of the complementary strand.

## *Origin*

In prokaryotes, initiation of DNA synthesis does not begin just anywhere along the molecule. Rather, replication begins at specific locations or sites referred to as replication origins.

*Escherichia coli*, like other prokaryotes, has a single origin composed of approximately 245 bp. At this site, initiation of synthesis occurs. Three stages or steps are needed: (1) origin recognition, (2) so-called melting of the duplex—the hydrogen bonds between the strands are broken, and (3) loading of the helicase enzyme.

In eukaryotes, yeast and animal cells, for example, the location and structure of origins have been more difficult to establish owing to both the more complex nature of the eukaryotic genome and the limitations of the assays used to detect the sites. It does appear, however, that, although initiation may occur at a number of different sites, specific origins may not be essential for eukaryotic DNA replication. Once initiation does occur, DNA synthesis in both prokaryotes and eukaryotes proceeds bidirectionally (in two directions) along a **replication fork.**

## *Proteins*

The DnaA protein (note the use of upper and lower case letters—this combination of letters is used to designate proteins needed for DNA synthesis) is the protein, in *E. coli*, that recognizes and binds the origin sequence—the initiation site for DNA replication.

A complex of proteins DnaB and DnaC, apparently guided by DnaA, then associates at the origin site to form a prepriming complex; DnaB has helicase activity and helps to separate strands.

As noted earlier, the two new strands of DNA are synthesized differently. The strand elongated by continuous synthesis is referred to as the **leading strand,** and the strand of Okazaki fragments is referred to as the **lagging strand.** Keep in mind that in both cases synthesis is in the $5' \longrightarrow 3'$ direction.

The Rep protein attaches to the template strand of DNA, upon which the leading strand of new DNA will be synthesized. The protein moves along the parental strand in the $3' \longrightarrow 5'$ direction (opposite the direction of new strand synthesis) and consumes two ATPs for each base pair separated.

SSB molecules are needed to keep the two strands of DNA separated once unwinding has occurred. These proteins attach to the parent strands and prevent the reformation of the base pair hydrogen bonds.

## *Nucleotides*

The synthesis of DNA requires four nucleotides in the deoxyribose triphosphate form and four in the ribose triphosphate form. Needed for DNA synthesis are deoxyadenine 5′-triphosphate, dexyguanosine 5′-triphosphate, deoxycytidine 5′-triphosphate, and deoxythymine 5′-triphosphate. These are abbreviated, respectively, as dATP, dGTP, dCTP, and dTTP (the *d* indicates "deoxy" and the *TP* indicates "triphosphate"). These are all high-energy forms of the nucleotides.

For the synthesis of the primer RNA, the nucleotides needed are adenosine 5′-triphosphate (ATP), guanosine 5′-triphosphate (GTP), cytidine 5′-triphosphate (CTP), and, in place of dTTP, RNA requires uridine 5′-triphosphate (UTP).

During aerobic respiration, glucose and fatty acids are oxidized. The electrons removed are shuttled to the electron transport system (in the cytochromes), which is coupled to a system of enzymes that eventually convert the energy stored in the electrons into ATP via a process called oxidative phosphorylation. Adenosine diphosphate (ADP) is phosphorylated to ATP during this process. So glucose and fatty acids indirectly supply the energy for DNA synthesis. The ATPs produced via oxidative phosphorylation are then used to produce the other triphosphate nucleotides.

To produce these other high-energy nucleotides, two different enzymatic reactions are required. The first, the production of the triphosphate forms, is a two-step reaction in which the specific nucleoside monophosphate kinase (e.g., adenylate kinase) converts the monophosphate to the diphosphate using ATP as the phosphate donor. For example, $AMP + ATP \longrightarrow 2\ ADP$ or $GMP + ATP \longrightarrow GDP + ADP$. Then, a second enzyme, a nucleoside diphosphate kinase, phosphorylates the diphosphates to the triphosphate forms, so $GDP + ATP \longrightarrow GTP + ADP$.

To produce the "deoxy" forms of the nucleotides requires still another group of enzymes, collectively called ribonucleotide reductases. Each reductase is specific for its corresponding ribonucleotide. The substrates for these enzymes are in the diphosphate forms; for example, ADP is converted into dADP, or CDP into dCDP. These are subsequently phosphorylated to the triphosphate forms for DNA synthesis.

The nucleotides of DNA synthesis are very important in three ways. They function

1. as the building blocks of DNA (and RNA)
2. as the information stored in DNA (more precisely, the sequence of nucleotides serves this function)
3. as the energy source (in the triphosphate forms) for synthesis

The first two functions were introduced in Chapter 1. As for the third function, remember that DNA synthesis is a building process. In thermodynamic terms, it is an endergonic reaction; that is, it requires energy. Also note, as we saw in Chapter 1, that the nucleosides appear in the nucleic acids as the monophosphates.

What drives the synthesis of DNA is a set of reactions common in the cell. An energy-requiring reaction is coupled to an energy-releasing reaction. Endergonic reactions are driven by exergonic reactions.

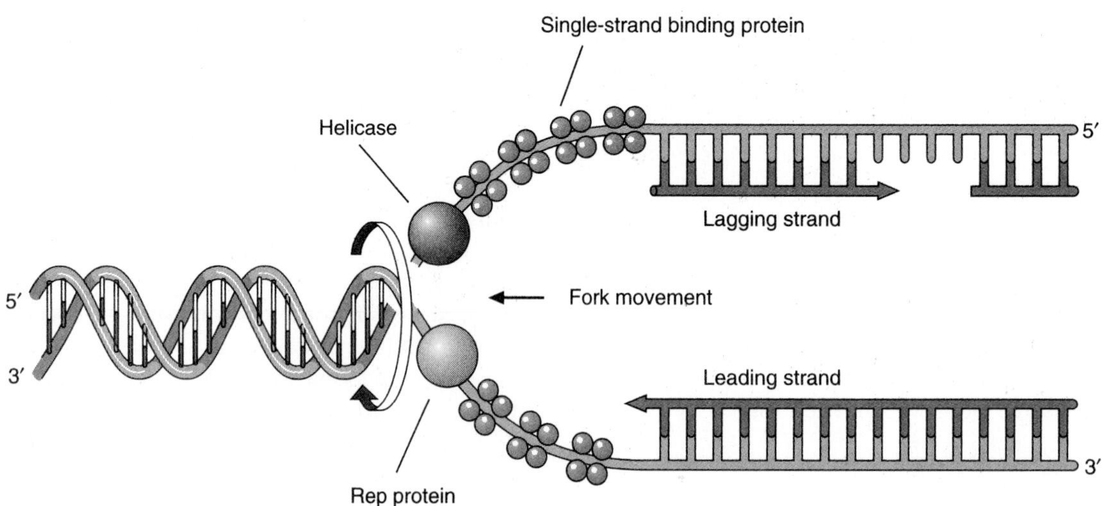


FIGURE 2.4
DNA is unwound as Rep protein binds to the leading strand template and helicase II (dnaB protein) binds to the lagging-strand template. Separated DNA strands are prevented from reannealing by the binding of SSBs.

So, during DNA synthesis, each of the triphosphate nucleotides is converted to the monophosphate form just as incorporation into the new DNA strand is about to take place. The release of the two phosphates (called pyrophosphates) releases energy that is used to incorporate the monophosphate nucleotide into the growing DNA strand: e.g., ATP ⟶ AMP + P~P + energy ⟶ DNA.

# Enzymes

## Gyrases

Before DNA synthesis can take place, the tightly coiled molecule must be "relaxed," uncoiled, and the double helix unwound. The relaxing and uncoiling of the molecule is essentially a release of tension along the DNA molecule just ahead of the point where synthesis will take place. And once DNA synthesis is completed, the daughter DNAs must be separated. All this activity is the result of enzymes called gyrases. The enzymes release tension by transient breaks of phosphodiester bonds. These breaks may occur as often as once in every 10 or so nucleotides.

## Helicases

Helicase activity is a property of DnaB protein and Rep protein, which sometimes together are referred to as the helicase enzymes. These proteins, along with the SSBs, separate the strands of the DNA molecule by breaking hydrogen bonds; the strands are then unwound, and attachment of the SSBs holds the strands apart so that each can act as a template (Figure 2.4). As these events occur, the replication forks are produced and bidirectional DNA synthesis takes place.

While the Rep protein moves along the parent strand acting as the template for leading strand synthesis (continuous synthesis), the DnaB protein apparently attaches to the parental strand, which acts as the template for the synthesis of the lagging strand (discontinuous synthesis). And as the Rep protein moves in a 3′ to 5′ direction, the DnaB moves in a 5′ to 3′ direction. So the DnaB protein is involved in Okazaki fragment synthesis.

## DNA-Directed RNA Polymerase and Primase

A somewhat strange aspect of DNA synthesis is that it first requires the synthesis of another nucleic acid, ribonucleic acid, or RNA. Like DNA, this second type of nucleic acid is made up of a sequence of nucleotides. But RNA differs from DNA in two important regards. First, RNA contains the pyrimidine uracil (see Figure 1.2) in place of the pyrimidine thymine. Second, instead of the sugar deoxyribose, RNA has ribose (see Figure 1.3). This versatile nucleic acid is not only an absolute requirement for DNA synthesis but, as we shall see, it can also function as genetic material for viruses (Chapter 3) and it is also an absolute requirement for the start of protein synthesis (Chapter 8).

DNA synthesis is initiated by RNA synthesis. A short piece of RNA, called primer RNA because it primes DNA synthesis, is needed before DNA chain formation can begin. The primer varies in length in different cell types, but it is on the order of 2–10 nucleotides. The leading strand of DNA synthesis is primed once, whereas the lagging strand, where discontinuous synthesis takes place, must be repeatedly primed. Each Okazaki fragment is begun with an RNA primer.

Primase and RNA polymerase catalyze the same reaction; so the primase is really a RNA polymerase. Both the primase and the RNA polymerase attach to the template strand for leading strand synthesis; primase alone attaches to the parent template strand for lagging strand synthesis.

RNA polymerases may be necessary to initiate DNA replication, since no polymerase discovered thus far has been found capable of chain initiation. It seems that DNA polymerases require at least one base pair in position before the enzymes can attach and extension of new strands can be begun. RNA polymerases, on the other hand, do not have this requirement. These enzymes can initiate synthesis and elongate chains—also only in the 5′ to 3′ direction.

RNA polymerases read the nucleotide sequences, one nucleotide at a time, of the parent molecule, and they base pair the appropriate nucleotide to the template parent strand, producing short sections of a hybrid molecule—RNA/DNA. Since it is RNA being synthesized, the triphosphates of the ribonucleotides, not of the deoxyribonucleotides, are needed. The latter will be needed for DNA synthesis.

Lastly, the polymerization of the primer RNA is directed by the DNA sequences, the template. In other words, the enzymes take their instructions from DNA with regard to the sequence of nucleotides to be used in constructing the primer.

Primase appears to be a special smaller monomeric molecule, whereas DNA-directed RNA polymerase is a much larger multimeric protein (Chapter 8).

## DNA-Directed DNA Polymerase III

DNA-directed DNA polymerase III is one of the two actual replicating enzymes. As its name implies, and like the RNA polymerases, it takes its instructions from the parent-template DNA strand. Using the RNA primer as an anchor and the triphosphate deoxyribonucleotides (dATP, dGTP, dCTP, and dTTP), DNA polymerase adds deoxyribonucleotides to the primer, thereby elongating the chain. But now, DNA is being produced.

We have a strand of nucleic acid composed of a small section of RNA and attached to it a growing section of DNA. As the enzyme moves along the parent strand, it "reads" each parent nucleotide and base pairs to each the appropriate new nucleotide, producing the adenine-thymine or guanine-cytosine pairings. Synthesis, (i.e., nucleotide addition to the growing DNA chain) is always in the 5′ ⟶ 3′ direction.

How are the correct base pairs produced, or, put another way, How does the enzyme "read" the template strand? Two factors appear to be important:

1. Enzymes are specific for substrates.
2. The correct pairing with the template makes the most energetically stable complex with the polymerase.

DNA polymerase III also guarantees that synthesis per se will happen by lowering the activation energy needed for synthesis, that synthesis will occur within the time frame required by the cell (minutes or hours), and that synthesis will result in the faithful reproduction of the parent molecule (errors are rare and correctable) and not in simply a joining of nucleotides in a random sequence. In the absence of DNA polymerase III, none of this would happen! The active enzyme, often called DNA polymerase III holoenzyme (meaning complete) is a multisubunit complex consisting of at least seven different protein molecules.

## DNA-Directed DNA Polymerase I

DNA-directed polymerase I is the second of the two DNA synthesizing enzymes. Sometimes called the Kornberg enzyme after its discoverer, polymerase I is essential for both the synthesis and the repair (Chapter 9) of DNA.

DNA polymerase I has, in fact, three enzyme activities:

1. DNA polymerization
2. depolymerization in a 5′ ⟶ 3′ direction
3. depolymerization in a 3′ ⟶ 5′ direction, the **proofreading** activity

The last two activities are referred to as **exonuclease** activity, which means that the enzyme is able to remove nucleotides, one at a time, from either the 5′ or the 3′ end of nucleic acid chains. In this chapter, we shall concern ourselves primarily with the polymerization and exonuclease activities in the 5′ ⟶ 3′ direction.

As we will see shortly, the lagging or discontinuous strand of newly synthesized DNA is produced in fragments of 1000–2000 nucleotides in prokaryotes and something on the order of 100–200 nucleotides in eukaryotes. Each of these fragments, the Okazaki fragments, is processed to remove the RNA primer and to replace the removed ribonucleotides with the appropriate deoxyribonucleotides of DNA (the same processing occurs on the leading strand but only once since there is only one primer). The Okazaki fragments are then joined to form a longer and longer strand of DNA.

As Okazaki fragments are being synthesized by DNA polymerase III, others are being processed by DNA polymerase I. The latter polymerase proofreads the growing chain as new nucleotides are being added. Should a base be attached to the 3′ end of a growing chain and it is not a correct base, the 3′ ⟶ 5′ exonuclease feature of polymerase I

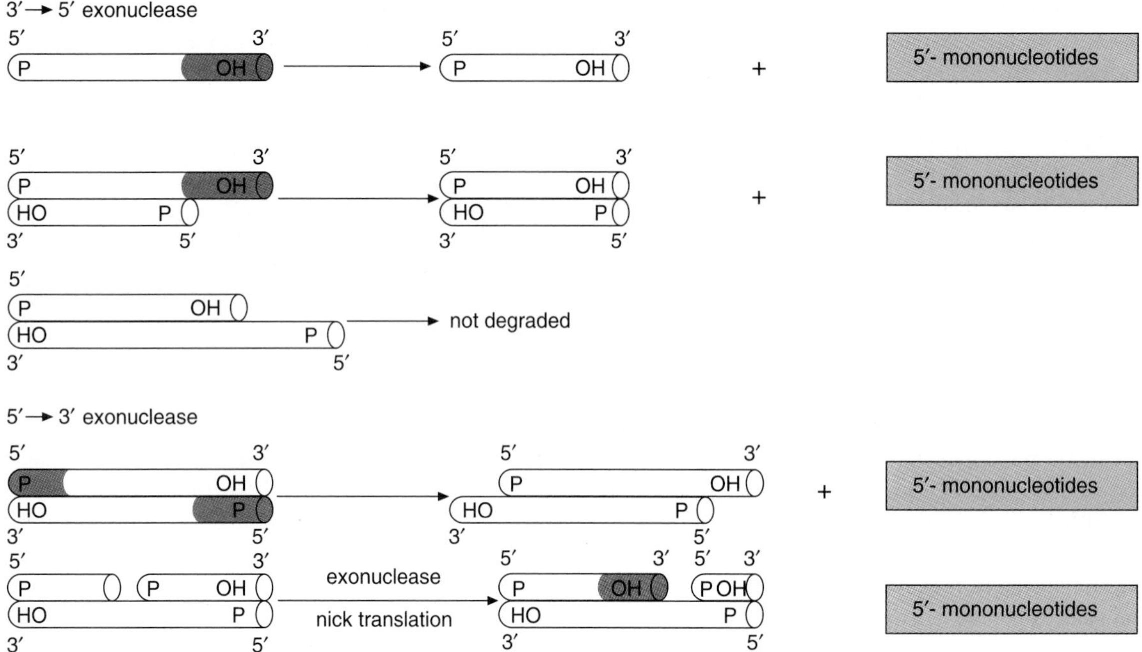


FIGURE 2.5
DNA polymerase I has both 3′ to 5′ and 5′ to 3′ exonuclease activity, and 5′ to 3′ polymerizing activity. It uses the latter two to remove RNA primers by simultaneously removing ribonucleotides from the 5′ end of a nick while it adds deoxyribonucleotides to the 3′ end of the nick. As a result, the nick moves down the chain in the 5′ to 3′ direction (nick translation).

would remove the unmatched base. This kind of proofreading activity helps to ensure a high degree of replication fidelity.

DNA polymerase I is also required for the removal of the RNA primer. The enzyme extends the 3′ end of one fragment by addition of deoxyribonucleotides while removing ribonucleotides (RNA) from the 5′ end of the following Okazaki fragment. This procedure is sometimes called **nick translation** (Figure 2.5).

Once these procedures are completed, two adjacent fragments of DNA are separated only by an unformed phosphodiester bond. This bond is formed by the last enzymes needed for DNA synthesis.

## *DNA Ligase*

The DNA ligases are responsible for connecting DNA segments during replication, repair, and recombination. All ligases join a 5′-phosphoryl group and a 3′-phosphoryl group on adjacent fragments, thereby sealing the nick (Figure 2.6). The reaction occurs in discrete steps and requires an energy source, either ATP (eukaryotes) or nicotinamide adenine dinucleotide, NAD (prokaryotes). The ligase works as follows:

1. The adenyl group of the ATP or NAD is transferred to the enzyme.
2. The enzyme then transfers the adenyl group to the 5′ phosphoryl terminus of the nick to form a pyrophosphate grouping. In effect, AMP is attached to the 5′-phosphoryl terminus.
3. DNA ligase then catalyzes the attachment of the 5′-phosphoryl group to the 3′ position of the adjacent fragment, and, in the process, splits out the AMP that was added in step 2.

# THE REPLICATION OF PROKARYOTIC DOUBLE-STRANDED DNA

The replication of *E. coli* DNA is a tightly controlled event occurring only once per cell division. The molecule is replicated at a rate of approximately 1500 nucleotides per second. The entire chromosome is duplicated in about 40 minutes. The exact nature of the triggering mechanism for DNA replication in both prokaryotes and eukaryotes is unknown. There are some clues, however. For example, in prokaryotes the availability of nutrients appears to be important; in eukaryotes, the presence or absence of growth factors may be critical.

Control of replication takes place at the stage of initiation, generally. What appears certain is the types of control needed:

1. The timing of replication must be closely coordinated with other physiological and biochemical events of cell metabolism and division.

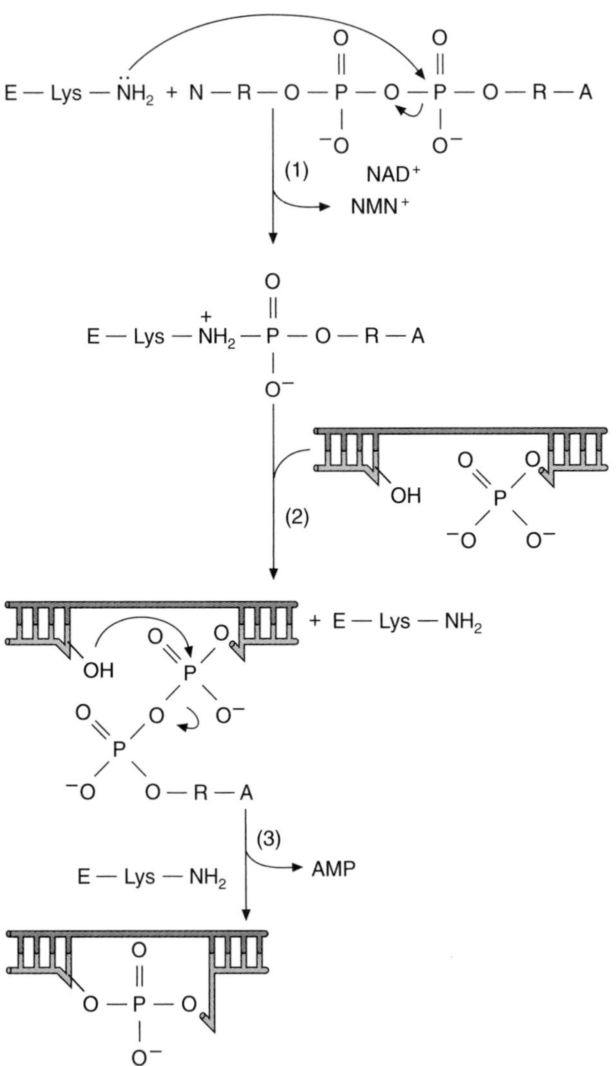


FIGURE 2.6
The reactions catalyzed by *E.coli* DNA ligase. In eukaryotic and T4 DNA ligases, $NAD^+$ is replaced by ATP so that $PP_i$ rather than $NMN^+$ is eliminated in the first reaction step.

2. The timing must be so regulated that duplication of DNA occurs only once per cell division.
3. The initiation of replication should occur only when conditions for cell growth are favorable. When conditions are not favorable, DNA replication ought to be inhibited. Inhibition should also occur if there is DNA damage or if a terminally differentiated stage has been reached.

In some prokaryotes, such as species of the bacterial genera *Bacillus* and *Clostridium*, the endospore might be considered a terminally differentiated stage. In eukaryotes, especially multicellular species such as human beings, the situation is somewhat more complicated. Different cells may or may not respond to replication signals. Nerve cells, in general, and the adult brain, in particular, are for all intents and purposes incapable of cell division and DNA replication. Other cells, however, such as liver and skin cells, though differentiated, are still capable of DNA synthesis and cell division.

## Semidiscontinuous Synthesis

Before dealing with the steps and mechanisms of replication, it would be useful to emphasize two important points already discussed:

1. The nucleotides are phosphorylated at the 5′ position.
2. Synthesis is, therefore, always in the 5′ ⟶ 3′ direction.

That is, each new nucleotide is added to the growing chain by joining the incoming nucleotide, via its phosphorylated 5′ position, to the unphosphorylated 3′ position of the last nucleotide of the chain. Another way to look at this is to remember that the growing, or elongating, strand of DNA is extended at its 3′ end only.

DNA replication occurs by **semidiscontinuous** processes. The two new strands, in other words, are produced in different ways. The term *semidiscontinuous* is not to be confused with the term *semiconservative*. The latter term denotes the role played by each strand of the parent DNA molecule.

## Continuous and Discontinuous Synthesis

The semidiscontinuous process has to do with the way in which each new strand is synthesized. In this process, one strand is elongated by the **continuous** additions of nucleotides to the 3′ end (after primer synthesis has taken place once) while the other new strand is produced by repeated synthesis of primer RNAs and short lengths of DNA (Okazaki fragments), which must eventually be joined by DNA ligase. This latter method is called **discontinuous** synthesis (Figure 2.7).

# SEMIDISCONTINUOUS REPLICATION OF DNA

The best-understood model of cellular double-stranded DNA replication is that of *E. coli*. The organism contains a single closed circular double-stranded molecule measuring some 1300 μm in length and made up of about $4.7 \times 10^6$ base pairs. Its replication rate is on the order of 1500 nucleotides per second. This macromolecule is contained in a cell measuring approximately 3 μm at the long axis.

To replicate a molecule of this length confined in so small a space poses enormous problems. And to compound

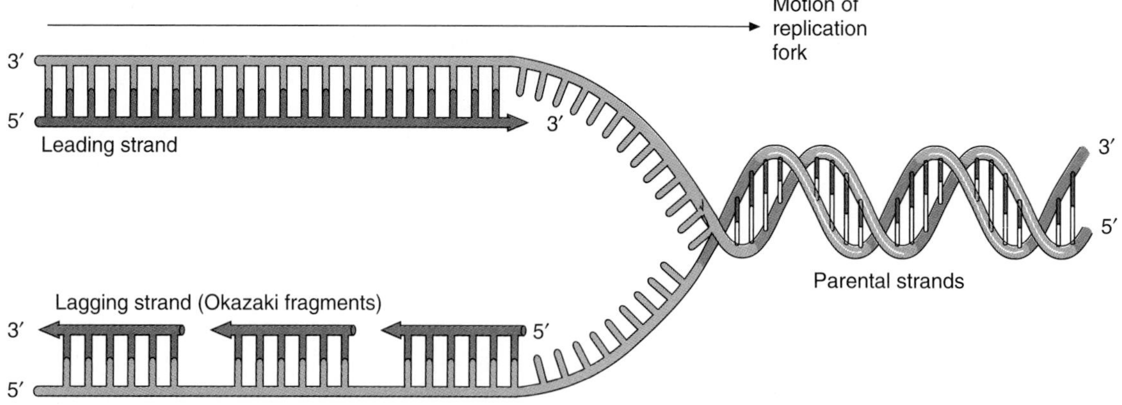


FIGURE 2.7
In DNA replication, both daughter strands (red) are synthesized in their 5′→ 3′ direction. The leading strand is synthesized continuously, whereas the lagging strand is synthesized discontinuously.

the difficulty, *E. coli*'s DNA must be tightly wound or packaged to fit the confines of the cell. As synthesis occurs, the molecule is unwound. Then, when replication is completed, each new double-stranded DNA molecule has to be rewound (Figure 2.8) and repackaged. Obviously, all this unwinding and rewinding must be done under strictly controlled conditions.

## Replication Initiation

To begin the replication process, an enzyme, the DNA gyrase, using ATP as an energy source, breaks the strands of DNA to allow controlled uncoiling of the molecule. This enzyme is also required to separate the two newly synthesized DNA double-stranded molecules and to recoil and repackage each in the daughter cells. Uncoiling of the DNA is required in order for assembly of the initiation complex, consisting of the Rep protein, helicase, and SSB, to occur.

The steps needed to initiate replication at the origin site, called oriC, are as follows:

1. DnaA protein binds at the oriC, causing local melting or breaking of hydrogen bonds between base pairs. About 40 bonds are ruptured. This is an energy-requiring process that consumes ATP.
2. The binding of DnaA then facilitates the binding of DnaB and DnaC proteins at the melted region. This completes formation of the prepriming complex.
3. The helicase enzymes, DnaB and Rep protein, along with SSB and gyrase then begin to uncoil and unwind and separate the strands of the parent DNA. This allows attachment of primase and RNA polymerase to the parent template for leading strand synthesis and of primase alone to the parent template for lagging strand synthesis.

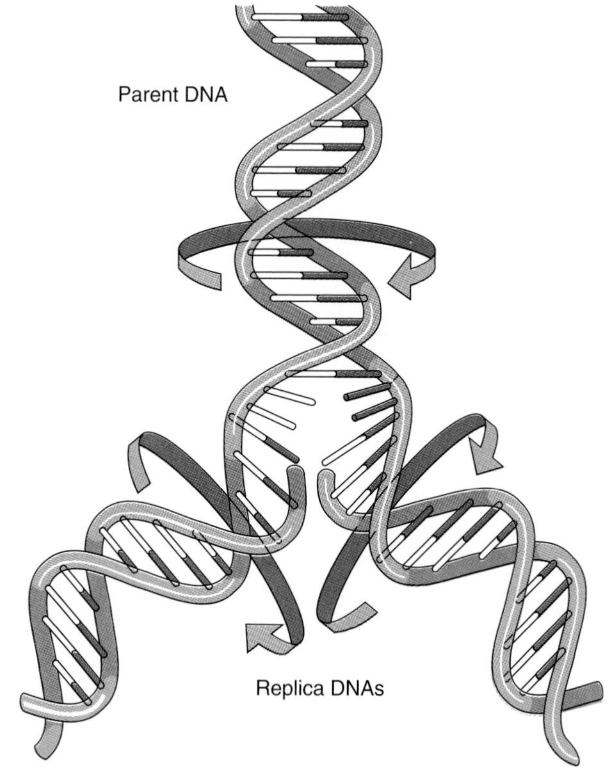


FIGURE 2.8
The replication of DNA.

At this point, it might be helpful to refer once again to Figure 2.4, the replication fork. Remember, the two strands of DNA are antiparallel; they run in opposite directions. Also, keep in mind that DNA synthesis is always in the 5′ to 3′ direction.

So from the initiation site, oriC, one parental strand runs in a 5′ ⟶ 3′ direction while the other strand is running in the 3′ ⟶ 5′ direction, for example:

3′ ⟶ 5′
*
5′ ⟶ 3′
⟵
direction of replication

If then initiation is at the asterisk and replication is to proceed from right to left, the bottom strand (5′ ⟶ 3′) will be the template from which the leading strand will be produced via continuous synthesis (5′ ⟶ 3′ synthesis). The top strand (3′ ⟶ 5′) will serve as the template for the synthesis of the lagging strand via discontinuous synthesis (also 5′ ⟶ 3′ synthesis).

### *Primosome Formation*

Now, once the helicase, Rep protein, and SSB proteins attach, primase can attach to form what is called the **primosome.** The primosome is the enzyme-protein complex within which primer RNA is produced to form the 5′ end of each Okazaki fragment during discontinuous synthesis. Primosomes appear not to be required for continuous synthesis.

### *Leading Strand Synthesis*

Since the leading strand is antiparallel to its template and the direction of its synthesis coincides with the overall direction of replication (see the above illustration), only one primer RNA need be made. DNA polymerase III can then take over and continuously add deoxyribonucleotides to the 3′ terminus of the lengthening strand.

### *Lagging Strand Synthesis*

Synthesis of the lagging strand, however, would also be antiparallel and, therefore, running in the direction opposite to the direction of replication. The template strand is running 3′ ⟶ 5′, so the new strand must run 5′ ⟶ 3′.

However, studies suggest that the lagging strand template loops back on itself, turning 180 degrees, so that, in effect, the synthesis of the lagging strand is now running in the same direction as that of the leading strand (Figure 2.9).

The sequence of events for lagging strand synthesis (Figure 2.9) is as follows:

1. RNA primer recently synthesized by the primosome.
2. The lagging strand behind the RNA–DNA duplex loops back 180 degrees into the DNA polymerase III complex, which then elongates the strand by synthesizing DNA.
3. Synthesis of DNA continues until the DNA polymerase has joined 1000–2000 deoxyribonucleotides and reaches the 5′ end of the previously made Okazaki fragment.
4. At this point, the DNA polymerase releases the lagging strand template. The approaching enzyme causes SSB proteins to be released also.
5. As DNA synthesis continues, more SSB proteins attach to the lagging strand template just behind the DNA polymerase.
6. Following attachment of SSB proteins, another primosome associates with the lagging strand template to begin another round of RNA primer synthesis and to repeat the cycle.
7. As the Okazaki fragments begin to accumulate (at least two must be present), DNA polymerase I, acting as the polymerase and the 5′ to 3′ exonuclease, starts to add deoxyribonucleotides to the 3′ terminus of one fragment while simultaneously removing ribonucleotides from the RNA primer (5′ end) of the fragment just ahead (see Figure 2.5). These two enzyme activities (polymerase and exonuclease) continue until all the primer RNA has been removed and the 3′ terminus of one fragment has reached the 5′ end of the DNA section just ahead. At this point, the two Okazaki fragments, now composed entirely of deoxribonucleotides, require only a phosphodiester bond to join them.
8. The enzyme DNA ligase now takes over. Using ATP as an energy source, the ligase synthesizes the phosphodiester bond by linking the 3′ end of one fragment to the 5′ end of the fragment ahead via the 5′ end's phosphoryl group.

These steps (1–8) are repeated until synthesis of both the leading and lagging strands of one replication fork contact the leading and lagging strands of the replication fork coming from the opposite direction around the circular chromosome; in this way bidirectional DNA synthesis is completed.

## BIDIRECTIONAL DNA SYNTHESIS

As indicated earlier, *E. coli* DNA replication and, for that matter, eukaryotic DNA replication, is bidirectional. That is, when the prepriming complex is formed at the origin, two replication forks are produced and move in opposite directions as synthesis proceeds. The events just described (steps 1–8) occur at both forks (Figure 2.9). Replication takes place in opposite directions around the closed circular chromosome until the two

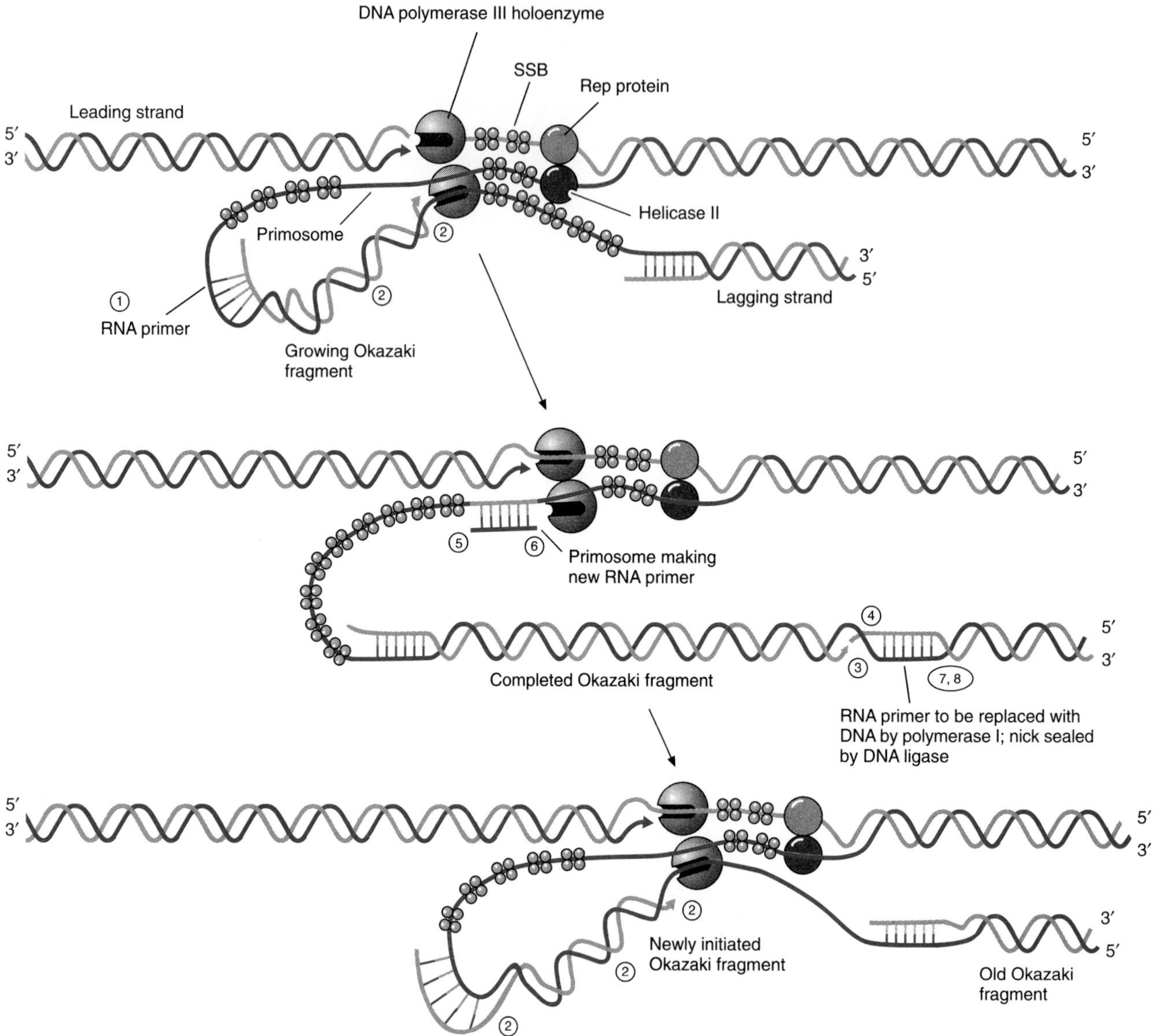


FIGURE 2.9
The lagging strand must be looped around or turned 180 degrees so that the DNA replisome (thought to contain two DNA polymerase III holoenzyme complexes) can add deoxyribonucleotides at the strand's 3′ end. Numbers in the diagram refer to steps outlined in the text.

forks merge, at which point duplication of the chromosome is completed.

Note in Figure 2.10 that the leading and lagging strands alternate around the bubble. The direction of replication, that is movement of the forks, determines which parent strand will act as the template for the leading and lagging strands.

As the forks move away from one another, the circular chromosome comes to resemble the Greek letter theta, so this type of replication is often referred to as theta replication and the replicating chromosome as the theta structure (Figure 2.11). Separation of the two double-stranded DNA molecules is then brought about by the enzyme DNA gyrase upon completion of replication.

## EUKARYOTIC DNA REPLICATION

Although not nearly as much is known about eukaryotic DNA replication as is known about the prokaryotic system, much information has accumulated during recent years. The data presently available appear to support the idea that the mechanism of eukaryotic DNA replication is very similar to the prokaryotic method. There are, however, some differences in how the two kinds of DNA are dupli-

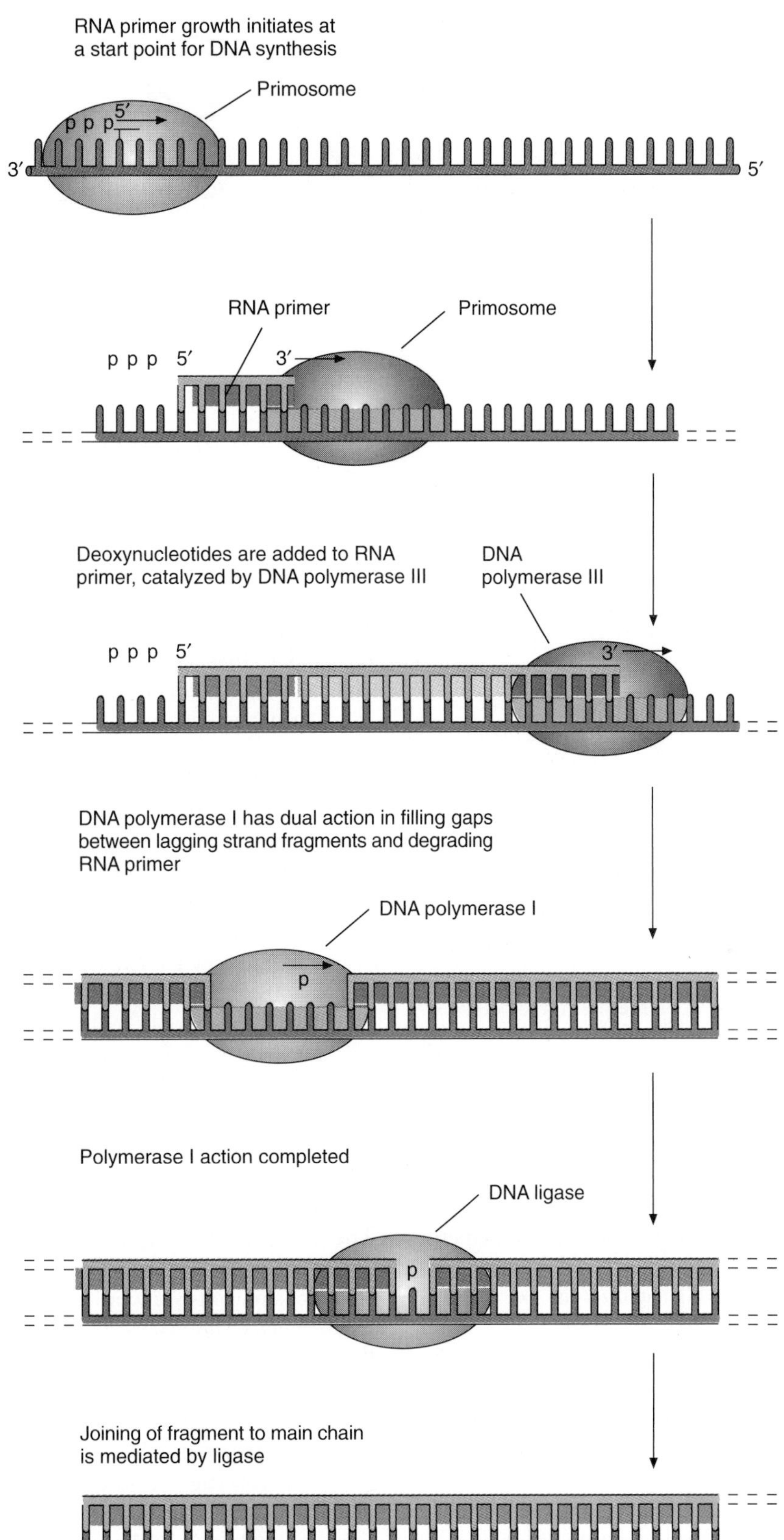


FIGURE 2.10
Detailed diagram of the initiation of DNA replication with an RNA primer and the subsequent removal of the primer by DNA polymerase I.

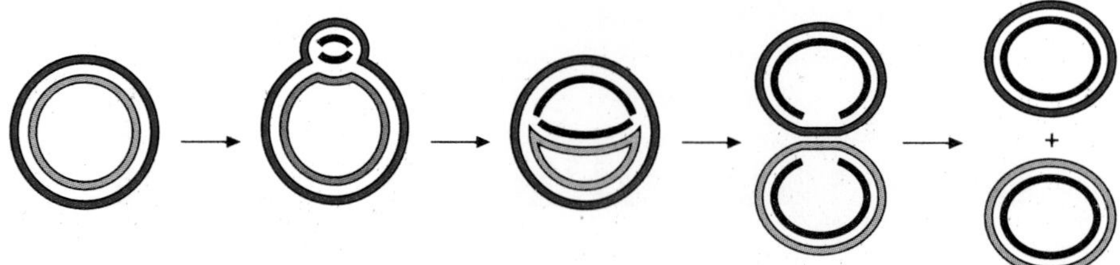


FIGURE 2.11
The theta mode of DNA replication in *Escherichia coli*.

cated. Rather than discuss these differences in detail, we shall simply list the major differences:

1. Eukaryotes have more and larger chromosomes.
2. Eukaryotic replication may require as much as 6–8 hours for completion versus the 40 minutes needed by *E. coli*.
3. There are multiple, rather than a single, replication origins along eukaryotic chromosomes. They are spaced about 20 kbp apart.
4. Bidirectional replication begins at each origin and continues until forks from adjacent origins merge.
5. Eukaryotic DNA replication is at the rate of about 10–100 nucleotides per second as opposed to the prokaryotic rate of about 1500 nucleotides per second.
6. The decision to replicate DNA is made in the $G_1$ phase of the cell cycle, during which time the concentrations of growth factors, the signal molecules, are increasing.
7. Eukaryotic DNA is duplicated during the S phase of the cell cycle.
8. At least five types of DNA polymerases have been found in eukaryotic cells.

## REPLICATION FIDELITY

All living things are dependent on the faithful duplication of their genetic information if their kind is to be perpetuated. Individuals of a species must pass on to the next generation the set of instructions that will ensure the survival of the species.

This passage of intact and correct information to the next generation is dependent on the DNA replication mechanism's degree of copying fidelity. On the other hand, changes must be possible because they are needed to cope with a changing environment. Genetic variation allows for what is called adaptive flexibility. But too many randomly occurring changes would lead to an organism's inability to deal with the current environment, so a high error rate is intolerable.

How, then, is the integrity of the genetic information maintained from one generation to the next? How do *E. coli* give rise to other *E. coli?* Roses to other roses? Liver cells to other liver cells? Human beings to other human beings?

In *E. coli*, the error rate is estimated to be about 1 error in 1 billion base pairs replicated, or about 1 error per 1000 cells per generation. This low rate of error results from many factors, some known and many, no doubt, still to be discovered. The fidelity of DNA replication in *E. coli* is checked at two levels during the duplication process. Each level has a different mechanism operating to control errors. The first mechanism is an error-avoidance system and the second is an error-correction system.

During replication, the error-avoidance system operates through two different enzyme activities. First, the DNA polymerase III selects the correct nucleotides for base pairing. This selection is the most important of the processes in guaranteeing the fidelity of replication. As noted earlier in this chapter, the correct pairing of the new nucleotide with the parent template nucleotide is the most energetically favorable. The second enzyme activity is that of proofreading by DNA polymerase I. This enzyme checks the most recently added nucleotide and removes it if an incorrect base pair would result. That is, if a Chargaff's base pair does not result from the addition, the new nucleotide at the 3′ end of the elongating strand is removed.

The second mechanism involves error correction and is referred to as the mismatch repair system. It is designed to catch errors missed by the first system of checks. This activity is also dependent on DNA polymerase I in *E. coli*. The ability of the enzyme (or enzymes) of this system to distinguish between the parent nucleotide and the new nucleotide is obviously of critical importance. Removal of the parent nucleotide instead of the new nucleotide and then the insertion into the parent strand of a nucleotide that base pairs with the new, but wrong, nucleotide would be the opposite of error correction; errors would instead be built into the parent DNA.

Some investigators have suggested as a possible answer to the problem of distinguishing parent from progeny the fact that, after the synthesis of new DNA, methylation of adenines takes place. Crucial to this operation is that there is a short time lag between synthesis and methylation. During this period, unmethylated, progeny DNA is checked and mismatches are repaired.

In eukaryotes, how parent strands are distinguished from newly synthesized strands is not known. In yeast and fruit flies, no methylation occurs; in other eukaryotes it is cytosine that is methylated. Investigators have suggested that the answer to this problem in both prokaryotes and eukaryotes lies in discontinuous synthesis. The gaps between Okazaki fragments are the marks of the new strand. How the leading strand would be checked remains a problem, however.

Another suggestion is that RNA primer synthesis itself may be another error-avoidance system. It appears that mismatching is most likely to occur as the first few nucleotides are assembled into a strand. But, since RNA primer will be removed and replaced by DNA polymerase I, errors made here can be corrected.

Finally, it should be noted that DNA repair is possible because of what might be called information redundancy. Each strand of double-stranded DNA can act as the template for the other strand. So, damage to either strand can be repaired by checking the nucleotides of the other, undamaged strand. In Chapter 9, we shall delve more deeply into DNA damage and repair.

## Summary

After Watson and Crick had discovered the structure of DNA and Meselson and Stahl had proved that the molecule replicates by a semiconservative method, there were years of intense effort to find the mechanism by which the molecule replicates itself faithfully. Chargaff's base pairing rule proved to be critical. Double-stranded DNA was shown to replicate in such a way that each strand acts as the template, or source of instruction, for the sequence of nucleotides of the other strand.

The mechanism of replication of double-stranded DNA was found to be a semidiscontinuous process. One strand is synthesized continuously in a 5′ to 3′ direction. The other strand is also synthesized in a 5′ to 3′ direction, but synthesis occurs discontinuously in sections called Okazaki fragments. These fragments are then joined to form a single unbroken strand.

A large number of proteins and enzymes are now known to be required for DNA synthesis. Proteins are needed to recognize the replication origin site and to stabilize the single strands as the two strands are unwound. Enzymes are needed for uncoiling, unwinding, polymerization, and to ensure duplication fidelity. The triphosphates of the nucleotides are required both as building blocks and sources of energy.

An interesting feature of DNA synthesis is the need for an RNA primer to which the deoxyribonucleotides can attach to begin the synthesis of DNA. The primer RNA is later removed and replaced by DNA via a special DNA polymerase. The synthesis of double-stranded DNA is bidirectional from the origin and occurs in two replication forks that move away from each other. Synthesis continues until the two replication forks moving around the circular prokaryotic chromosome meet and merge. Prokaryotes have been found to have a single replication origin, whereas eukaryotes have multiple origin sites.

A critical condition of DNA duplication is that the process operate with a very high degree of fidelity. How this fidelity is achieved is not completely understood. A number of cellular activities that contribute to duplication fidelity have been found. The new DNA strands are checked at at least two levels. The first is called the error-avoidance system. It operates to prevent mistakes from being built into the new strand at the very beginning. This system is dependent on two enzymes: DNA polymerase III and DNA polymerase I. The second level corrects errors that may have escaped the first check. This second checking system is called error correction and is also dependent on the activity of the enzyme DNA polymerase I. How these enzymes are able to distinguish between parental and progeny DNA is a problem still under study.

In *E. coli*, duplication fidelity is efficient enough so that errors occur only once in a billion base pairs replicated.

## Study Questions

1. What did Watson and Crick mean by "specific pairing"?
2. Briefly explain the semiconservative model of DNA replication.
3. Describe the Meselson-Stahl experimental protocol.
4. Show how the data from the Meselson-Stahl experiments confirmed the semiconservative model.
5. Describe the role of each of the following in prokaryotic DNA replication: triphosphate nucleotides, template DNA, primase, RNA primer, DNA polymerases I and III, and DNA ligase.
6. Describe how leading and lagging strand synthesis are similar and different.
7. What are Okazaki fragments?
8. What is the direction of DNA synthesis?
9. How can both new strands of DNA be synthesized in the same direction even though the template strands are running antiparallel?
10. Show the direction of synthesis of the new strands at each replication fork.
11. List at least five major differences between eukaryotic and prokayotic DNA synthesis.
12. Briefly describe how error avoidance and error correction ensure DNA replication fidelity.

## Readings and References


Cairns, J. 1963. The bacterial chromosome and its manner of replication as seen by autoradiography. *J. Mol. Biol.* 6:208–13.

Delbruck, M., and G. S. Stent. 1957. Page 699 in *The chemical basis of heredity*, ed. W. D. McElroy, and B. Glass. Baltimore, Md.: John Hopkins Press.

Echols, H., and M. F. Goodman. 1991. Fidelity mechanisms in DNA replication. *Annu. Rev. Biochem.* 60:477–511.

Kelman, Z., and M. O'Donnell. 1995. DNA polymerase III holoenzyme: Structure and function of a chromosomal replicating machine. *Annu. Rev. Biochem.* 64:171–200.

Kornberg, A. 1960. Biologic synthesis of deoxyribonucleic acid. *Science* 131:1503–08.

Kornberg, A., and T. A. Baker. 1992. *DNA replication*. 2d ed. New York: W.H. Freeman.

Loeb, L. A., and T. A. Kunkel. 1982. Fidelity of DNA synthesis. *Annu. Rev. Biochem.* 51:429–57.

Marians, K. J. 1992. Prokaryotic DNA replication. *Annu. Rev. Biochem.* 61:673–719.

Meselson, M., and F. W. Stahl. 1958. The replication of DNA in *Escherichia coli*. *Proc. Natl. Acad. Sci. USA* 7:671–82.

Ogawa, T., and T. Okazaki. 1980. Discontinuous DNA replication. *Annu. Rev. Biochem.* 49:421–57.

Okazaki, T., et al. 1968. Mechanism of DNA chain growth. I. Possible discontinuity and unusual secondary structure of newly synthesized chains. *Proc. Natl. Acad. Sci. USA* 59:598–605.

Radman, M., and R. Wagner. 1988. The high fidelity of DNA duplication. *Sci. Am.* 259 (August):40–46.

Singer, M., and P. Berg. 1991. *Genes and genomes*. Mill Valley, Calif.: University Science Books.

Voet, D., and J. G. Voet. 1995. *Biochemistry*. 2d ed. New York: John Wiley & Sons.

Wang, T. S.-F. 1991. Eukaryotic DNA polymerases. *Annu. Rev. Biochem.* 60:513–51.

Watson, J. D., and F. H. C. Crick. 1953. Genetical implications of the structure of deoxyribonucleic acid. *Nature* 171:964–67.

Watson, J. D., and F. H. C. Crick. 1953. Molecular structure of nucleic acids: A structure for deoxyribose nucleic acid. *Nature* 171:737–38.

Wilkins, M. H. F., A. R. Stokes, and H. R. Wilson. 1953. Molecular structure of deoxypentose nucleic acids. *Nature* 171:738–40.

CHAPTER THREE

# THE REPLICATION OF OTHER NUCLEIC ACIDS

## OUTLINE



## CHAPTER OBJECTIVES

*This chapter will discuss:*

- How single-stranded DNA is replicated: ØX174
- The synthesis of single-stranded RNA: tobacco mosaic virus
- How RNA can act as a template for DNA synthesis: retroviruses
- How the insertion of foreign DNA into cellular DNA is controlled: lambda virus and lysogeny

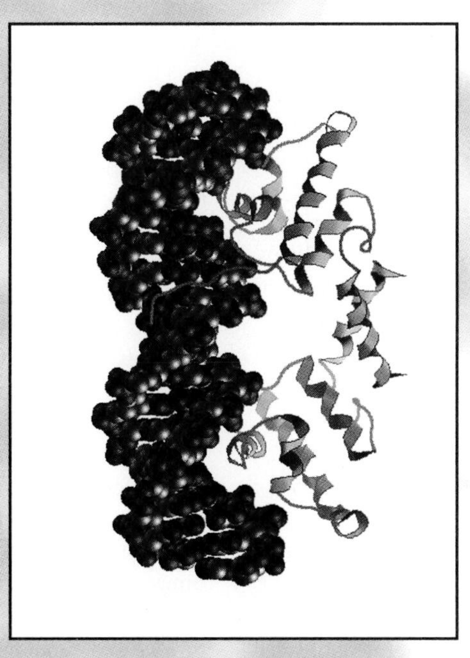

## INTRODUCTION

All cellular forms of life studied thus far contain double-stranded DNA as the genetic material. The key word here is *cellular*. This word includes the prokaryotes as well as the eukaryotes. There is, however, another life form, a noncellular form, namely, the viruses, whose genetic material can be different.

Of the four possible types of **nucleic acids** functioning as genetic material, all have been found in one virus or another. There are, for example, viruses containing double- or single-stranded DNA and others whose genome is double- or single-stranded RNA. The nucleic acid may also be linear or circular. For any particular virus, however, only one type of nucleic acid serves as genetic material.

Because viruses have different forms and types of nucleic acids, they must have different cellular mechanisms for the synthesis of their genomes. So viruses provide us an opportunity to examine the replication of nucleic acids other than double-stranded DNA. We shall also see that foreign DNA or RNA can be processed in unusual ways so as to become a part of a host's genome. Viruses have played an important role in our understanding of the behavior of newly acquired genetic nucleic acids inserted into a host. Understanding this process has been indispensable in genetic engineering studies (Chapter 11).

Viruses range in size from the minute satellite tobacco necrosis virus (STNV), which has a single gene, to the relatively large pox viruses, whose genomes contain something on the order of 240 genes.

Viruses are unusual because all of them are obligate intracellular parasites. In other words, all viruses are absolutely dependent on living, functioning cells for reproduction. The reason is that, as a group, viruses lack enzymes and energy-harnessing systems, or, for that matter, any independent metabolism; they are incapable of producing ATP or proteins.

Viruses are able to infect all cellular life forms, from bacteria to plants to animals. But each virus is host-specific, and some, such as HIV and polio, can be tissue-specific as well.

The problem facing viruses is that they must replicate within cells whose enzyme systems are designed to replicate only double-stranded DNA not single-stranded DNA or any form of RNA. In this chapter, we shall restrict ourselves to those viruses about which quite a bit is known and that have unusual forms and methods of replicating those other nucleic acids:

1. ØX174: a circular, single-stranded DNA molecule
2. tobacco mosaic virus (TMV): single-stranded, linear RNA
3. human immunodeficiency virus (HIV): HIV's genome consists of a single-stranded linear molecule of RNA, but the organism carries two copies of this molecule along with two copies of a special enzyme, reverse transcriptase. This virus can incorporate its genome into the host cell's genome.
4. Lambda: This virus's genome is a linear double-stranded DNA molecule that circularizes upon infection of its host cell. This organism is also capable of integrating its nucleic acid into the host cell's DNA.

All viruses must go through a number of stages or steps to replicate themselves. Duplication of the genome is only one of those steps. In brief, the stages of viral replication are:

1. Attachment or adsorption of the virus to the host cell.
2. Penetration of the entire virus into an animal or plant host cell or of the virus nucleic acid into a bacteria host cell.
3. Alteration of the host cell's biochemical machinery to allow for virus production.
4. Replication of viral nucleic acid.
5. Protein synthesis of virus coat, etc.
6. Assembly of the various components of the virion into complete particles.
7. Release from the cell of the mature virus progeny.

## ØX174

Among the smallest viruses known are the icosahedral DNA viruses of bacteria. (Bacterial viruses are also called **bacteriophage** or simply **phage**.) The term *icosahedral* refers to the geometry of the phage. These viruses have a protein coat, called the capsid, surrounding the DNA. The capsid is composed of 20 equilateral faces and 12 corners. Many of the spherical eukaryotic viruses and bacteriophage are icosahedral.

ØX174 has a genome consisting of 5386 nucleotides as a single-stranded circular DNA molecule (Figure 3.1). It, therefore, carries a limited amount of information in its DNA. The genome encodes information for 11 proteins. For replication, ØX174 is dependent on the host cell's DNA-replicating machinery. The host cell in this case is *E. coli*.

ØX174 has been extensively studied for a number of reasons. Its genome contains overlapping genes (Chapter 4); that is, genetic information or nucleotide sequences for different proteins overlap (see gene A* contained in gene A and gene E contained in gene D). This phenomenon has also been found in eukaryotic organisms (see Chapter 10). This organism's genome was also the first to be completely sequenced (by F. Sanger in 1977). And the study of this virus's replicating strategy has been a useful model for the study of how the chromosome of its host, *E. coli*, 1000 times the size of the viral genome, is replicated. And, of

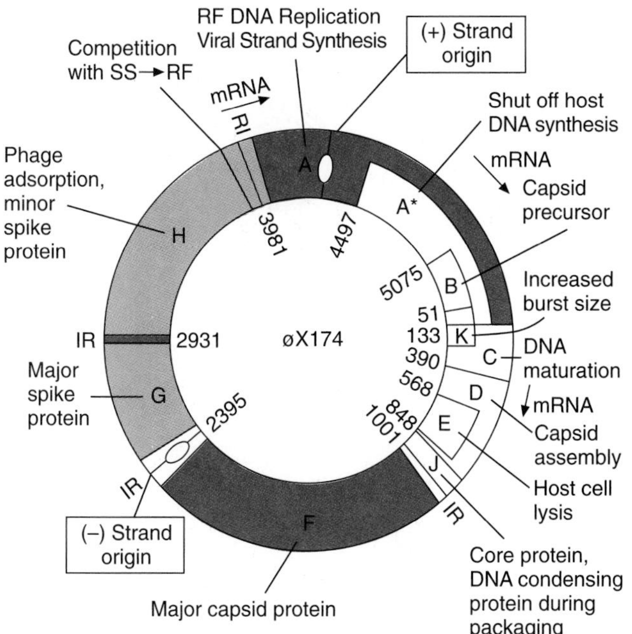


FIGURE 3.1
Genetic map of phage ØX174 with suggested functions of gene products. Letters around the rim indicate known genes and their function; numbers on the inside of the circle indicate distance in base pairs from a reference point called position 1. IR = intergenic region; arrow = mRNA promoter and direction of transcription.

course, the study of the bacterial replication method has aided the study of eukaryotic chromosome replication, including those of human beings.

Upon adsorption to *E. coli*, ØX174 injects its DNA into the cell. This injected DNA is designated the positive (+) strand. The (+) strand encodes the genetic information of the phage in its sequence of nucleotides. The protein product of phage gene A* produced in the host cell then inhibits DNA synthesis within 20–30 minutes after infection, possibly by cleaving single-stranded regions at host replication forks. This activity may be required to release the limited number of host replication enzymes (e.g., primosomes) now needed by the phage.

Keep in mind, the genetic information of the virus is in the sequence of nucleotides designated the positive (+) strand; this is the strand that must be replicated. We know that DNA replication requires a template. If the (+) strand is the template, what would be produced would be a complement of the (+) strand (Chargaff's base pairing). This complement would have a sequence of nucleotides different from that of the original, or parent, strand. The new strand is designated the negative (–) strand. So, in effect, what happens is that a complement of the (+) strand, a (–) strand, is synthesized, and then a complement of the (–) strand is produced. This final product would be the same as the (+) strand and would have the original sequence of nucleotides.

ØX174 must now go through three stages to complete the duplication of its genome, the (+) strand (Figure 3.2).

1. Conversion of viral single-stranded DNA (SS) to a duplex replicative form (RF) consisting of a (+) and a (–) strand: viral SS → parental RF.
2. The multiplication of RF by way of **rolling-circle replication:** parental RF → progeny RF.
3. Synthesis of (+) strands using the (–) strands of the RF's as templates via rolling-circle replication: RF → SS.

Synthesis of the (–) strand to produce RF's is carried out in six steps, and the procedure is entirely dependent on host enzymes and proteins (Figure 3.2). The conversion to a duplex DNA molecule is analogous to lagging, or discontinuous, strand synthesis discussed in Chapter 2.

The six steps needed for SS → RF conversion are as follows:

1. Prepriming: Attachment of SSB proteins to ØX174 DNA after injection into the cell and uncoiling of viral DNA by host gyrase. This is followed by association of proteins *n*, *n′*, and *n″* and then proteins *i*, DnaB, and DnaC to form a prepriming complex.
2. Priming: The enzyme primase binds to form the primosome.
3. Elongation: Propelled by *n′*-catalyzed ATP hydrolysis, the primosome moves in a 5′ to 3′ direction, displacing SSB proteins as it moves and synthesizes RNA primer. This process requires DnaB protein. The initiation of primer RNA synthesis occurs at random sites. As primer RNA synthesis is completed, DNA polymerase III elongates the strand using deoxribonucleotides to produce DNA and Okazaki fragments.
4. Gap filling: DNA polymerase I then removes primer RNA, replacing it with DNA.
5. Ligation: DNA ligase now joins the individual fragments via phosphodiester bonds.
6. Supercoiling: DNA gyrase then coils the new duplex DNA composed of a (+) and a (–) strand to produce an RF.

We now have a template for (+) strand synthesis, the (–) strand of the RF. The primosome remains attached to the RF, where it will take part in (+) strand synthesis. The time required to complete an individual RF unit is about one minute.

The second stage of ØX174 DNA replication, parental RF → progeny RF, is much simpler than the first stage. It requires two additional proteins: the product of gene A (gpA), encoded in the viral genome, is needed to initiate and terminate the process; DNA helicase, the product of *E. coli*'s *rep* gene, is needed here for unwinding the RF unit. The two proteins help to produce a

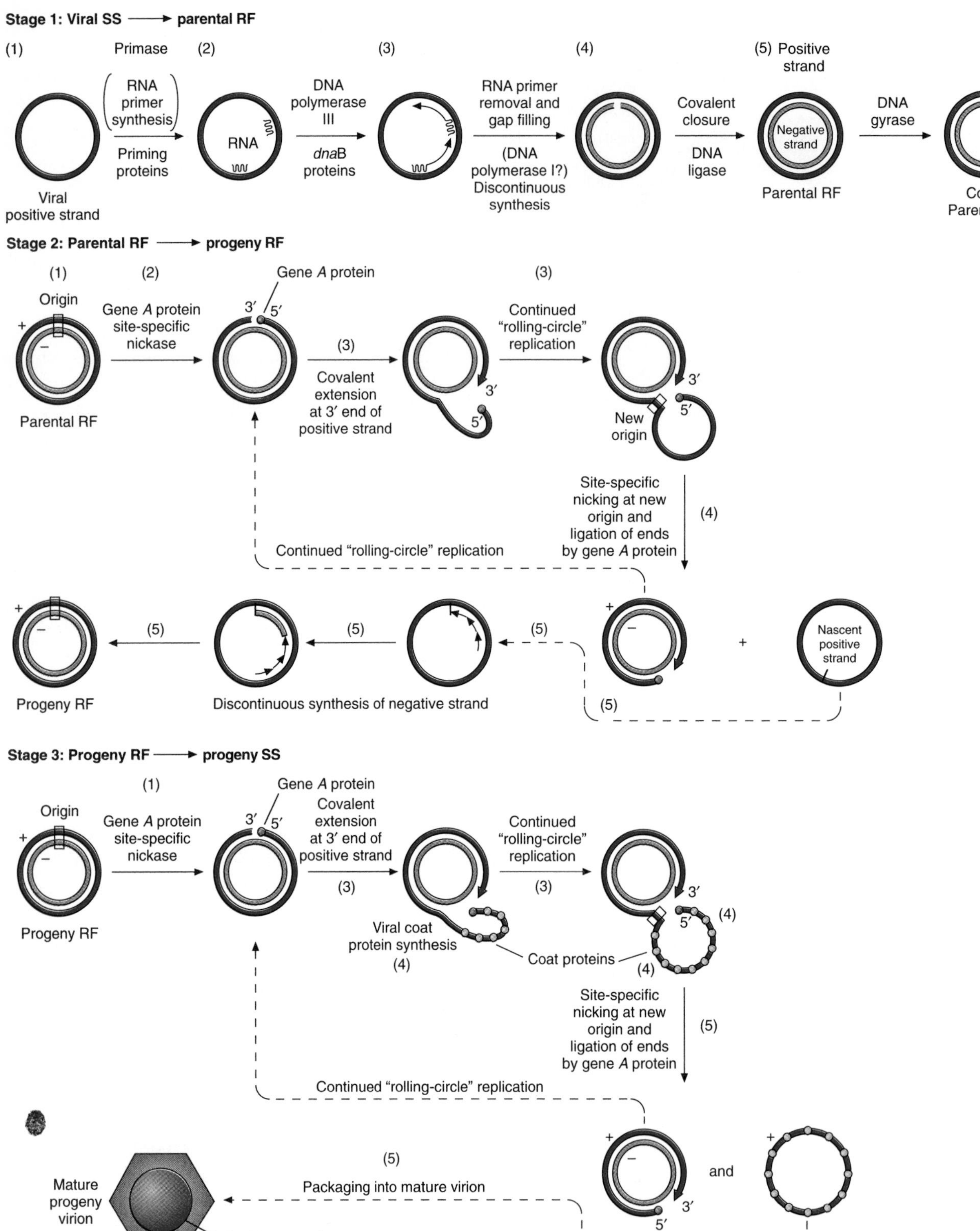


FIGURE 3.2
Duplication of the ØX174 genome requires three stages.

single-stranded, circular (+) molecule using the rolling-circle mechanism (Figure 3.2).

DNA helicase, gpA, SSB proteins, and DNA polymerase III are all that are needed to synthesize viral (+) circles of DNA. The process is as follows:

1. DNA helicase and SSB proteins attach to the duplex, unwinding the parental RF and stabilizing the separated strands.
2. gpA protein nicks the (+) strand at a specific site and remains attached to the 5′ end of the (+) strand produced by the nick.
3. The 5′ end of the (+) strand is displaced from the duplex as DNA polymerase III adds deoxribonucleotides at the 3′ end of the nick using the (–) strand as the template. Synthesis of this (+) strand is continuous and moves around the (–) strand template while the 5′ end curls back on itself. As more DNA is made, more of the original (+) strand DNA is peeled from the duplex.
4. Once the nick-specific site has been reached and duplicated, gpA protein cleaves the DNA at this point and simultaneously closes the circle of the peeled-away DNA by forming a phosphodiester bond. In effect, what has happened is that the synthesis of new DNA along the template (–) strand causes the original (+) strand to be peeled away. In other words, in the duplex DNA, or RF, a new (+) strand displaces the older (+) strand.
5. The displaced (+) strand is now used as the template to produce a (–) strand via discontinuous DNA synthesis, as happened just after the viral DNA was injected into the cell (stage 1 is repeated).

In this way, the original RF unit is replicated, perhaps 35 times, to produce a population of these duplexes from which (+) strands can be made. Parental RF → progeny RF synthesis may require up to 20 minutes. At this point, 20–25 minutes after infection, both RF multiplication and host cell DNA synthesis have stopped.

The last stage in the replication of ØX174, progeny RF → progeny SS, is the synthesis of the (+) strands alone. Using the population of duplex (–) strands as the templates and the rolling-circle mechanism, as in the parental RF → progeny RF second stage, new (+) strands are made. This is leading strand, or continuous, DNA synthesis. The newly produced (+) strands are then packaged into progeny virus particles.

The steps needed for progeny RF → progeny SS are these:

1. gpA protein, assisted by the primosome, which is still attached to the duplex, again nicks the (+) strand at the specific site to produce 3′ and 5′ ends. gpA protein remains attached to the (+) strand's 5′ end.
2. DNA helicase, the host enzyme, attaches to the (–) strand at the origin, or nick site. Along with the primosome and SSB proteins it begins to unwind the duplex as the primosome synthesizes primer RNA beginning at the 3′ end of the template (–) strand.
3. When primer RNA synthesis is completed, DNA polymerase III adds deoxyribonucleotides to the primer.
4. As the (+) strand is stripped away from the duplex in rolling-circle replication, viral coat proteins, produced by the host cell's protein-synthesizing machinery, begin to attach to the newly released DNA. This attachment of proteins prevents the (+) strands from acting as the templates for (–) strand synthesis.
5. gpA protein finally cleaves the stripped-away (+) DNA strand at the newly made origin site and closes the circle with a phosphodiester bond. Last, at this stage, more coat proteins are added to the (+) strand to complete the phage particle.

This stage of replication, progeny RF → progeny SS, may require something like 20–30 minutes. In the end, a single ØX174 particle infecting a single *E. coli* cell will give rise to 35 duplexes (RFs) and 500 SS progeny viruses in about one hour. And, since none of the parental DNA shows up in the progeny viruses, this is a conservative method of DNA synthesis.

## TOBACCO MOSAIC VIRUS (TMV)

Tobacco mosaic virus (TMV) is of interest because it is the first known example of the self-assembly of a biological structure. And it is of historical significance because it was the first virus to be crystallized (Stanley 1935).

TMV is a plant virus whose genome is a single-stranded RNA molecule consisting of 6390 nucleotides. The small genome of this organism contains information encoded in four genes. The RNA is wound within a protein coat of about 2130 identical protein subunits. The RNA and proteins are arranged in a right-handed helix. Each protein subunit binds three nucleotides of the RNA.

The replication of this RNA virus (see Figure 3.3) requires the synthesis of a complementary strand, as in the case of ØX174, and the participation of specific enzymes of the class RNA-directed RNA polymerase, which are often referred to as **replicases.** The information for the enzyme is encoded in the viral genome.

Synthesis of RNA, like that of DNA, is always in the 5′ to 3′ direction. So synthesis of viral RNA begins at the 3′ end of the parental template. Interestingly, there are no error-correcting mechanisms for viral single-stranded RNA synthesis; this fact may explain their

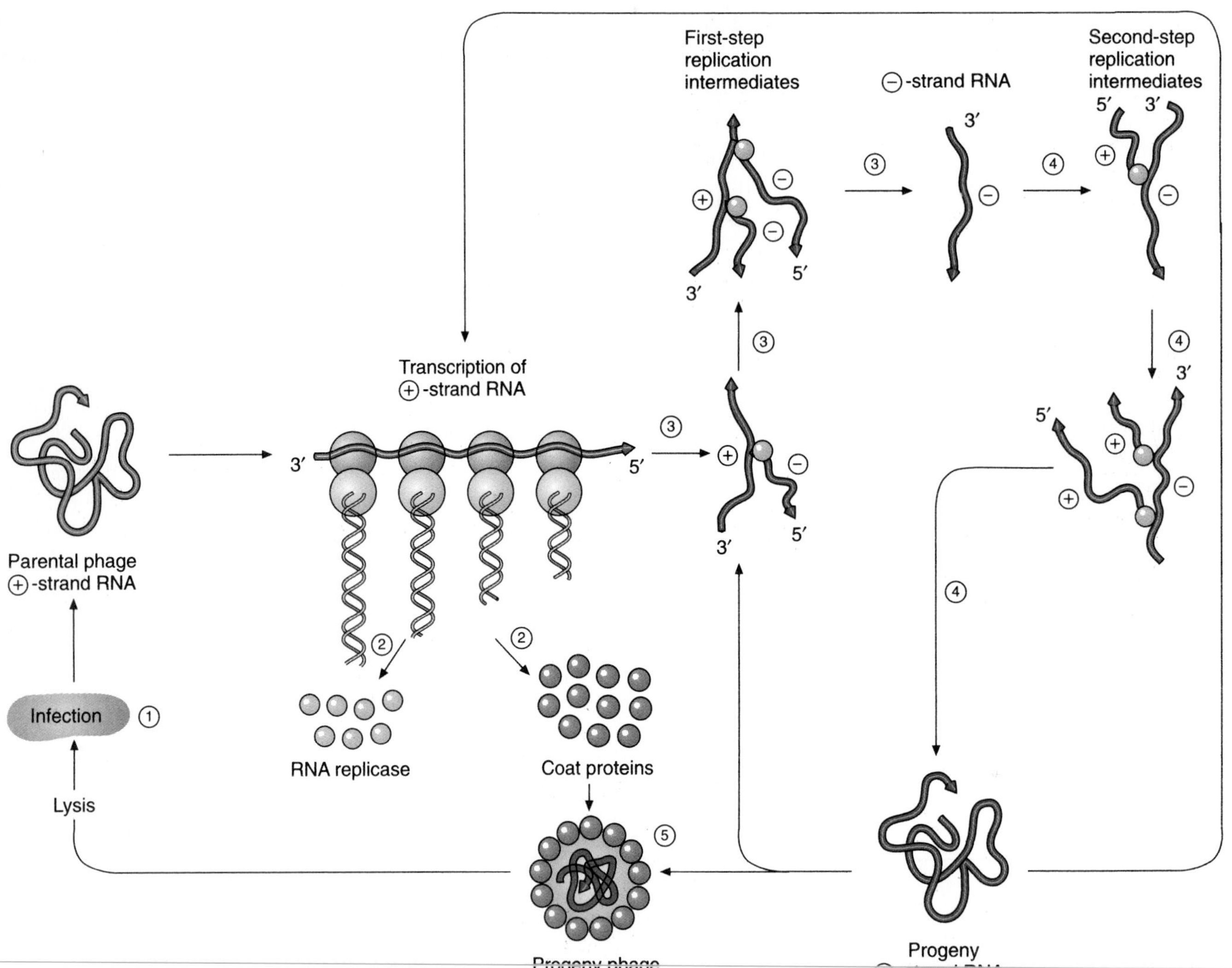


FIGURE 3.3
The steps of TMV replication following infection of a tobacco leaf and release of the TMV (+) strand. Numbers in the diagram refer to steps outlined in the text. Note that in the first step (+) strands serve as templates for the synthesis of (–) strands; in the second step (–) strands serve as templates for the synthesis of (+) strands.

rapid rate of mutation. (Keep this in mind when we discuss HIV replication in the next section.)

Once infection of a tobacco leaf cell has taken place and the (+) strand of TMV is released, replication can begin. Remember, this is RNA synthesis taking place, so ribonucleotides, *not* deoxyribonucleotides, are used. But the ribonucleotides must still be in the triphosphate form so that they can also act as an energy source.

The steps in TMV replication are (Figure 3.3):

1. A tobacco plant cell is infected.
2. Direct translation of the viral RNA results in the synthesis of a number of copies of replicase and some coat proteins.
3. Replicase produces complementary strands (–) of RNA using the parent (+) strand as the template. Synthesis begins at the 3′ end of the template, so the new strand (–) is made in a 5′ to 3′ direction. No duplex RNAs are formed.
4. The (–) strands are then used by the replicases as the templates for the synthesis of (+) strands. Synthesis now begins at the 3′ end of the (–) strands so that the new (+) strands are produced in the 5′ to 3′ direction. These (+) strands have a ribonucleotide sequence, absent mistakes, identical to the parent-infecting RNA.
5. The early coat proteins (step 2) form what is called a protein disc. This disc is thought to recognize a particular segment of the (+) strand RNA molecule, which has folded back on itself. As the 3′ end elongates, more proteins are added to the top of the disc at the fold in the molecule; as more protein is added to the disc, the 5′ end of the RNA is pulled up through the growing

coat. So when the virus is completed, the 3′ terminus is at one end and the 5′ terminus is at the other end.

The completed virus is a helical array of proteins surrounding the RNA genome. The virus can now burst the plant cell and infect adjacent cells to repeat the process (steps 1–5).

## A RETROVIRUS: HIV

**Retroviruses** are a group of animal viruses named for their "backward" (*retro*) way of replicating their nucleic acids. They were given this name because they violate the central dogma of molecular biology (Chapter 1), which says that information flows from DNA to RNA and not the other way round.

One of these retroviruses is the human immunodeficiency virus, or HIV. HIV attacks the CD4 helper lymphocytes that are crucial in the stimulation of an immune response. The result of HIV infection is very often the disease AIDS (acquired immunodeficiency syndrome). This virus attacks the very cells whose function it is to assist in mounting a defense against viruses and other substances foreign to the host (Chapter 10). Other retroviruses were the first viruses shown to cause cancer.

Since the immune system is, in effect, the court of last resort where body defenses are concerned, its destruction or serious impairment can have serious, and sometimes lethal, consequences. In the United States, the Centers for Disease Control and Prevention estimates that the deaths each year from AIDS number in the thousands, and worldwide the disease claims tens of thousands of lives. The rate at which new infections occur worldwide is in the hundreds of thousands.

Like all retroviruses, HIV contains linear, single-stranded RNA as its genome. In fact, each HIV particle contains two identical copies of the RNA bound together at their 5′ ends by a molecule of cellular transfer RNA (tRNA; see Chapter 8). Figure 3.4a is a representation of the structure of HIV.

The viral genome has three genes (Figure 3.4b) whose products and functions are known: *gag* codes for internal structural proteins, *env* codes for envelope proteins, and *pol* codes for two enzymes critical for replication. The two critical enzymes are reverse transcriptase (discovered independently by David Baltimore and Howard M. Temin in 1970) and integrase. There are about half a dozen other genes whose roles in the virus's life cycle are not well understood.

Each of the RNA molecules has associated with it a copy of the enzyme reverse transcriptase. This enzyme is so named because it uses the genomic RNA as the template to synthesize DNA. (Remember, in the cell, information flow is from DNA to RNA, so this is a reversal of the flow of information.) The synthesis of DNA from RNA is the first step in HIV replication. (Incidentally, the drug AZT—a thymine analog used to treat AIDS patients—interferes with the synthesis of this DNA.)

Replication of HIV requires six steps (Figure 3.5):

1. Virus attaches to receptor sites on the CD4 lymphocytes.
2. Single-stranded DNA is synthesized by reverse transcriptase, using one or the other of the single-stranded RNA molecules as the template. The linear single-stranded DNA is then converted to a linear double helix by the same enzyme.
3. The enzyme integrase then incorporates the DNA duplex into the host cell chromosome, apparently into any chromosome. At this point, the virus is referred to as a **provirus.** (Keep in mind, HIV's genome is RNA *not* DNA.)
4. Viral RNA is produced by cellular RNA polymerases using the provirus DNA as the template. These RNA molecules serve both as the source of information for the synthesis of viral structural proteins and enzymes and as the viral progeny genomes.
5. Viral progeny RNA is encapsulated into the core proteins and core shell proteins (see Figure 3.4).
6. Viral particles are then budded through the cell's plasma membrane, where the virus acquires its lipid envelope (see Figure 3.4).

Synthesis of the duplex DNA occurs in the cell's cytoplasm. The DNA molecule is transported from the cytoplasm to the cell's nucleus, where integration into the cell's chromosome takes place. It is also in the cell's nucleus that viral progeny RNA synthesis occurs. The viral genome is then transported to the cytoplasm, where the cell's protein-synthesizing machinery is located and where viral proteins are made in accordance with the instructions supplied by the viral RNA.

The unique feature of the retroviruses is their ability to produce DNA from RNA. They are the only known organisms capable of doing so. The enzyme **reverse transcriptase,** which brings about this conversion, is really an **RNA-directed DNA polymerase** with three enzyme activities (remember, DNA polymerase I also has three enzyme activities):

1. Reverse transcriptase produces a single-stranded DNA molecule using RNA as the template. This is an RNA-directed DNA polymerase activity.
2. The enzyme synthesizes the second strand of DNA to form a double-stranded DNA molecule. This is a DNA-directed DNA polymerase activity.
3. The molecule has what is termed ribonuclease H activity: as we have seen, DNA synthesis requires an RNA primer. In this case, the primer is the cellular tRNA bound to the viral genomic RNA. Once DNA

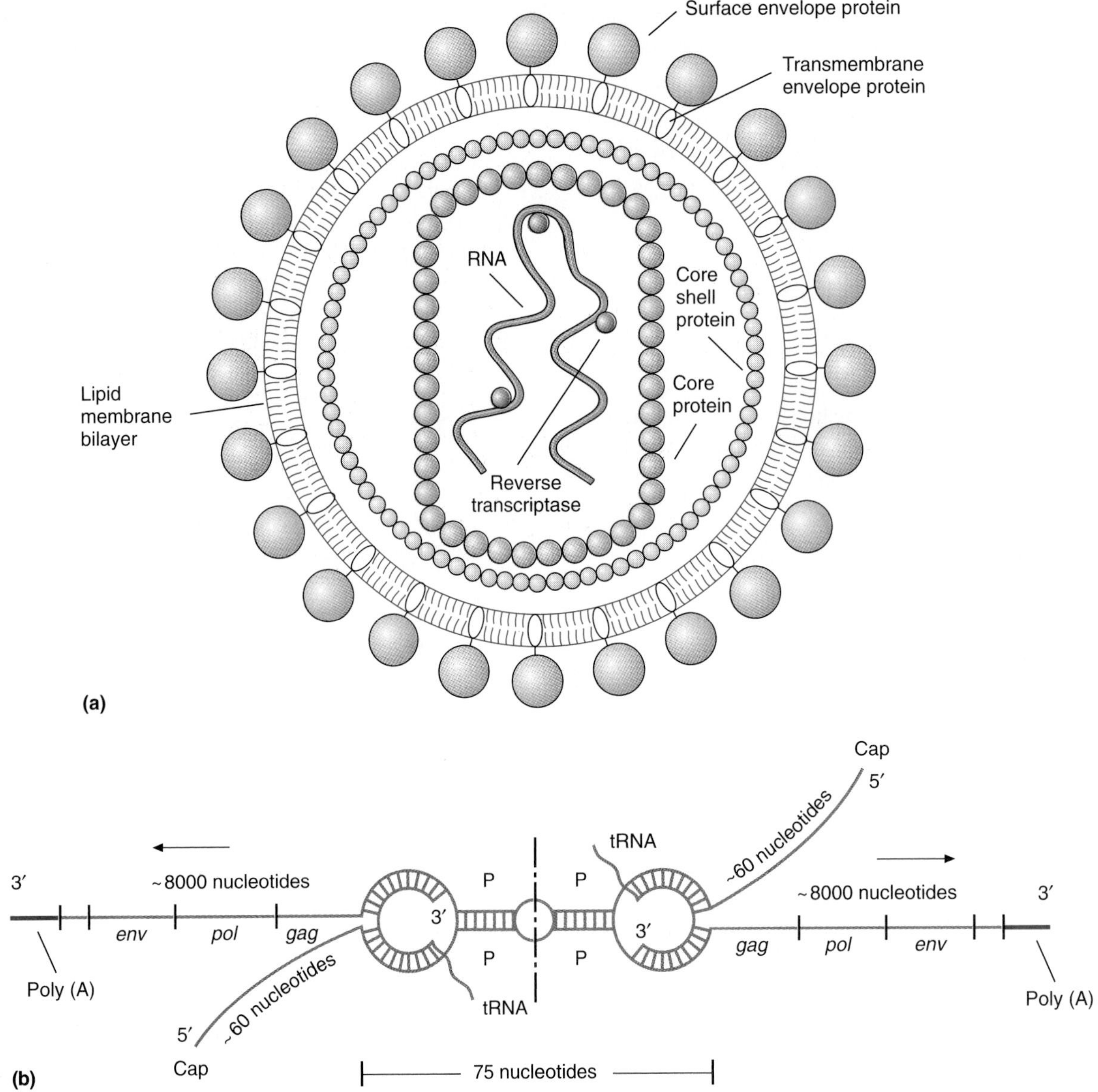


FIGURE 3.4
Retrovirus structure and function. (a) Structure of a retrovirus. (b) Structure of a retrovirus genome, showing the manner by which the two identical viral RNA molecules are held together.

synthesis begins and the primer is no longer needed, it is removed by ribonuclease H activity. The letter H is from the word *hybrid*. The hybrid is the RNA-DNA formed by reverse transcriptase.

The retrovirus group includes, besides HIV, the Rous sarcoma virus, feline leukemia virus, and mouse mammary tumor virus. All these organisms have the ability to maintain themselves in the provirus state almost indefinitely. In this state, no disease symptoms are obvious or clinically manifested. The viral DNA is replicated along with the host cell's DNA before cell division. That is why diseases such as AIDS have long latent periods.

An intriguing and critical question, therefore, is: What triggers the synthesis of viral RNA? In other words, what signal does HIV, for example, recognize and respond to when it is in the provirus state? What causes the decoding of the HIV DNA "genes" to begin viral replication and release from the cell so that the virus can go on to infect other lymphocytes, thus bringing about AIDS?

Viral reproduction is not an inevitable consequence of infection. How this quiescent state is brought about and maintained has been studied most intensively in the prokaryotic *E. coli* infected by lambda phage, our next example of nucleic acid synthesis.

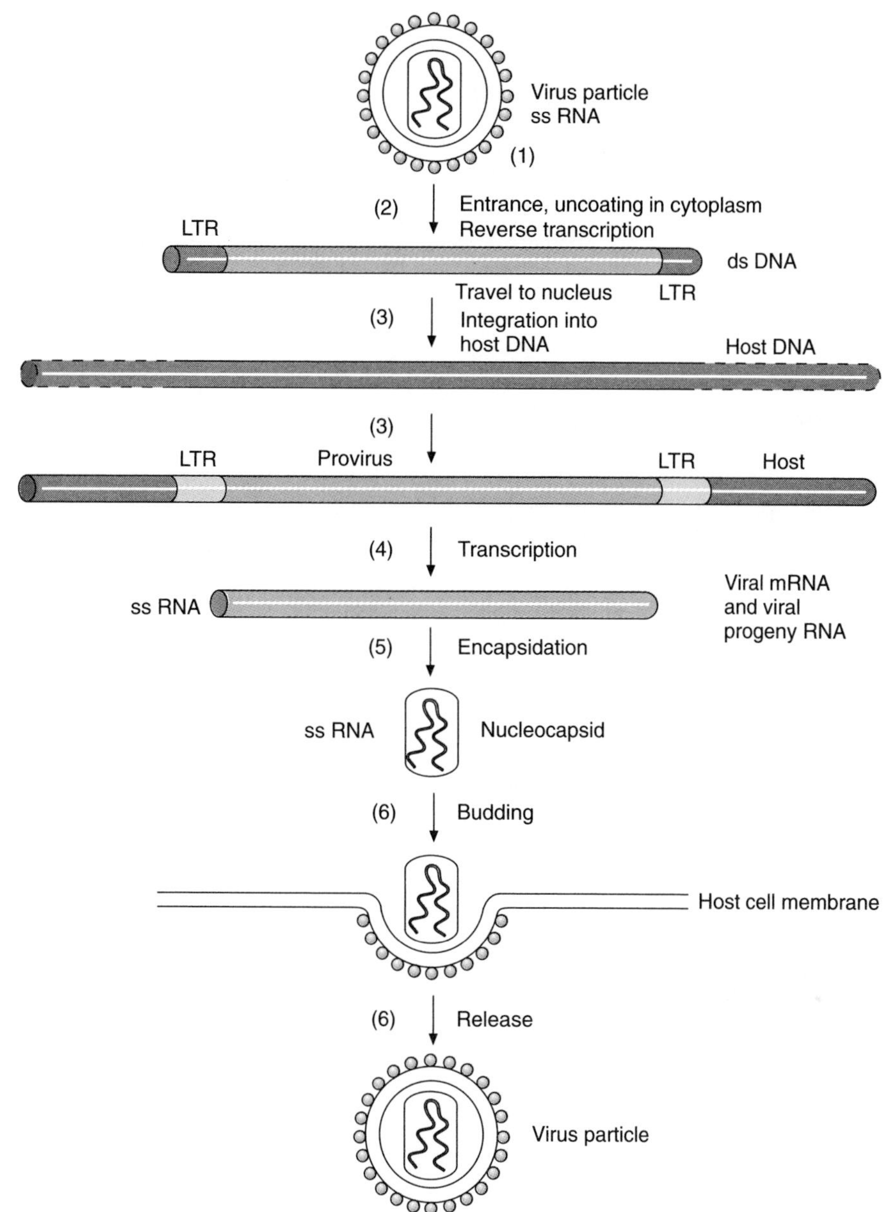


FIGURE 3.5
Replication process of retrovirus. LTR = long terminal repeats.

## LAMBDA PHAGE AND LYSOGENY

With some exceptions, bacteriophage and animal and plant viruses have a lytic life cycle only. That is, after reproducing themselves, most viruses kill the host cell in the process of being released from the cell. These viruses are often described as virulent viruses.

There is, for some viruses (lambda and HIV, for example), another possibility. These organisms may integrate their genetic material (or a substitute such as HIV's DNA) into the host cell's genome. The virus may then remain there, being passed from one generation of cells to the next as the cell reproduces, until some event, mostly unknown at present, causes the viral nucleic acid to begin reproducing itself.

Bacterial viruses incorporated into the host cell genome are called **prophage;** this stage is analogous to animal virus's provirus state. Bacterial viruses capable of becoming prophage are referred to as **temperate viruses** or **temperate phage.** The bacteria containing prophage are said to be in a state of lysogeny. The bacterial cells are **lysogenic:** capable of being lysed.

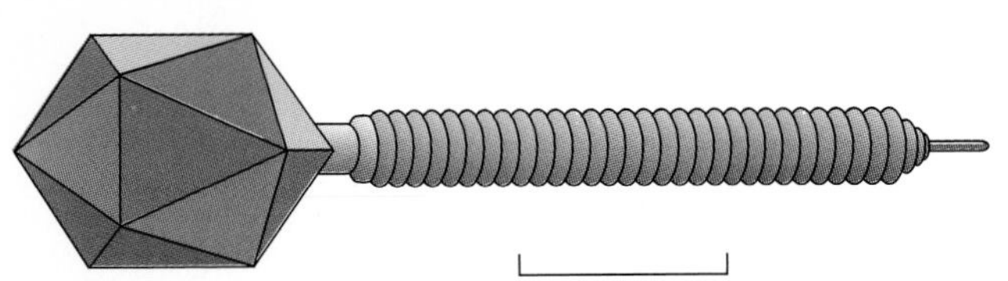


FIGURE 3.6
A sketch of bacteriophage λ.

What has intrigued scientists for years is the control mechanisms by which phages such as lambda, and by extension animal viruses, initiate and maintain the prophage state. The importance of understanding these mechanisms goes way beyond just an interest in viral reproduction. For example, in lambda phage infection of *E. coli,* we have one of the best examples of a genetic regulatory system. The information obtained studying this phage has contributed in no small measure to our understanding of the control of development in higher organisms.

Lambda phage is a medium-sized virus consisting of an icosahedral head and a flexible tail that ends in a single thin fiber (Figure 3.6). The organism's DNA is double-stranded and linear, has 48,502 base pairs and about 50 genes. The linear DNA molecule has complementary single-stranded 5′ ends of about 12 nucleotides. These ends are referred to as **cohesive ends,** or **cos sites** (more on these ends in Chapter 11). After injection of the viral linear DNA into *E. coli,* the molecule is circularized at its 5′ cos sites, and the enzyme DNA ligase covalently closes the circle by forming phosphodiester bonds. DNA gyrase then supercoils the molecule.

Because of the extreme complexity of the control systems of lambda phage reproduction—particularly control of the prophage state—no attempt at a comprehensive explanation will be given here. For our purposes, an overview is all that is required.

Upon infection of *E. coli,* lambda may follow one of two pathways. One pathway results in production of progeny phage and lysis of the host cell. The other pathway results in lambda's becoming a prophage, in which case the cell enters the lysogenic state.

Which of these two possibilities occurs depends on a number of factors, most prominent of which are the concentration of nutrients in the medium and the degree of multiplicity of infection—i.e., the degree to which the number of phage exceeds the number of bacteria in the culture.

When the supply of nutrients is low, bacterial metabolism is affected to the point where the cells may begin to degrade their own proteins and RNAs. The phage's fate is obviously dependent on an actively metabolizing cell since the viruses are obligate intracellular parasites. Should the phage life cycle be interrupted owing to a lack of nutrients, the virus is doomed and the cell dies.

In the case of a high multiplicity of infection, absent lysogeny, all bacteria in a population would be infected and eventually destroyed as the phage are released from the cells. Should a lysogenic state develop (the virus becomes a prophage), however, the bacterial cell cannot be reinfected by a phage of the same type—the cell is said to be immune. But even after many cell generations, a prophage can initiate a lytic cycle; this process is called **induction.** Let's now look at the lytic cycle and lysogeny.

## The Lytic Cycle

In the lytic cycle, viral DNA replication proceeds via the rolling-circle and theta methods (Chapter 2) and is at first dependent on host RNA polymerases, which are used to decode viral genes. The lytic process follows these steps (Figure 3.7):

1. Injection of viral linear DNA into the host cell.
2. Circularization of the viral DNA by base pairing of the complementary 5′ cos sites and closure of the circle by phosphodiester bond formation (Figure 3.8).
3. Using host cell enzymes, DNA replication at first follows the bidirectional theta model and later switches to the rolling-circle mechanism exclusively (see Figure 3.2). The theta mode increases the number of DNA templates available for the decoding of genes and for further replication. The rolling-circle mode produces the DNA for progeny phage. In either case, double-stranded DNA is produced. In the rolling-circle method, the DNA being peeled from the template is cleaved at the cos sites and then packaged into phage heads. The tails are assembled and attached to the heads. But unlike TMV, these viruses are not self-assembling; some host proteins are required.
4. By enzymatic action, the host cell is lysed and phage particles are released. A single phage infecting a cell results in the production of approximately 100 progeny.

## Lysogeny

Given a low supply of nutrients or a high multiplicity of infection or both, it would be to the phage's advantage to lie dormant until conditions improve. The dormant state for the virus is the prophage state in which the cell becomes lysogenic. Viral DNA is integrated into the host's chromosome, and, at the same time, those viral genes needed for viral reproduction and cell lysis are repressed.

Insertion of lambda DNA into host DNA is site-specific and occurs between the *E. coli* genes for galactose metabolism and those for biotin synthesis. The site is called the lambda attachment site and has been given the genetic designation *att*B. There is also a specific site on the viral genome designated *att*P (Figure 3.8).

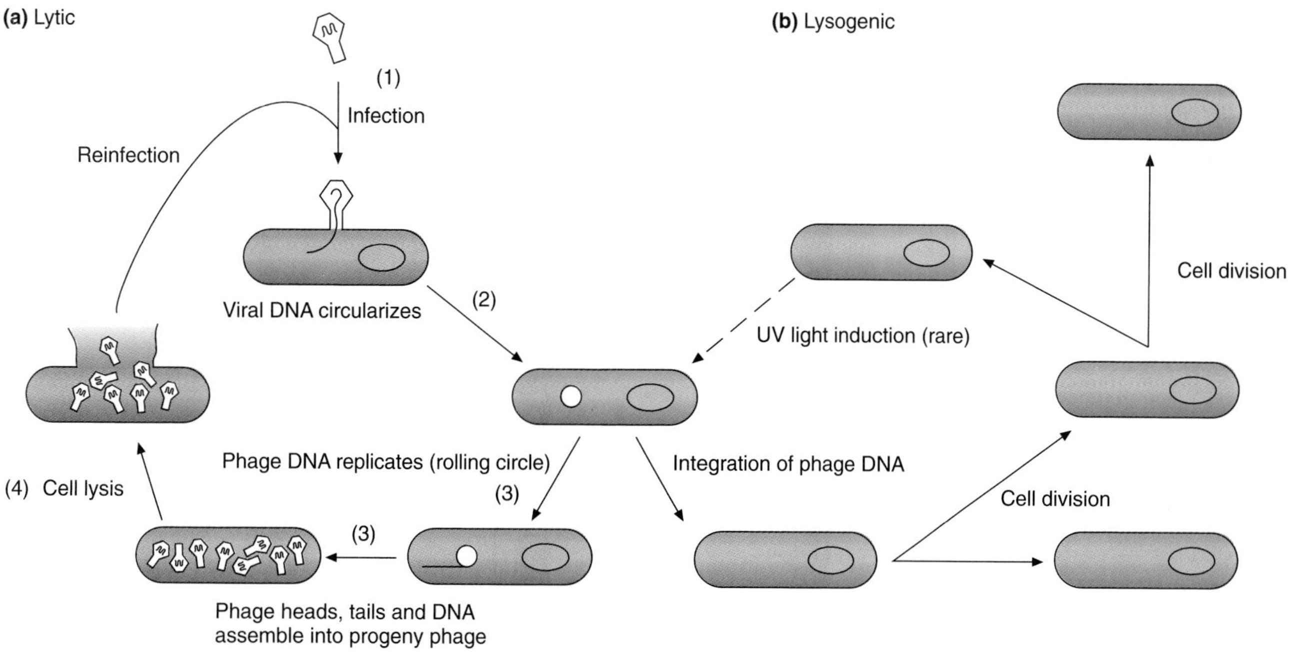


FIGURE 3.7
Lytic versus lysogenic infection by phage λ. (a) Lytic infection; (b) lysogenic infection.

The insertion process is referred to as integration. This process requires a phage-coded enzyme, integrase, acting in combination with a host protein called integration host factor. Although integration is not an energy-requiring process, it is not readily reversible. To reverse the process requires another enzyme called **excisionase,** along with integrase and integration host factor.

The choice between lysogeny and lytic reproduction by lambda is critically dependent on the concentration of a very unstable gene product, a protein called cII. This protein stimulates the production of two other proteins: One is cI protein (also called the repressor protein, or simply the repressor), which shuts down the expression of practically all phage genes, thereby limiting the gene products needed for reproduction and the enzyme integrase. The second is Cro protein, which stimulates RNA polymerase to decode the lytic reproduction genes and prevents RNA polymerase from decoding the repressor gene.

Figure 3.9 is a simplified representation of how the molecular switch for lysogeny and lytic reproduction works.

Three protein-binding sites are located at the region of lambda's DNA called the right operator, which separates the genes needed for lysogeny from those needed for lytic reproduction ($O_R1$, $O_R2$, $O_R3$ in Figure 3.9). To the right of these binding sites are the genes for virus lytic reproduction. To the left of the sites are the genes for cI repressor protein and prophage production—lysogeny.

Upon infection of a cell, and if conditions are conducive (nutrient supplies are low), cII protein and then cI protein and integrase are produced. Two molecules of the repressor form a dimer and bind to site 1. Blockage at this site does two things: (1) it prevents RNA polymerase from decoding the lytic reproduction genes to the right of site 1, and (2) it facilitates the binding of repressor protein dimer to site 2. Repressor binding at site 2 then stimulates RNA polymerase to bind at site 3 and to move to the left to decode repressor protein genes. In other words, repressor protein (cI) stimulates its own production. Repression of lytic reproduction proteins brings about the prophage state and lysogeny, the incorporation by integrase of lambda DNA into the host DNA.

As mentioned above, cII is a very unstable protein. It is preferentially degraded by host enzymes. The cII protein is, however, needed to begin production of the repressor protein. Once this production begins, cII is no longer needed since cI can stimulate its own synthesis. Repressor protein, then, is both a positive and a negative control molecule: it stimulates its own production and inhibits production of proteins needed for lytic reproduction of the phage.

Having introduced the cII protein, we can now explain how a poor nutritional state and a high multiplicity of infection bring about prophage development and lysogeny. In a poor nutritional state, *E. coli* produces quantities of a signal molecule, cyclic adenosine monophosphate, or cAMP (Chapter 7). The synthesis of one of the enzymes that degrades cII is cAMP-dependent. It appears that at high cAMP levels the synthesis of the degrading enzyme is inhibited. So as the cell's nutritional state worsens, more and more cAMP is produced and less enzyme is

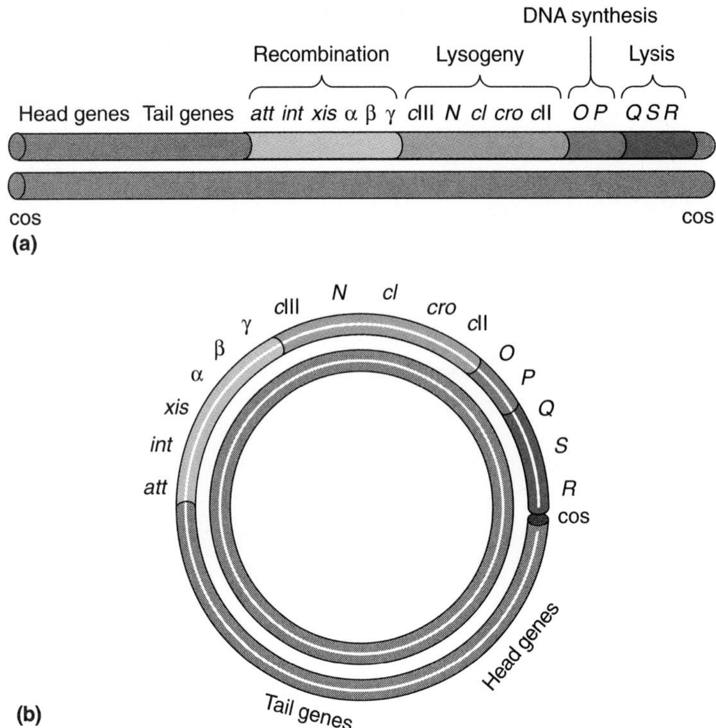


FIGURE 3.8
Genetic map of phage λ. (a) The map is shown in linear form, as the DNA exists in the phage particles; the cohesive ends (cos) are at the ends of the map. The genes are grouped primarily according to function. (b) The map is shown in circular form, as it exists in the host cell during a lytic infection after annealing of the cohesive ends.

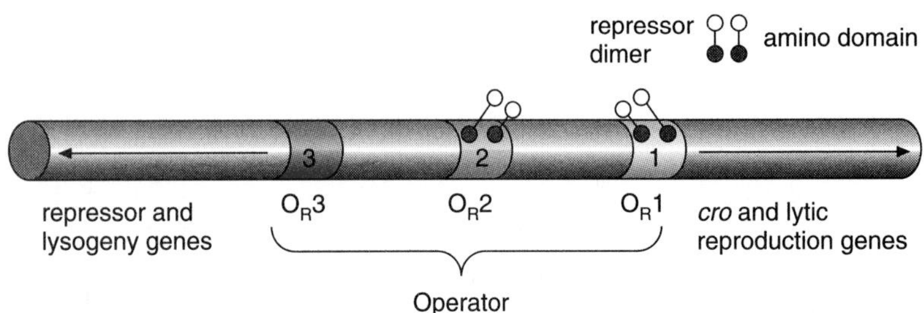


FIGURE 3.9
Separation of genes for lysogeny and lytic reproduction by the lambda right operator. Repressor binds $O_R1$ 10 times tighter than either $O_R2$ or $O_R3$. Binding of $O_R1$ stimulates the binding of $O_R2$. When both $O_R1$ and $O_R2$ are bound, transcription of *cro* and reproduction genes is blocked. RMA polymerase cannot bind to DNA. But RNA polymerase binding at $O_R3$-promoter is stimulated, and transcription of repressor and lysogeny genes occurs.

available to degrade cII. Thus the lysogenic state is initiated and maintained via the synthesis of cI repressor protein as described above.

Conditions of high multiplicity of infection allow for many viral cII genes to be decoded since there are more molecules of viral DNA in the host cell. Large quantities of the cII protein are thus produced before the genes for lytic reproduction can be activated. These genes are inactivated by the cII stimulation of the cI gene and the resulting production of repressor protein.

Should the lysogenic cell be subjected to conditions that result in DNA damage or inhibition of DNA synthesis (for example, by UV or x-irradiation, thymine starvation, or the presence of nitrogen mustard) phage induction will occur; a lytic cycle of reproduction will begin. For this to happen, however, requires inactivation of the repressor protein (Figure 3.10).

Host DNA damage initiates what is called the SOS response (Chapter 9). This response results in the decoding of a number of host genes. One of these genes codes for the en-

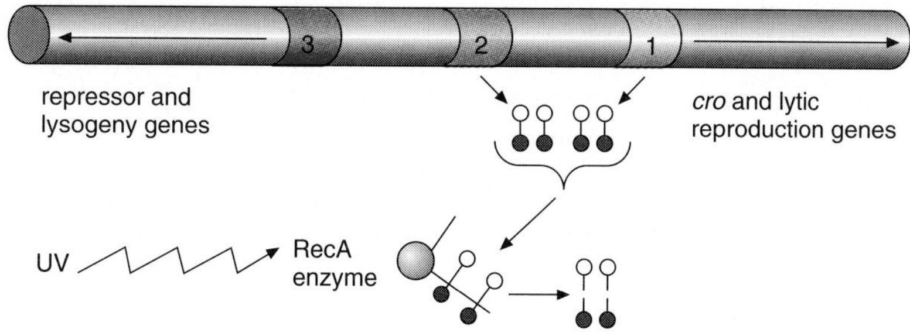


FIGURE 3.10
UV light stimulates the SOS response and RecA production. RecA leaves the repressor timers, freeing $O_R1$ and $O_R2$, allowing RNA polymerase to transcribe the *cro* and lytic reproduction genes.

zyme RecA, which destroys lambda repressor protein. This destruction inactivates the repressor gene, cI (remember, the repressor protein stimulates its own synthesis) and activates a new set of genes. Among these genes is one called the *cro* gene (exactly what the letters *cro* stand for is lost; some have suggested they stand for "control of repressor and other things"). The product of this *cro* gene is a protein that is also a positive and a negative control molecule. The *cro* protein stimulates the decoding of the genes needed for lytic reproduction (genes to the right of the right operator) and suppresses the decoding of the lambda repressor gene (to the left of the right operator). *Cro* frees sites 1 and 2, allowing RNA polymerase to decode the lytic reproduction genes, and it binds to site 3, preventing RNA polymerase from decoding the repressor gene.

Under these conditions, the prophage dissociates from the bacterial DNA, using its integrase and excisionase enzymes along with integration host factor, and freely replicates itself. Upon lysis of the host cell, the new phage particles infect new cells, and once again the decision to integrate or replicate must be made.

A study of the natural way by which lambda phage is released from *E. coli*'s chromosome may shed some light on the mechanisms by which other viruses, particularly animal viruses, are released from chromosomes. This knowledge is sure to lead to a better understanding of how viruses such as HIV are released from their host cell chromosomes. In the case of HIV, this release results in the death of the CD4 lymphocytes and the onset of AIDS.

## SUMMARY

Unlike cellular life forms, viruses may have as genetic material types and forms of nucleic acids other than double-stranded DNA, such as single-stranded DNA or single- or double-stranded RNA. Because viruses lack the necessary enzyme systems to harness energy or to synthesize nucleic acids or proteins, they are obligate intracellular parasites. The problem for viruses, however, is that cells are designed to replicate only double-stranded DNA and no RNA.

Viruses, therefore, have had to evolve strategies to take advantage of the cell's energy-harnessing abilities and biosynthesizing systems. As a group, these organisms have developed a number of ways to circumvent their shortcomings and the cell's limited capabilities.

Although all cells synthesize DNA using the semiconservative method, single-stranded viruses such as ØX174 use a conservative method of replication in which parental strands of DNA do not appear in the progeny. This virus produces an intermediate double-stranded replicative form during reproduction from which, using a rolling-circle mechanism, the original nucleotide sequence is reproduced. As the template, the (–) strand of the double-stranded replicative form is used. The virus can make as many as 500 copies of itself in about 1 hour using a combination of viral and cellular proteins and enzymes.

Tobacco mosaic virus has as its genetic material a linear single-stranded RNA molecule. Upon infecting a tobacco plant cell, TMV's RNA, (+) strand, is used as the template to synthesize a number of (–) strands by the viral encoded enzyme replicase. However, unlike ØX174, it makes no intermediate duplex or replicative form. Instead, replicase uses the (–) strands as templates to produce the parental nucleotide sequence in a (+) strand. The newly formed linear RNA molecules are then packaged into protein coats and released from the cell.

The retroviruses, animal viruses such as HIV, have devised yet another method of replicating their RNA in cells. These organisms use a viral enzyme, reverse transcriptase, to convert their single-stranded RNA genome into double-stranded DNA. This DNA molecule is then incorporated into the host's chromosome, using the enzyme integrase, which is coded for by the viral RNA.

HIV is now a provirus. As the host replicates its chromosomes before mitosis and cell division, the provirus is also replicated. New viral particles are produced when this new DNA, the provirus, is transcribed into (+) strands of viral RNA.

Among the bacterial viruses are some that are capable of both a lytic cycle, in which viral reproduction and destruction of the host cell occur, and a nonlytic phase, in which neither reproduction nor cell destruction occurs. Instead, in the latter case, much like HIV, these bacterial viruses can become a part of the host cell's chromosome, existing as prophage. During this time, the cell is said to be lysogenic. One of these viruses is lambda, a phage that infects *E. coli*. In a complex series of events, lambda's linear double-stranded DNA molecule is first circularized upon entering the cell, and then the organism enters either a lytic or a nonlytic cycle.

If a lytic cycle is to occur, lambda will replicate its DNA via the rolling-circle and theta mechanisms, both of which rely on host enzymes. Depending on the host cell's physiological condition or the number of viruses infecting the cell, lambda may, instead, become a prophage. This process of incorporation into the host's chromosome is site-specific for both the host and the virus and requires a phage enzyme, integrase, and a host protein, integration host factor. Then, as in the case of HIV proviruses, when *E. coli* divides, it replicates both its own DNA and the incorporated viral DNA, the prophage.

Should conditions that damage host DNA come about, prophage are released from host chromosomes to begin a lytic cycle of reproduction.

## Study Questions

1. List the stages of viral replication.
2. What is the role of the replicative form (RF) in ØX174 replication?
3. To replicate ØX174, why is the (–) strand required?
4. Why is the (–) strand of viral replication not viral genetic information?
5. Briefly describe the steps needed to synthesize the ØX174 (+) circle of DNA.
6. List the stages of TMV replication.
7. How does TMV replication differ from ØX174 replication?
8. How does the replication of retroviruses, such as HIV, differ from the replication of other RNA viruses and from that of DNA viruses?
9. What do proviruses and prophage have in common?
10. Why are the replications of ØX174, TMV, and HIV considered conservative replication?
11. What does the enzyme reverse transcriptase do, and why is this unusual? What is another name for this enzyme?
12. What are the two pathways lambda phage can take, and how do these pathways differ from one another?
13. Explain the conditions under which *E. coli* might become a lysogenic cell.
14. Briefly describe how lambda incorporates its DNA into the host's DNA.
15. Explain the process by which lambda prophage is released from the host chromosome to begin a lytic cycle.

## Readings and References


Alberts, B., D. Brav, J. Lewis, M. Raff, K. Roberts, and J. D. Watson. 1989. *Molecular biology of the cell*. 2d ed. New York: Garland.

Brock, T. D., and M. T. Madigan. 1991. *Biology of microorganisms*. 6th ed. Englewood Cliffs, N.J.: Prentice Hall.

Fraenkel-Conrat, H., and B. Singer. 1957. Virus reconstitution. II. Combination of protein and nucleic acid from different strains. *Biochim. Biophys. Acta* 24:540–48.

Haseltine, W., and F. Wong-Staal. 1988. The molecular biology of the AIDS virus. *Sci. Am.* 259 (October):52–62.

Kornberg, A., and T. A. Baker. 1992. *DNA Replication*. 2d ed. New York: W.H. Freeman.

Murialdo, H. 1991. Bacteriophage lambda DNA maturation and packaging. *Annu. Rev. Biochem.* 60:125–53.

Ptashne, M. 1992. *A genetic switch*. 2d ed. Cambridge, Mass.: Blackwell Scientific.

Ptashne, M., A. D. Johnson, and C. O. Pabo. 1982. A genetic switch in a bacterial virus. *Sci. Am.* 247:128–40.

Ptashne, M., A. Jeffrey, A. D. Johnson, R. Maurer, B. J. Meyer, C. O. Pabo, T. M. Roberts, and R. T. Sauer. 1980. How lambda repressor and *cro* work. *Cell* 19:1–19.

Stanley, W. M. 1935. *Science* 81:644.

Vaishnav, Y. N., and F. Wong-Staal. 1991. The biochemistry of AIDS. *Annu. Rev. Biochem.* 60:577–630.

Voet, D., and J. G. Voet. 1990. *Biochemistry*. New York: John Wiley & Sons.

CHAPTER FOUR

# THE PROTEINS

## OUTLINE



## CHAPTER OBJECTIVES

*This chapter will discuss:*

- The abundance and importance of proteins in cellular metabolism
- The functional classification of proteins
- Examples of the functions of proteins

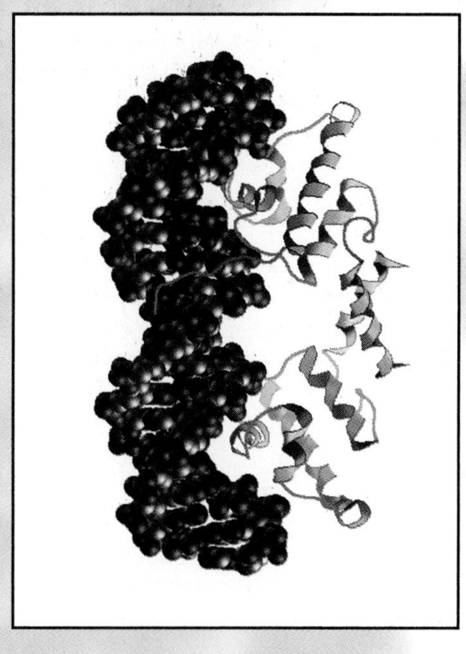

# INTRODUCTION

The DNA of all cells contains a variety of information (Chapter 5). There are control sequences, start and stop signals, attenuators and enhancers, and information for RNAs and proteins.

By far, however, information for the synthesis of proteins accounts for most of the information stored in the genes. Although DNA is the "master molecule" of the cell, it is actually a static molecule. That is, it really does very little apart from storing information. Rather, things are done to it either for its replication (Chapter 2) or for purposes of extracting the information it holds. For most of the cell cycle—that is, during the time when DNA synthesis or cell division is not occurring—extraction of information is a continuous process affecting DNA. Information extraction is absolutely vital to the well-being of the cell. This information retrieval is primarily for the purpose of protein synthesis.

It is in the form of proteins that most of the information contained in genes is manifested in the cell. This point cannot be overemphasized. *Genes encode the information that determines the structure of proteins. In turn, the protein's structure determines its biological activity. And through the biological activity of proteins the cell's physiological role is determined.*

Any changes in DNA, such as mutations (Chapter 9), affect the cell through changes in protein structure. Ultimately, the cell will be affected, because altered protein structure alters protein biological activity. So, the cell's **genotype**—its store of information—determines the cell's **phenotype**—the expression of the genotype via proteins.

# FUNCTIONAL GROUPINGS OF PROTEINS

Proteins are far and away the most abundant organic molecules in the cell; they account for more than 50% of the dry weight (excluding water) of the cell. Why are the proteins so abundant and vital to the cell, and how did they come to attain this importance? This chapter is not intended as anywhere near an exhaustive examination of the roles proteins play in the life of the cell. Rather, it is intended to give some idea of the great variety of proteins and functions in the physiological activities of the cell.

Each cell type in a multicellular organism such as a human being will have a particular part to play in maintaining the functioning and well-being of the whole organism. The part played by any cell type is primarily determined by its complement of proteins. And the proteins present in any given cell type are dependent on which genes of the cell are active and which are not.

Proteins are crucial to the cell because they are the cell's workhorses. They serve in a number of capacities, and for each capacity, for example enzymes, there are literally dozens and dozens of different protein molecules, each with a unique structure and function.

Proteins are capable of assuming so many roles because their structures can vary enormously. The variation in structure results from the variation in the amino acid sequence of the molecule. Two molecules with an identical number of amino acids but with different sequences would have different structures and, therefore, different biological activities. In other words, the variety of protein structures possible is almost limitless and, therefore, the variety of possible functions is correspondingly great. To simplify matters, we can classify proteins by their biological activity or *function*, as in Table 4.1. We can now take a brief look at some examples of the various functional groupings of proteins found among higher organisms such as human beings.

# CARRIER PROTEINS

**Carrier proteins** are located in the blood, lymph, and cells. Their role is to move other molecules within and between cells or between the bloodstream and the lymphatic system. So, for example, those proteins listed in Table 4.1 are responsible for the transfer of fatty acids (albumin), oxygen (hemoglobin), cholesterol (lipoproteins), and iron (transferrin).

Hemoglobin is one of the most exhaustively studied proteins. The molecule is actually composed of two parts. The heme portion, which is responsible for the red color of blood, is a porphyrin derivative consisting of four pyrrole rings surrounding an iron atom, to which oxygen binds (Figure 4.1). (This is the same heme found in cytochromes and in myoglobin, the oxygen transporter of muscle). The globin portion is a four-chain protein composed of two identical alpha chains and two identical beta chains. Each protein chain binds one heme molecule.

An important feature of hemoglobin is that it binds four molecules of $O_2$ cooperatively. That is, the binding of one molecule of oxygen facilitates the binding of each successive oxygen molecule. In other words, as oxygen binds, the affinity of the hemoglobin for the next oxygen molecule increases, so that the fourth oxygen binds with an affinity of 100 times that of the first oxygen molecule. The binding of oxygen molecules to hemoglobin causes extensive quaternary structural (Chapter 8) changes in the protein portion of the hemoglobin molecule. These changes are necessary for cooperative binding of oxygen.

The ability of the heme to bind oxygen *reversibly* is a function of the globin protein. The protein not only makes oxygen binding and release possible, but it also makes the cooperative binding possible. In fact, alone, heme cannot

**table 4.1**

**A Classification of Proteins by Biological Activity or Function**

| Type of protein | Biological activity or function | Examples |
|---|---|---|
| Carrier proteins | Move molecules within and between cells or between the bloodstream and the lymphatic system | Albumin; hemoglobin; lipoproteins; transferrin |
| Enzymes | Organic catalysts that increase rate of chemical reactions without themselves being consumed | Alcohol dehydrogenase; hexokinase; hyaluronidase |
| G-proteins | Transduce signals from outside the cell to the inside by stimulating production of second messengers | Transducin; $G_s$; $\alpha_i$ |
| Information molecules | Hormones and some other proteins that, when functioning, cause a particular metabolic or physiological activity in their target cells | Insulin; glucagon; luteinizing hormone; prolactin |
| Muscle proteins | Interaction of actin and myosin causes muscle contraction | Actin; myosin |
| Protection molecules | Protect against microbial invasion and toxins | Antibodies; complement; interferons; interleukins |
| Receptor proteins | Transmembrane proteins that relay information transmitted by hormones and neurotransmitters and that are needed for endocytosis | Insulin and adrenalin receptors on cell surfaces |
| Regulatory proteins | Control gene and cellular activities | Intercellular hormone receptors; repressors and corepressors |
| Structural proteins | Provide shape, internal organization, and movement to cells and their components | Cytochromes; cytoskeleton; histones; ribosomes |
| Miscellaneous proteins | Channel proteins, which allow movement of ions across membranes, and special proteins involved in cell-to-cell communication and transport | Adhesion proteins; chloride ion channel proteins |

bind oxygen reversibly. (Incidentally, CO, NO, and $H_2S$ are toxic because they bind to the heme of hemoglobin, myoglobin, and cytochromes where oxygen normally does but with a greater affinity than does oxygen.)

## ENZYMES

**Enzymes** are organic catalysts. These molecules increase the rate of chemical reactions without themselves being consumed in the process. At the completion of the reaction, the enzyme is regenerated in its original form to begin the reaction over again. Compared with the number of substrate molecules used, the number of enzyme molecules needed is extremely small. All biological reactions in which a molecule is changed from one form to another, from substrate to product, requires a protein enzyme (two exceptions are discussed below). Hemoglobin, therefore, does not qualify as an enzyme—substrate and product are the same, oxygen.

Until recently, all known enzymes were proteins. But, largely as the result of the recent work of Thomas Cech and Sidney Altman, we now have examples of two other types of molecules that have enzyme activity.

Cech has shown that, in the ciliated protozoan *Tetrahymena thermophilia*, RNA can have catalytic properties. In this organism, the newly produced **ribosomal RNA (rRNA)** is autocatalytic (Chapter 10). That is, in the absence of protein, the ribosomal RNA can cut and splice itself, releasing a 413-nucleotide sequence from the RNA in the process. The RNA is said to be self-splicing. The process is a series of three reactions resembling the splicing of messenger RNA (Chapter 10).

In the bacterium *Escherichia coli*, the enzyme RNaseP is composed of a 377-nucleotide RNA component and a much smaller protein subunit. Altman demonstrated that

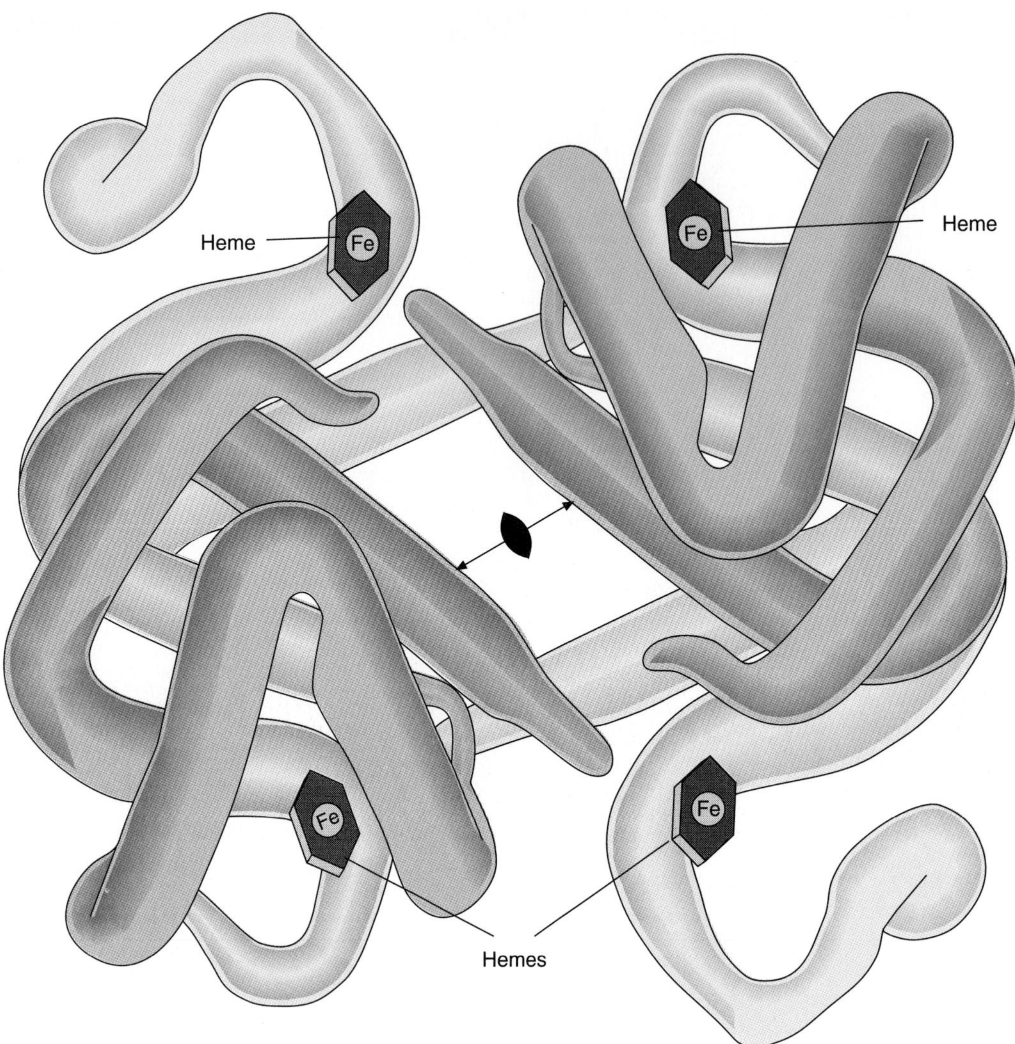


FIGURE 4.1
Hemoglobin. Note the heme molecules, one for each peptide chain.

it is the RNA of this hybrid molecule that is the catalytic site. RNaseP is needed for the processing of both ribosomal and transfer RNAs. All these RNAs with catalytic properties have been given the name **ribozymes.**

Over 1000 protein enzymes are known, and each is capable of carrying out a unique chemical transformation. This is possible because proteins can assume any of thousands of conformations, depending on their amino acid sequences. Their particular structures give proteins unique chemical abilities. Each of the hundreds upon hundreds of reactions required to keep the cells, and thus the body, running smoothly requires a specific enzyme to carry out a specific chemical transformation.

For example, a different enzyme is needed for each of the more than 25 reactions necessary to convert a single molecule of glucose into a usable form of energy—ATP. In some cases, one reaction may require more than one enzyme.

Enzymes are also needed for the synthesis of all other proteins, including other enzymes. They are used to convert vitamins into coenzymes (molecules needed by some enzymes to carry out their reactions). For example, they convert panthothenic acid into coenzyme A, a vital component of the system of enzymes that converts glucose into ATP.

It is an enzyme, alcohol dehydrogenase, that allows us to detoxify alcohol so that we can enjoy a cold beer or glass of wine without becoming staggeringly drunk. Another enzyme, called hyaluronidase, stored in the acrosome, or head, of sperm, is released when the sperm encounters the egg. This enzyme, along with others, digests the outer protective layers of the egg, allowing the sperm to penetrate.

Genetic defects known as "inborn errors of metabolism" are the result of the absence of an enzyme or the presence of a defective one.

## G-PROTEINS

A family of proteins known as **G-proteins** function as transducers of signals from outside the cell (see Information Molecules, below) to its inside. The name derives from the fact that they use GTP as an energy source. They transduce signals by stimulating the production of so-called second messengers. G-proteins are a link between receptors on the cell's surface and biochemical pathways within the cell. Different G-proteins are used by different hormonal and neurotransmitter receptors. Some signal molecules are adrenaline, glucagon, and insulin (i.e., hormones) and acetylcholine (i.e., a neurotransmitter). Glucose metabolism, motor neurons, and neuromuscular junctions, which are regulated by hormones and acetylcholine, respectively, would be affected if their G-proteins were absent or defective.

The G-proteins, which consist of three protein chains, are located beneath the cell's surface receptors, inside the cell. When a receptor is stimulated by a hormone (the first messenger), the receptor activates the alpha chain of the G-protein by enzymatically catalyzing the exchange of a GTP for a GDP at the alpha chain. This chain then separates from the other two and activates a so-called effector (the second messenger), such as cAMP (Chapter 7), which in turn activates an enzyme in a particular pathway to begin a series of biochemical reactions. In this way, the hormone has set in motion, from outside the cell, specific chemical reactions within the cell. Once the activation of the effector is completed the alpha chain with its GDP attached rejoins the other two chains and the process can be repeated—GTP exchanging for GDP, etc.

## INFORMATION MOLECULES

Information molecules include not only the many hormones but also such proteins as ubiquitin and neuropeptides. A common characteristic of the many informational proteins is that when functioning they each cause the cell to respond in a particular way. Thus, information has to be imparted to the cell. Many of the neuropeptides affect brain activity, but just how is still an enigma. We shall confine our discussion, therefore, to the hormones and ubiquitin, about which much more is known.

The hormones can be divided into what are often called the polypeptide or polypeptide-like hormones, such as insulin (a polypeptide), and the steroid hormones, such as the sex hormones. Here, too, we shall limit ourselves to a discussion of the polypeptide-like hormones.

Hormones can be regarded as informational molecules, or chemical messengers, because, upon reaching their target cells, they cause a variety of metabolic or physiological activities to occur. Hormones are transported by the bloodstream throughout the body. They affect their target cells because only the target cells have the appropriate receptor proteins on their membrane surfaces to recognize the particular hormones. As we saw in the previous section, the hormone's message is transferred into the cell via the G-proteins, which then activate the second messenger, often cAMP.

The receptors (more about them later) are proteins with a high affinity for the proper hormone. It does not take much effort to imagine the havoc that would result if hormone receptors were not specific; the cell would be continually responding to wrong signals. However, upon binding to the proper receptor protein, the hormone causes internal cellular activities to begin.

Let's take a look at the events triggered by a drop in the blood sugar level. The normal sugar level is maintained between 80 and 120 mg/100 ml. When a drop below about 80 mg/100 ml occurs, the alpha cells of the pancreas respond by releasing the polypeptide hormone glucagon into the bloodstream. At the target cells, the liver cells in particular, glucagon binds its receptor protein on the outer surface of the cell's membrane. The binding, via a G-protein, causes the synthesis of cAMP, the second messenger, along the inner surface of the cell membrane. The cAMP, in turn, is responsible for activating a series of enzymes, in a process often referred to as the cascade effect. The end result of the cascade effect is to get liver cells to release glucose from its storage form, glycogen, and to release the glucose into the bloodstream to bring the blood glucose level back to normal.

Ubiquitin is a protein whose information is of a completely different kind. This 76-amino acid protein is the most highly conserved in nature. As its name implies, it is ubiquitous (and abundant) in eukaryotes. In organisms as different as fish, toads, human beings, and the fruit fly *Drosophila* the molecule is identical. Between human beings and yeast the difference is only 3 amino acids.

Often, proteins that are to be degraded are marked by being covalently bound to ubiquitin first. So the information ubiquitin passes on to cellular enzymes is that any other protein associated with it is to be degraded. Degradation is an energy-dependent (ATP) process requiring a not-well-understood protein complex called "ubiquitin-conjugated degrading enzyme." Ubiquitin also serves to modify other proteins such as the histones H2A and H2B and a number of cell surface receptors. In this way, it is thought to influence the other proteins' functions.

## MUSCLE PROTEINS

It is the interplay of the muscle proteins actin and myosin together with ATP as the energy source that makes muscle contraction possible. Actin is a protein that stimulates the hydrolysis of ATP, thereby causing the cyclic association and dissociation of actin and myosin that we recognize as muscle contraction. The ATP is actually hydrolyzed by myosin, but myosin requires the participation of actin to bind the ATP before its hydrolysis. In fact, the name *actin* was given to this protein because of its capacity to "activate" ATP in the presence of myosin.

## PROTECTION MOLECULES

Animal fluids, primarily blood, contain a variety of proteins whose functions are to protect the body against microbial (bacterial, viral, protozoan, etc.) invasion or against the toxins of some of these microorganisms, which themselves are proteins in many cases (such as tetanus, botulinum, and diphtheria toxins).

In human beings, for example, saliva, perspiration, and even tears contain an enzyme called lysozyme, which can digest the cell walls of a number of bacteria, including species of *Staphylococcus* and *Streptococcus*. Properdin is a protein found in serum. It is a part of a rather complicated series of reactions involving enzyme reactions and protein-protein interactions. The reactions eventually lead to the association of a group of proteins, known as the membrane attack complex, that attaches to bacterial cell membranes and causes their disruption, killing the bacterial cell in the process.

The interferons are a group of several related proteins produced by many cell types of the human body in response to viral infection. Interferons, as the name suggests, interfere with viral multiplication and also stimulate cells of the immune system to respond to the infection.

By far, the most important of the protection molecules are components of and are synthesized by the **immune system** (Chapter 10) of vertebrates including human beings. Among these protection molecules are the specially produced proteins known as **antibodies,** also called **immunoglobulins.** Other special proteins are the complement group proteins, the receptors on immune system cells, and the interleukins. Together, these proteins are designed to inactivate or destroy microorganisms and to neutralize their toxins. Antibodies and complement can destroy cells or their toxins. The interleukins are chemical messengers, or informational molecules, needed for communication between and among the cells of the immune system. Vaccinations are given to stimulate the immune system for protection of the individual.

The human immune system is capable of making literally thousands of unique antibodies on demand (i.e., vaccination). These proteins are so specialized in function that they will react only with the molecule that caused their production in the first place; molecules that stimulate the immune system to produce antibodies are called **antigens.** Antigens may be encountered either naturally by infection or artificially by vaccination. So antibodies result from either natural or artificial stimulation of the immune system.

The importance of the immune system cannot be overstated. It is the court of last resort. Should it fail, an individual is doomed. The disease AIDS is fatal because it disrupts the immune system (hence "acquired *immunodeficiency* syndrome"); the immune system itself is under attack and eventually is destroyed or made ineffective. Unfortunately, there is no backup to the immune system.

There is a second immune system that protects by means of specialized cells that, of course, are specialized because of their special proteins. This is what is called the cellular immune system. And here, too, besides the cells' activities, communication between cells is via protein molecules. This is the system responsible, in large measure, for the rejection of organ transplants—heart, lung, liver. The cellular immune system does not distinguish between harmful invasion of the body by microorganisms or their toxins and helpful invasion by heart cells for example.

Basically, the immune systems distinguish the self from the nonself and operate on the principle that what is not "self" is dangerous to the body and must be destroyed.

## RECEPTOR PROTEINS

**Receptor proteins** are found on the cell's surface and extend through the plasma membrane to emerge on the inner side of the membrane within the cytosol. Receptors are transmembrane proteins. They are critical conduits for relaying information transmitted by hormones and neurotransmitters such as insulin, glucogan, adrenaline, and acetylcholine to the interior of the cell. These proteins are designed (via their amino acid sequences) to be soluble in an aqueous environment—the internal and external environment of the cell—and in a nonaqueous environment—the lipid membrane of the cell. They are also capable of two activities. On the outside of the cell, receptors recognize and react with signal molecules (e.g., hormones), and on the inside of the cell, they cause the activation of the G-proteins and subsequent synthesis of a second messenger (e.g., cAMP). Depending on the signal received and the receptor receiving the signal, the cell then alters its metabolism accordingly. For each hormone and neurotransmitter there are different receptors. The combination of different receptors on the cell's surface then determines the range or variety of signal molecules to which the particular cell can respond.

Receptor proteins are also needed for the process of endocytosis by which cells engulf and bring in materials from the outside. There are special receptors for this process. So, for example, cells can take up such substances as cholesterol embedded in lipoproteins and iron carried by the protein transferrin.

## REGULATORY PROTEINS

Among the **regulatory proteins** are those that regulate the activity of genes and those that regulate cellular components. We shall be most concerned with gene regulation in this text (Chapter 7), but just make mention of other systems of regulation.

For example, calmodulin modulates the calcium level in cells, a critical process in controlling the $Ca^{2+}$ pump at the plasma membrane and in such cellular activities as glycogen metabolism, regulation of energy metabolism via its effect on the citric acid cycle, muscle contraction, release of neurotransmitters, and as a second messenger for some hormones.

Regulator proteins are needed to activate and deactivate genes in an orderly manner so that all the proteins and RNAs required by the cell are available on demand. We shall deal in greater detail with the regulatory role of these proteins in Chapter 7. For now, we need only say that these control proteins are vital to the process of transcription, during which genes are decoded and mRNA is synthesized. The cell is continually turning genes on and off in response to signals originating from both within and without.

## STRUCTURAL PROTEINS

**Structural proteins** are necessary to provide shape, internal organization, and movement to the cell and its various parts or components. For example, the cytoskeleton proteins are required to maintain the cell's shape. Histones (Chapter 10), small basic proteins, are needed to package DNA. Other structural proteins are necessary to maintain the conformations of such crucial structures as ribosomes, on which proteins are made (Chapter 8), and cytochromes, which pass electrons to oxygen during aerobic respiration and play a vital role in ATP synthesis.

The major cytoskeleton proteins are actin filaments, intermediate filaments, and microtubules. The interaction of accessory proteins and the cytoskeleton proteins allows for the movement of cells in, for example, muscle contraction, which depends on the interaction of actin filaments and the thick filaments known as myosin. Likewise, the beating of the cilia of the respiratory tract, an important cleansing activity, and the ability of sperm to swim, using the beating of their tails to propel them toward the waiting egg, are dependent on the interaction of accessory proteins and cytoskeleton proteins.

The histones are a group of five basic proteins containing a very high proportion of the positively charged amino acids lysine and arginine. The five proteins, given the designations H1, H2A, H2B, H3, and H4, are used to negate the negative charge on DNA, thereby facilitating packaging. Like ubiquitin, during the course of evolution, these proteins have been highly conserved. The nucleotide sequences of their genes, particularly of H4, and, therefore, the amino acid sequences of the molecules vary little from the relatively simplest organism to the more advanced organism. For example, although the pea and the cow diverged some 1.2 billion years ago, the difference between their H4 proteins is two amino acids out of a total of 102. Such stability of amino acid sequence over more than a billion years of evolution suggests a critical role for these proteins—a role that precludes changes. The proteins are well designed for their biological activity—the packaging of DNA.

Ribosomes (more on these structures in Chapter 10) are large aggregates of two types of molecules: proteins and RNA. Two subunits, one large and one small, make up one ribosome. Each subunit is composed of approximately 60% RNA and 40% protein. The larger of the two subunits contains 50 proteins, and the smaller unit 33 proteins. That is for the eukaryotic ribosome. The prokaryotic ribosome is smaller overall; it has 32 proteins in its larger subunit and about 21 proteins in the smaller one.

Ribosomes are required for protein synthesis during the process called translation. Translation is the stage of protein synthesis during which information derived from the gene, and now in the form of the nucleotide sequence of messenger RNA, is "read" and the appropriate sequence of amino acids is constructed, producing a protein molecule. In the absence of ribosomes, proteins could not be made and, consequently, the cell could not exist as we know it.

Cytochromes are molecules consisting of an iron atom embedded in a heme group, which is in turn associated with proteins, as in the hemoglobin molecule discussed earlier. The protein portion of this complex, like the histones, has been highly conserved throughout the course of evolution. The structures of cytochromes are so well suited to their biological activities that very little change in the protein molecules has been tolerated over billions of years of evolution. For example, the proteins of the cytochromes of bacteria are very similar to those of humans.

Cytochromes function in a number of systems—from photosynthesis to aerobic and anaerobic respiration, detoxification pathways, and omega fatty acid oxidation. During aerobic respiration, human breathing, cytochromes in a series are used to transport electrons from oxidized food stuffs, primarily glucose and fatty acids, to oxygen. The energy stored in the electrons is used in the synthesis of the high-energy compound ATP in a process called oxidative phosphorylation.

The cytochromes of this cytochrome system—also known as the cytochrome chain, the respiratory system or chain, and the electron transport system or chain—alternately accept and pass on to the next cytochrome electrons that are ultimately passed to oxygen to produce the water we exhale when we breathe. These are oxidation-reduction reactions, and that is the reason we need oxygen—to get rid of electrons released from food stuffs. (Incidentally, this oxidation of food stuffs to generate energy or, for that matter, to lose weight, is commonly referred to as burning up sugar or fats.)

## MISCELLANEOUS PROTEINS

The miscellaneous protein category is used only to describe some proteins that do not fit neatly into any of the other categories. But their roles in the lives of cells or multicellular organisms are no less critical.

Channel proteins, for example, are involved in the selective permeabilities of cell and organelle (e.g., mitochondria) membranes to particular ions. They enable critical ions such as sodium, potassium, and chloride to be moved in the proper directions at the appropriate times. For example, in the transmission of nerve impulses, nerve membranes must be alternately polarized and depolarized. The transport of glucose into human cells is dependent on a $Na^+$-$K^+$ pump, which is, in turn, dependent on the movement of the ions through channels.

It has recently been discovered that the disease cystic fibrosis (Chapter 12) results from a defective gene encoding a protein located in the chloride ion channels of lung and other cells. An inability to move chloride ions in the proper direction because of the defective gene's protein product causes a buildup of lung fluid, which, at present, leads to death at a very early age. Efforts to cure this malady using genetic engineering techniques are currently under way (Chapter 11).

Special proteins are also important for what are called gap junctions. These are cell-to-cell channels important to intercellular communication and to the passage of inorganic ions and most metabolites, such as sugar, amino acids, and nucleotides from one cell to another. Larger molecules such as proteins, nucleic acids, and polysaccharides cannot pass through these junctions.

## CONCLUSION

To bring this chapter to an end, it needs to be said that proteins are normally found associated with other types of molecules and rarely as pure proteins, especially in eukaryotes. These other molecules are important to the functioning of the proteins because they help to stabilize the proper structure or conformation. Proteins are commonly found in association with carbohydrates—to produce glycoproteins. Many receptor proteins are of this type. Other proteins are complexed with lipids (fats)—to produce lipoproteins, the carriers of cholesterol, for example. And, last, proteins, particularly enzymes, may require metal ions, such as $Mg^{2+}$, $Mn^{2+}$, and $K^+$, for maximum activity and efficiency; these ions are often referred to as cofactors of enzymes.

## SUMMARY

This chapter is intended as a brief look at proteins. In particular, it surveys the variety of types and functions of proteins and of their importance to the activities of the cell.

Proteins are the most abundant organic constituents of the cell. Together, the various proteins make up more than 50% of the dry weight of the cell. This abundance is a reflection of the number and importance of the roles proteins play in cellular metabolism. The vast majority of the information encoded in the DNA molecule is for the synthesis of proteins.

The ability of proteins to function in so many roles is due to the tremendous structural variations proteins can assume. This structural variation is the result of the enormous number of combinations that can be gotten by combining 20 different amino acids. The greater the number of amino acids that go into making up proteins the greater the possible structural variations and so the greater the functional capabilities.

Proteins can be grouped into at least 10 functional classes. In each class or group, there are dozens upon dozens of different molecular types, each of which is needed to perform a specific function for the cell. The protein molecules are distributed throughout the cell from the plasma membrane to the cytoplasm and from the nucleoplasm to cell organelles.

Without proteins, the cell could not exist. If too many proteins were defective, the cell would cease to function and would die.

## STUDY QUESTIONS

1. Explain why proteins are so vital to the cell.
2. How is it possible for proteins to play such a wide variety of roles in the cell?
3. List the classes or groups of proteins.
4. Give an example from each class of protein listed in the answer to question 2.
5. Explain the role each example given in the answer to question 3 has in the activities of the cell.
6. Give an example of a nonprotein enzyme and explain its role in the cell.

## Readings and References

Alberts, B., D. Bray, J. Lewis, M. Raff, K. Roberts, and J. D. Watson. 1994. *Molecular biology of the cell*. 3rd ed. New York: Garland.

Baer, M., and S. Altman. 1985. A catalytic RNA and its gene from *Salmonella typhimurium*. *Science* 228:999–1002.

Lehninger, A. L. 1975. *Biochemistry*. 2d ed. San Francisco: Worth.

Stryer, L. 1995. *Biochemistry*. 4th ed. W.H. Freeman.

Voet, D., and J. G. Voet. 1995. *Biochemistry*. 2d ed. New York: John Wiley & Sons.

Weaver, R. F., and P. W. Hedrick. 1997. *Genetics*. 3rd ed. Dubuque, Iowa: Wm. C. Brown.

Zaug, A. J., P. J. Grabowski, and T. R. Cech. 1983. Autocatalytic cyclization of an excised intervening sequence RNA is a clearage-ligation reaction. *Nature* 301:578–83.

CHAPTER FIVE

# THE NATURE OF THE GENE: THE REPOSITORY OF INFORMATION

## OUTLINE



## CHAPTER OBJECTIVES

*This chapter will discuss:*

- The gene and the information it contains
- The nature of structural genes and regulatory genes
- Genes with different end products: proteins and RNAs

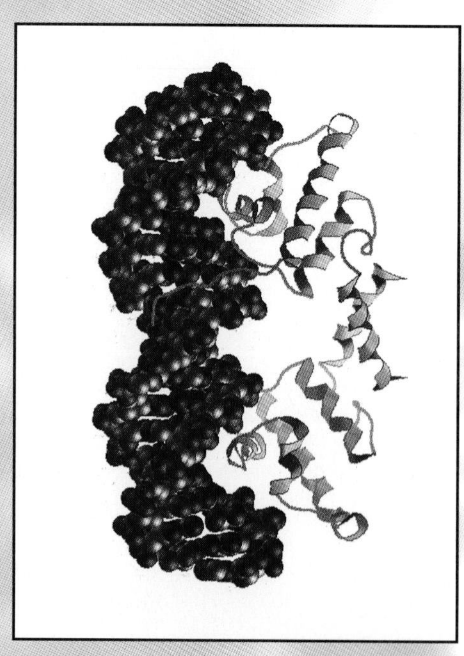

## INTRODUCTION

Arguably, humans have had an intuitive understanding of genetics at least since the time it was first noticed that plants, animals, and human beings had a remarkable similarity to their parents. No doubt, by the time the domestication of plants and, especially, of animals began, farmers knew that progeny with particular characteristics could be gotten if the right two individuals were mated. They recognized the occurrence of inheritance.

But, of course, not until Mendel and Miescher began their work was there a systematic, scientific study of inheritance. And not until the 1940s and 1950s was the material of inheritance finally and conclusively identified and the mechanism of inheritance glimpsed. The search for the elusive gene was coming into its own. Genes have been a central part of living organisms since the very beginning of life on earth (Figure 5.1).

The questions being asked today concerning the gene are: What exactly is the nature of the gene? How does the gene work? What information of inheritance does the gene contain? Are all genes basically alike, or are there different kinds of genes?

In this chapter, we will look at the answers to some of those questions, answers that have been discovered over the past four or so decades. In subsequent chapters we will look at the answers to other questions, such as How are genes activated, deactivated, and controlled (Chapter 7)? We begin this chapter by defining the gene as we currently understand it.

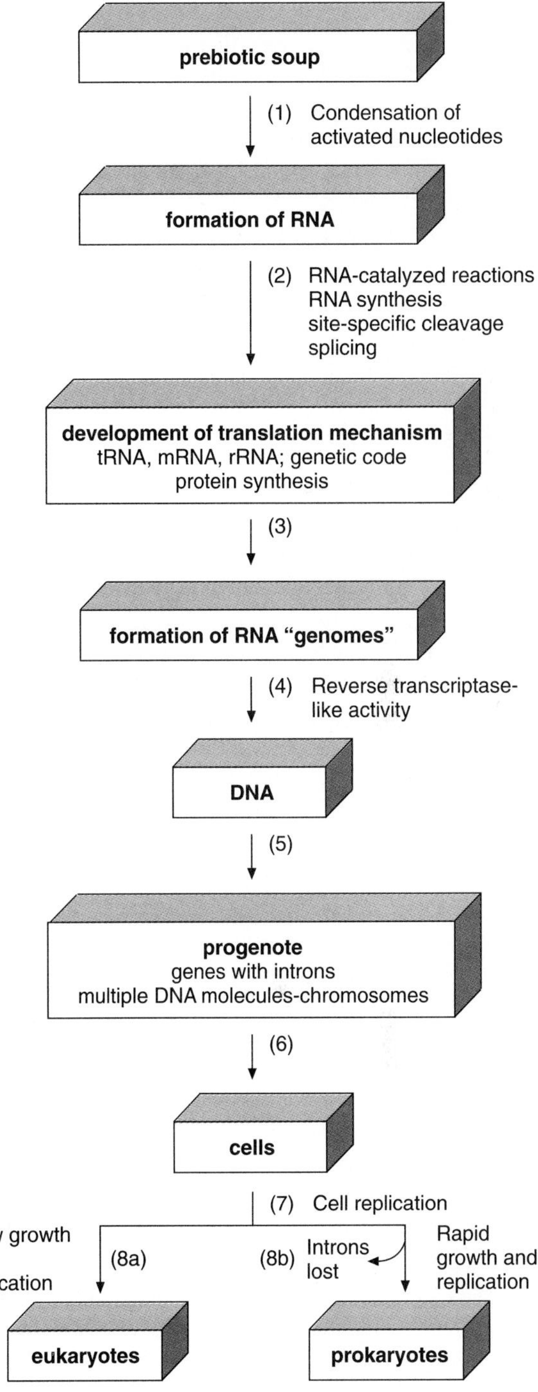


FIGURE 5.1
A speculative scheme for the early molecular and cellular events in the evolution of genetic systems.

## A GENE DEFINED

A short description of a **gene** is that it is a unit of inheritance that contains the information (a linear sequence of nucleotides) for a polypeptide or for an RNA molecule. A **polypeptide** is a linear arrangement of amino acids joined one to another by a carbon-to-nitrogen bond known as a **peptide bond** (Chapter 8). The term *polypeptide*, then, refers to the many peptide bonds in the molecule. A polypeptide chain may then associate with other polypeptides to form a biologically active protein molecule. Polypeptides are very often biologically inactive. The terms *polypeptide* and *protein* are frequently used interchangeably, but it is important to understand that proteins are often composed of more than a single chain.

Genes can be conveniently assigned to one of two broad functional categories: structural genes and regulatory genes. It is the role of the end product of these genes that distinguishes structural and regulatory genes.

1. **Structural genes** code for polypeptides or RNAs needed for the normal metabolic activities of the cell—e.g., enzymes, structural proteins, and receptors (Chapter 4). So a structural gene is an amount of DNA, or RNA in some viruses, that contains the information for one complete, mature transfer RNA (tRNA), ribosomal RNA (rRNA), or polypeptide.
2. **Regulatory genes** code for polypeptides that form proteins whose function is to control the expression of structural genes. With regard to their makeup, these genes are like structural genes.

The gene usually occupies a particular location within the chromosome, has a specific effect on the organism's morphology or physiology, can be mutated (i.e., changed), and can recombine with other genes.

An operational definition of the gene used throughout the text is as follows: a gene is a linear sequence of nucleotides that codes for a polypeptide or for an RNA molecule. Note the idea of a linear sequence of nucleotides. In other words, the information in a gene is the sequence of nucleotides along one DNA or RNA strand; it does not pertain to the relation of nucleotides across the two strands of the DNA double helix.

An important point to keep in mind is that the gene per se does not make, manufacture, or produce anything. It is a store of information. It is acted upon; it does not initiate any action. It is for all intents and purposes at the mercy of the conditions within the cell and in the cell's immediate environment.

The complete set of genes of an organism, what is called the genetic constitution of the organism, is its **genotype.** The physical or chemical manifestation or expression of the genotype is the **phenotype** (i.e., the cell's morphology and physiology). For example, the genes that control eye color in animals, including human beings, are a part of the animal's genotype. The actual eye color (blue, brown, hazel) is the physical or chemical manifestation or expression of those genes, the phenotype.

If a particular characteristic, such as brown eye color, is a part of an organism's phenotype—that is, if it is expressed—it can be said that the individual has the gene for that characteristic. If, however, a particular characteristic is not expressed, one cannot conclude that the gene is absent. Again, using eye color as the example, we can say that a brown-eyed individual has genes for brown eyes, but we cannot say that the individual does not have genes for blue eyes. Because gene expression can be repressed, as happens in bacteria (Chapter 7) and in higher organisms such as animals, one gene may be recessive (not expressed; e.g., blue eyes) while another is dominant (expressed; e.g., brown eyes).

Genes may be located on either strand of the double-stranded DNA. But, regardless of which strand contains a particular gene, all genes are read in a 5′ to 3′ direction, and the strand containing the particular gene is referred to as the coding strand (more on this subject in Chapter 8).

## GENE SIZE

The sizes of genes have been estimated using various techniques. A direct method is simply to divide the number of nucleotides in a chromosome by the number of genes known to exist for the organism. Another, but indirect, method is use of genetic mapping and cross-over frequencies. This method depends on the assumption that crossing-over frequencies are about equal along an entire chromosome. The problem is that a high frequency of crossing-over can be interpreted to mean that the distance between the genes in question is greater than may in fact be the case since the farther apart genes are the more likely it is that crossing-over will occur. On the other hand, genes that show a low frequency of crossing-over may be seen as being closer together than they really are; these genes may cross over at a low frequency for some unknown reason. So estimates of genome size and of numbers of genes may be larger or smaller than is actually the case. The results, however, regardless of the method used, are comparable; the estimate is that an average prokaryotic gene contains somewhere between 900 and 1500 base pairs. The average eukaryotic gene is more difficult to determine, but genes on the order of several hundred thousand nucleotides have been discovered in the human genome.

## THE C-VALUE PARADOX

It is not an unreasonable assumption that the more complex an organism, morphologically and physiologically, the more information is required and, therefore, the more DNA (or RNA in the case of viruses) is needed to store the information. The experimental evidence bears this assumption out. For example, ØX174 has a chromosome that is 1.8 µm long and contains approximately a dozen genes. Lambda phage has almost 10 times as much DNA and some 50 genes. *Escherichia coli* is estimated to have, perhaps, 4700 genes in a DNA molecule of about 1360 µm (that is something like a thousand times as long as the cell). And the DNA of human diploid cells, if placed end to end, would measure some 2 m, or about 6 ft, and contain an estimated 50,000–100,000 genes. Individual human chromosomes contain from 48 million to 240 million base pairs; the total number of base pairs in human diploid cells is estimated at 6 billion. The DNA from these individual chromosomes is measured at from 1.5 to 8.5 cm in length.

There are, however, some anomalies in these data. The anomalies are referred to as the **C-value paradox** (C for "content"). The **C-value** is the amount of DNA (in picograms) in the haploid genome of a species. As noted above, there is a correlation between DNA content and complexity. But, paradoxically, for some organisms, there seems to be no correlation. Although the minimum C-value for eukaryotic species does increase as complexity increases, some organisms have much more DNA than appears to be required.

For example, the range of DNA for mammals is between 2 and 3 pg (picograms: 1 pg = $10^{-12}$ g, or 1 trillionth of a gram). But, for the amphibia, the range is between 1 and 100 pg; some flowering plants, the lily, for example, have 100 times the DNA content of a human cell, as does the lungfish.

The current explanation for this phenomenon is that many organisms, including human beings, simply have more DNA than they need. The extra DNA is thought to be noncoding DNA. That is, a good deal of the cell's DNA does not code for protein, RNA, or regulatory information to control gene activity. This statement, of course, begs the questions: What is its function? What is the reason for all this noncoding DNA? Why have organisms accumulated so much more DNA than they apparently need for protein and RNA synthesis?

Just as the amount of DNA varies from species to species so too does the diploid number of chromosomes. For example, human beings have 46 chromosomes, dogs have 78, the fruit fly has 8, the hermit crab has 250, and the potato has 48. The most complex organisms do not necessarily have the most chromosomes.

## THE INFORMATION IN GENES

In Chapter 1, we noted that one of the characteristics of DNA, functioning as genetic material, is to be the repository of information for the cell. And, further, we noted that the information is found in linear portions of the DNA called the genes. In particular, the information resides in the sequence of nucleotides. What is important here are two facts:

1. In the nucleotides of DNA, it is the bases, not the sugar-phosphates, that are the information.
2. It is the linear sequence of the bases that is critical as information. (Remember, the words *bases* and *nucleotides* are routinely used interchangeably in this context.)

An analogy can be made to language. Letters and sequences of letters make up words, which are information. Different letter combinations make up different words (information), and words in various combinations make available different kinds of information.

Genes are composed of "words" made up of "letters," the nucleotides represented by their first letters: A, T, C, and G. In different combinations, these letters make up different words. And in different combinations, those words give different information to the cell. In the gene, a word is always of three letters, no more, no less. And the gene itself can be understood as a "sentence" composed of these three-letter words. In Chapter 6, we shall delve further into this information system.

For the moment, however, we need to know that genes contain the information, the nucleotide sequences. These sequences tell the cell either how to arrange amino acids so that an almost infinite variety of proteins can be made or how to arrange other nucleotides to construct tRNA or rRNA. (Since, however, the vast majority of genes are required for protein synthesis, we shall concentrate on that function.)

The sequence of nucleotides in genes provides two kinds of information during protein synthesis. The first is the particular amino acid that the three-letter words in a gene specify. The second is the position each amino acid will occupy in the protein. These pieces of information—which amino acid to use and its position in the protein—determine what is called the primary structure of the polypeptide, its linear arrangement of amino acids. The primary structure of the polypeptide ultimately determines the biological activity of the protein that contains the polypeptide. The primary structure determines the secondary, tertiary, and quaternary structures (that is, the physical shape of the molecule), and the polypeptide's shape, in turn, determines how the molecule will associate with other polypeptides (see Chapter 8). And, depending on whether or not the primary, secondary, tertiary, and quaternary structures are correct, the resulting protein may or may not have biological activity. The key phrase here is *biological activity*. Two examples of the results of badly folded proteins are sickle-cell anemia and the neurodegenerative disease amyotrophic lateral sclerosis, "Lou Gehrig's disease."

Located usually at the 5′ end of the gene are **regulatory sequences.** These are nucleotide sequences that respond to chemical signals originating from inside and outside the cell. As conditions change, internally or externally, or both, the cell responds to these chemical signals via their interactions with regulatory sequences; these interactions activate or inactivate structural genes. In this way, the types, amounts, and rate of synthesis of proteins are controlled (Chapter 7). These proteins may be needed inside the cell or on its surface, or they may be exported from the cell (e.g., the exoenzymes of bacteria).

Regulatory sequences have no end products. Their information is to provide a recognition sequence for signal molecules and, in consequence of the interaction of the signal molecule and the regulatory sequence, to control the activity of structural genes. The regulatory sequences go by names that suggest the function of each type of sequence: **promoters, operators, attenuators,** and **enhancers.**

Let us now take a closer look at both structural genes and regulatory sequences.

## STRUCTURAL GENES

The structural genes of various organisms, from viruses to prokaryotes to eukaryotes, differ with regard to their informational makeup and even with regard to their stability at a particular location within the DNA molecule. For example, there are genes whose information is contained in a continuous linear arrangement of nucleotides, as is commonly the case with prokaryotes. There are genes whose information is interrupted by noncoding sequences of nucleotides, as generally happens in eukaryotes. Other genes have information that can be combined with information from still other genes to create a third gene, as in the virus

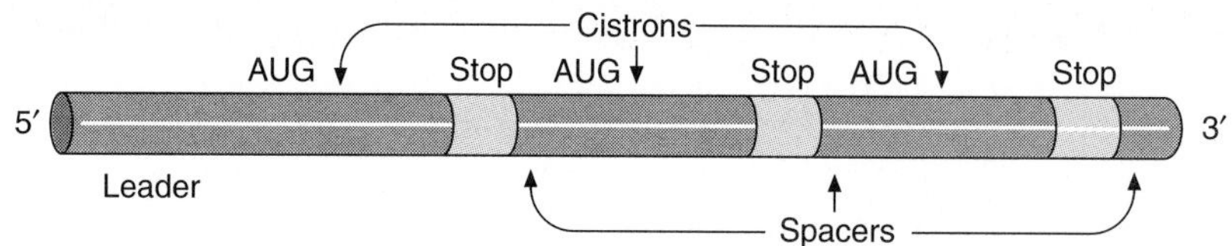


FIGURE 5.2
Arrangement of cistrons (black) and untranslated regions (red) in a typical polycistronic mRNA molecule.

ØX174 and in human beings. There are even genes that can relocate to different positions on a given chromosome or to another chromosome altogether, as happens in both prokaryotes and eukaryotes.

# Prokaryotic Genes

Most prokaryotic genes are composed of nucleotide sequences that are not interrupted by noncoding sequences; though, in some *Eubacteria* and *Archaebacteria*, noncoding sequences have been found. In general, however, all the nucleotides found in prokaryotic genes are bits of information (letters or words) used to construct a protein or an RNA molecule.

The structural genes of prokaryotes are most often activated in clusters, which are often referred to as polycistronic transcription units. That is, the transcription unit contains information for the construction of more than one polypeptide. The transcription units, which are **mRNA** molecules (Figure 5.2), have the following nucleotide arrangements starting at the gene's 5′ end: a leader sequence, a start sequence, a nucleotide sequence for the polypeptide, a stop sequence, and a spacer sequence of from 5 to 20 nucleotides to separate one gene from the next.

All these sequences are repeated for each gene decoded in a polycistronic transcription unit except for the leader sequence, which appears only once in each polycistronic transcription unit. In prokaryotes, then, mRNA carries more than one message; mRNA has information for more than one polypeptide.

# Eukaryotic Genes

In contrast to prokaryotes, eukaryotes generally use **monocistronic transcription units.** The genes of these organisms are not activated in clusters but rather as single entities. Messenger RNA carries information for only a single polypeptide.

The eukaryotic gene is most often an arrangement of sequences similar to that of the prokaryotic gene (leader, start, etc.), but these genes also have coding and noncoding sequences of nucleotides. Consequently, the eukaryotic gene is sometimes called a split gene; the information is dispersed along the DNA molecule. There are some rare exceptions to this arrangement. For example, the genes that encode the histone proteins needed to package eukaryotic DNA (Chapter 10) and the genes needed for the synthesis of alpha and beta interferons, proteins that help protect against viral infections, lack noncoding sequences.

# Exons and Introns

In eukaryotic genes, the nucleotide sequences that specify amino acids (the coding sequences) are called **exons,** from the word "*ex*pressed." The noncoding nucleotide sequences are called **introns,** for "*int*ervening" sequences. Table 5.1 gives some examples of exons and introns.

Introns vary in size, number, location, and nucleotide sequence from one gene to another. Often the total length of the introns of a gene exceeds the total length of the exons of that gene by anywhere from 2 to 10 times and more. The number of introns appears to increase as the size of the gene increases.

Interestingly, introns do not occur randomly along a gene's length. Rather, they occur at particular locations. For example, they occur adjacent to coding sequences that specify the amino acids of special structural or functional domains of the complete protein. And, in multicellular organisms, regardless of the cell in which a gene is found, from germ line cells to somatic cells, split genes retain the same structure; so the introns are invariant features of those genes.

As we shall see in Chapter 10, during the process of protein synthesis in eukaryotes, introns are removed from the mRNA molecule soon after the gene has been decoded and mRNA has been produced.

# Overlapping Genes

In the very small DNA viruses such as ØX174 and the very small bacteriophage MS2, F2, and QB, along with the animal virus simian virus 40 (SV40), the genetic material is much too small to hold all the information needed to code for all the proteins they require. The genomes of these organisms are composed of fewer than 5400 nucleotides.

The phage ØX174, for example, needs 11 proteins that have a total of about 2300 amino acids. Since each amino acid is coded for by 3 nucleotides, the length of

table 5.1

CODING (EXONS) AND NONCODING (INTRONS)

| | | EXONS | INTRONS | |
|---|---|---|---|---|
| GENE PRODUCT | ORGANISM | (TOTAL bp) | NUMBER | TOTAL bp |
| Adenosine deaminase | Human | 1500 | 11 | 30,000 |
| Apolipoprotein B | Human | 14,000 | `28 | 29,000 |
| β-globin | Mouse | 432 | 2 | 762 |
| Cytochrome b | Yeast (mitochondria) | 2200 | 6 | 5100 |
| Dihydrofolate reductase | Mouse | 568 | 5 | 31,500 |
| Erythropoietin | Human | 582 | 4 | 1562 |
| Factor VIII | Human | 9000 | 25 | 177,000 |
| Fibroin (silk) | Silkworm | 18,000 | 1 | 970 |
| Hypoxanthine phosphoribosyl transferase | Mouse | 1307 | 8 | 32,000 |
| α-interferon | Human | 600 | 0 | 0 |
| Low-density lipoprotein receptor | Human | 5100 | 17 | 40,000 |
| Phaseolin | French bean | 1263 | 5 | 515 |
| Thyroglobulin | Human | 8500 | >40 | 100,000 |
| $tRNA^{Tyr}$ | Yeast | 76 | 1 | 14 |
| Uricase subunit | Soybean | 300 | 7 | 4500 |
| Vitellogenin | Toad | 6300 | 33 | 20,000 |
| Zein | Maize | 700 | 0 | 0 |

ØX174 DNA should be at minimum 6900 nucleotides not counting start, stop sequences and the like.

These organisms have solved the problem by developing what are called **overlapping genes.** See Figure 3.1 for a genetic map of ØX174 illustrating the overlapping feature of this genome. As can be seen from the map, the nucleotide sequences for some genes are completely contained within the sequences of other genes. Note, for example, that gene B is included in the sequence of gene A, and gene K is positioned within genes A and C. This strategy allows a larger amount of information to be contained within a particular length of DNA than would be the case if each gene occupied a discrete length of the genome. In other words, contiguous genes require more DNA than do overlapping genes.

Overlapping genes have been discovered even in eukaryotes. The gene for human factor VIII (a blood-clotting protein) is found on the X chromosome and has about 186,000 bp. Nested within the factor VIII gene is another protein gene. And, within the thyroid hormone receptor gene, there is another gene related to the thyroid hormone receptor gene. Other examples of genes within genes have been discovered (Chapter 10).

With overlapping genes, then, we have overlapping information. But each gene, whether overlapping or not, has its own start, or initiation, sequence that distinguishes one set of instructions from another. Although this arrangement is a very efficient use of a limited amount of DNA or RNA, there are problems. A mutation in an overlapping region can affect not one but two genes and consequently more than one protein or RNA.

## Transposons

In 1951, Barbara McClintock reported the discovery of transposable elements, or movable genes (now called **transposons**), in maize. Since then literally dozens of other such jumping genes have been discovered in both prokaryotes and eukaryotes, although higher-organism transposons are not as well characterized as those of prokaryotes. Transposons are capable of causing rearrangements of nucleotide sequences, because their moving about may create new sequences via deletions, inversions, or the change to a new location. The rearrangement may also change the function of existing sequences by placing them under the control of a new regulatory sequence. Structurally, transposons are nucleotide sequences immediately flanked on either end by inverted repeat sequences that are themselves flanked by direct repeat sequences, for example:

5′ ATGCA/CCGTAA[GGTTCCATTT]AATGCC/ATGCA 3′
3′ TACGT/GGCATT[CCAAGGTAAA]TTACGG/TACGT 5′

We have here in a 5′ to 3′ reading direction a transposon in brackets [GGTTCCATTT] that is flanked by an inverted repeat sequence of CCGTAA inverted to AATGCC, which is then flanked by a direct repeat sequence of ATGCA.

Simple transposons code only for the genes needed for their transposition; more complex elements may carry genes for such things as drug resistance and sugar fermentation. Some of these transposons share certain general features. They encode information for their own transposition to other chromosomal sites via the enzyme transposase and for other structural and regulatory sequences. Some elements can insert themselves into specific target sequences on new chromosomes, and some elements can insert themselves at any of a number of target sites. The so-called P elements of *Drosophila* cause sterility. The Ac and Ds elements of maize can alter the level of gene activity, can disrupt normal gene control at specific developmental times or in specific tissues, and can cause mutations resulting in the production of biologically inactive proteins.

Some scientists are now suggesting that these movable genes may play a role in evolution. There is speculation that these jumping genes are even capable of transferring genetic information from one species to another. One group of investigators reported some limited evidence that a mite had transmitted a P element from one species of fruit fly to another during the past century. Although no direct demonstration of the phenomenon is available, there is now much discussion regarding experiments to investigate this possibility.

## REGULATORY GENES

**Regulatory genes** are referred to by a variety of other terms, including regulatory sequences, regulation units, regulatory regions, control units, control sequences, and control regions. And, depending on their function vis-à-vis the structural gene to be influenced, they are also classified as promoters, operators, attenuators, or enhancers.

These sequences (genes) may occur at one end of a structural gene, as promoters and operators do (the 5′ end); between promoter-operator sequences and their structural genes, as attenuators do; or at some distance from the gene-decoding start site, as is the case with enhancers. Some regulatory sequences have even been found in the introns of some genes.

The terms *upstream* and *downstream* are often used in describing the position of sequences, with the reference point being the particular sequence under discussion. **Upstream** means in the 5′ direction with regard to the reference sequence; **downstream** means in the 3′ direction with regard to the reference sequence (Figure 5.3). So we can refer to promoters and operators as being upstream of the structural gene start site and to particular enhancers as being either upstream or downstream of the structural gene start site.

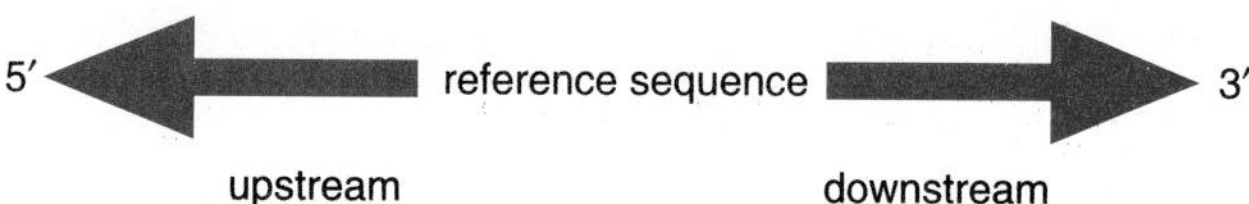


FIGURE 5.3
Conventionally, upstream and downstream are to the left and right of the reference sequence, corresponding to the usual way in which mRNA is written: 5′ → 3′.

Except for attenuators, a common feature of all regulatory sequences is that they interact with protein molecules or protein molecules coupled to nonprotein molecules (Chapter 7, Components of the *lac* Operon: the repressor protein and CAP protein, respectively). Attenuators bind ribosomes.

Here we shall take a brief look at the regulatory sequences mentioned above—promoters, operators, attenuators, and enhancers. In Chapters 7, 8, and 10, the functions of the sequences will be discussed further.

### Promoters

**Promoters** are DNA nucleotide sequences recognized by the DNA-directed RNA polymerases as their attachment sites. Promoters are located upstream, or to the 5′ end of the structural gene. The DNA sequences range from about 20 to 200 bp.

Common to many bacterial promoters is that they have two identical or nearly identical sequences. The first is a TATAAT sequence. David Pribnow, in 1975, reported its discovery, so it is often called the **Pribnow box.** Sometimes it is called the **–10 region** to indicate that the sequence begins 10 nucleotides upstream from the gene start site. The second sequence, TTGACA, is called the **–35 region,** also named in reference to its position vis-à-vis the structural gene's start site. The Pribnow box is the attachment point of the sigma unit of the RNA polymerase, and the –35 sequence is another attachment site for the enzyme.

In eukaryotes, a sequence analogous to the Pribnow box, called the **Hogness box** (after its discoverer, David S. Hogness) is a TATAAA sequence. This region lies 19–27 bp upstream of the eukaryotic gene's start site.

Promoters may bind proteins other than the RNA polymerases (Chapter 7). These other proteins bind before the polymerase and are required to allow binding of the RNA polymerase itself. These proteins, then, exert a positive, or stimulatory, control on RNA polymerase activity and consequently on gene decoding or transcription.

### Operators

**Operators** are nucleotide sequences that lie between the promoter and the structural gene. They are the regions of

DNA to which repressor proteins bind (Chapter 7) and thereby prevent transcription. Repressor proteins have a very high affinity for operator sequences. Repression of transcription is accomplished by the repressor protein's attaching to the operator sequence downstream of the promoter sequence (the point of attachment of the RNA polymerase). The enzyme must pass the operator sequence to reach the structural gene's start site. The repressor protein bound to the operator physically prevents this passage, and, as a result, transcription by the polymerase cannot occur.

### Attenuators

The **attenuator** sequences are found in bacterial gene clusters that code for enzymes involved in amino acid biosynthesis (Chapter 7). Attenuators are located within so-called leader sequences, a unit of about 162 nucleotide pairs situated between the promoter-operator region and the first structural gene start site of the cluster. Attenuation has a 10-fold effect on transcription. As the level of an amino acid in the cell rises and falls, attenuation adjusts the level of transcription to accommodate the changing levels of the amino acid. High concentrations of the amino acid result in low levels of transcription of the structural genes, and low concentrations of the amino acid result in high levels of transcription. Attenuation proceeds independently of repression; the two phenomena are not dependent on each other. Attenuation results in the premature termination of transcription of the structural genes.

### Enhancers

Enhancers were first discovered in the animal DNA virus SV40. These control elements are located near the virus's replication origin. Enhancers have now been found in eukaryotic cells and in RNA viruses as well. The function of **enhancer** sequences appears to be to increase the number of RNA polymerase molecules transcribing a structural gene. The enhancers seem not to have particular positions relative to the initiation site of the structural gene as the promoter-operators do. In fact, they have been found some distance upstream and downstream of the start site of the genes they control.

Enhancers are activated by the binding of sequence-specific proteins. One explanation of how they work is that the binding of the sequence-specific protein causes introns to loop out (see Chapter 7, Looping). This looping brings the enhancer closer to the RNA polymerase binding site—the promoter. Then, in some way, the attachment of the polymerase to the promoter is stimulated, perhaps because the promoter is now more accessible to the enzyme.

Enhancers have been implicated in tissue-specific regulation and temporal regulation during the development phase of growth. Some investigators (e.g., Lewin) have argued that enhancers may, in fact, be a promoter variant: enhancers and promoters have sequences in common, both are required for the regulation of transcription, and both can bind sequence-specific proteins.

All cellular enhancers discovered thus far have been found to be required for selective gene expression in specific tissues.

## GENES FOR PROTEINS AND RNAS

In 1909, the English physician Archibald Garrod published his book *Inborn Errors of Metabolism,* in which he noted that a number of human diseases followed the rules of Mendelian inheritance. He also took a leap of faith and connected the disorders to a deficiency or absence of particular enzymes. Garrod then speculated that the genetic material (genes) was what controlled the production of enzymes. But it was not until 1941, when a paper by George W. Beadle and Edward L. Tatum appeared, that the connection between enzymes and genes was experimentally confirmed. In their paper, Beadle and Tatum postulated the "one gene–one enzyme" theory.

Since then much work in genetics and protein chemistry has led to a refinement of this theory. Proteins, as noted earlier, are most often composed of chains, or subunits, called polypeptides. We now know that genes code for polypeptides rather than for complete multichain proteins. We have, then, a new postulate: "**one gene**–(codes for) **one polypeptide.**" Most often, more than one gene is needed to construct a protein. It is important to repeat here that there is little *chemical* difference between a polypeptide and a protein. Both are made of amino acids joined by peptide bonds; *polypeptide* refers to the many peptide bonds in the molecule.

There is, however, often *biological* difference between the two types of molecules. Proteins have biological activity, but polypeptides may not. Again, the example of human hemoglobin (Chapter 4) can be used. This oxygen-carrying protein is composed of four polypeptides, two identical copies of two different polypeptides. The polypeptides lack biological activity (i.e., oxygen-carrying capacity), but the complete protein has it. In this case, two genes are required to synthesize the hemoglobin molecule.

So our definition of a gene, with regard to proteins, is: a gene is a sequence of nucleotides that codes for a polypeptide. A polypeptide is synthesized via two processes already mentioned: transcription, the decoding of the gene, and translation, the arranging of amino acids in a linear sequence in accordance with the instructions decoded from the gene (see Chapter 8, Protein Synthesis).

Our earlier description of a gene included the idea that there are also genes for RNA molecules. All cellular life forms contain genes needed to synthesize **ribosomal RNA** (rRNA) and **transfer RNA** (tRNA). The first of these RNAs is found in ribosomes, the structures on which polypeptides are produced during translation. The second type of RNA, tRNA, consists of the molecules that carry amino acids to the ribosomes and that also "read" the set of instructions in the mRNA, which, in turn, determines the amino acid sequence of the polypeptide (Chapter 8).

In eukaryotes, but not in prokaryotes, the rRNA genes are located in one or a few specific chromosomes in places called nucleolar organizer regions. These regions may contain anywhere from a hundred to several thousand genes, depending on the species.

For tRNAs, there are something on the order of 60 different genes. But their organization in both eukaryotes and prokaryotes is poorly understood.

A major difference between eukaryotes and prokaryotes is that, whereas the prokaryotes have a single RNA polymerase to transcribe protein and RNA genes, eukaryotes have three RNA polymerases; one polymerase transcribes rRNA genes, a second polymerase transcribes tRNA genes, and a third polymerase is needed to transcribe the protein genes. We shall have more to say about these polymerases in Chapter 10.

## Summary

A gene, whether composed of DNA or RNA, is a linear sequence of nucleotides that may code for a polypeptide, tRNA, or rRNA. The gene is the unit of inheritance.

It is the sequence of bases in the nucleic acids that is the information; in this regard, the sugar-phosphate groups play no part. The genes may be located on either strand of a double-helix DNA or on RNA.

Genes may be assigned to one of two functional categories: structural genes or regulatory genes. The first encodes information for the cell's metabolic proteins and RNAs; the second encodes information for proteins that regulate the activity of structural genes.

The information in the genes is stored in sequences of three nucleotides (bases), each of which codes for both the type and position of amino acids in the polypeptide. In the case of regulatory sequences, the sequence is critical for recognition by signal molecules. These sequences can be classified on the basis of their roles during the process of transcription. Promoters and operators, located at the 5′ end of prokaryotic structural gene clusters, are sites of attachment of RNA polymerase (promoters) and repressor protein (operators). Other types of regulatory sequences are required to control the level or rate of transcription. Two such sequences are the attenuators found in prokaryotes and the enhancers found in eukaryotes. Attenuators affect the transcription of genes coding for enzymes needed for amino acid biosynthesis. Enhancers appear to affect the binding of RNA polymerase to eukaryotic promoters.

Prokaryotic structural genes are usually continuous informational sequences of nucleotides. Eukaryotic genes, on the other hand, are usually split. Eukaryotic genes contain noncoding sequences interspersed among coding sequences. The noncoding sequences are called introns (intervening sequences), and the coding sequences are called exons (expressed sequences).

Some viruses and eukaryotes have been found to contain nucleic acids in which there are overlapping informational nucleotide sequences so that more information can be stored in a molecule of a given length than would otherwise be the case. For example, using overlapping genes, the DNA of ØX174 is able to hold at least three more genes than it could if all its genes were contiguous.

Although most genes occupy a more or less stable position within the DNA molecule, there is a group of genes for which mobility is the normal state of affairs. These are the transposable elements, or transposons, found in diverse life forms, both prokaryotic and eukaryotic.

Besides protein structural genes, all cells also have RNA structural genes. These genes are needed for the production of ribosomal RNAs (rRNAs) and transfer RNAs (tRNAs). In eukaryotic cells, the hundreds of rRNA genes are located in clusters in the nucleolar organizer regions. There are at least 60 different genes for the various tRNAs needed by the cell, though the organization of these genes is not completely understood.

To transcribe or decode the protein and RNA genes, prokaryotes make do with a single RNA polymerase. Eukaryotes, in contrast, require a special RNA polymerase for each type of gene.

## Study Questions

1. What is the chemical composition of the gene?
2. What are the two types of genes?
3. Briefly describe the role of the end product of each type of gene.
4. Give a one-sentence description of a gene.
5. How is the information stored in a gene?
6. How do structural genes and regulatory sequences differ?
7. What kinds of information regarding amino acids is stored in the genes needed for protein synthesis?
8. What structural features distinguish eukaryotic genes from prokaryotic genes?

9. Briefly describe each of the following: overlapping genes, split genes, and transposons.
10. Describe the role of each of the following in gene expression: promoters, operators, attenuators, and enhancers.

## Readings and References

Beadle, G. W., and E. L. Tatum. 1941. Genetic control of biochemical reactions in *Neurospora*. *Proc. Natl. Acad. Sci. USA* 27:499–506.

King, R. C., and W. D. Stansfield. 1990. *A dictionary of genetics*. New York: Oxford University Press.

Lewin, B. 1997. *Genes VI*. New York: Oxford University Press.

Oliver, S. G., and J. M. Ward. 1985. *A dictionary of genetic engineering*. Cambridge: Cambridge University Press.

Singer, M., and P. Berg. 1991. *Genes and genomes*. Mill Valley, Calif.: University Science Books.

Travis, J. 1992. Possible evolutionary role explored for "jumping genes." *Science* 257:884.

Voet, D., and J. G. Voet. 1995. *Biochemistry*. 2d ed. New York: John Wiley & Sons.

Watson, J. D. 1976. *Molecular biology of the gene*. 3rd ed. Menlo Park, Calif.: W.A. Benjamin.

CHAPTER SIX

# THE GENETIC CODE

## OUTLINE



## CHAPTER OBJECTIVES

*This chapter will discuss:*

- The characteristics of the genetic code
- How the codons of DNA and RNA are different and how they are used
- How a redundant genetic code operates
- Some variations of the genetic code
- The functions of some special codons
- Some terms and phrases used in discussions of the genetic code

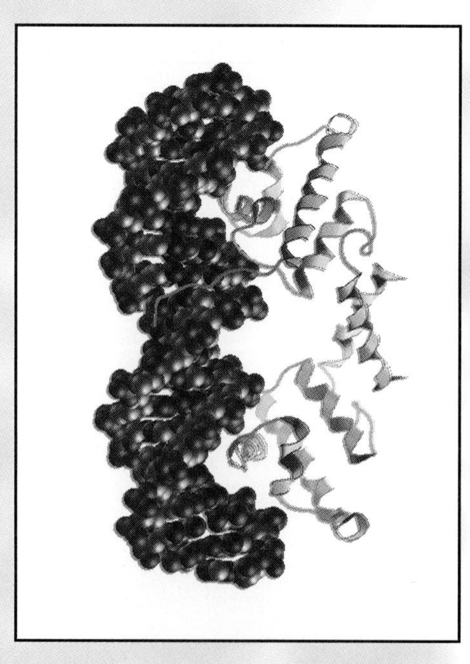

## INTRODUCTION

The genetic information of all organisms—from viruses to human beings—is found in one or the other nucleic acid, DNA or RNA. For all cellular life forms, the information is in the DNA molecule. The question is: In what form is the information stored? We have alluded to the answer several times by saying that the information is in the sequence of bases, which code for polypeptides (or RNA). What we need to know now is how the sequences of nucleotides are related to the storage of discrete bits of information. In other words, How is the seemingly monotonous repetition of four nucleotides parceled into unique portions of genetic information? How is the gene constructed so as to contain information?

In the 40 or so years since scientists began trying to unravel what has come to be called the genetic code, we have learned that the code has a number of general features and that the code directs or instructs the cell's protein-synthesizing machinery to insert a particular amino acid into a particular position (Chapter 5) of the desired polypeptide in accordance with the instructions found in DNA.

There is, then, a relationship between the nucleotide sequence of a gene and the amino acid sequence of a polypeptide. That relationship is what we call the genetic code. So we have two unrelated chemical species interacting via the genetic code. **Nucleotides** serve as the coding species for **amino acids.**

However, as noted in Chapter 5, the gene is a static unit. It does not do anything during protein synthesis. Things are done to it. So the genetic information must be translated into something that the cell can use. That something is a protein molecule.

It is the array of protein molecules found in the cell that gives cells their unique characters and abilities to function in particular ways—as a brain cell or liver cell, a bacterium, a plant cell, and so on. In effect, the protein is the molecule that is the physical manifestation of genetic information.

In reality, the genetic code is a part of the DNA molecule (it is the genes contained therein). But normally, in discussions of the code, instructions, or information of protein synthesis, we refer to what appears in mRNA as the code. The reason is that, when it was discovered that this species of RNA delivers the instructions (message) from gene to protein-synthesizing machinery, the code was deciphered using mRNA.

## GENERAL FEATURES

A careful look at the genetic code reveals a number of general characteristics.

1. The code is a linear sequence of nucleotides. That is, the information stored in the genes is in the sequence of nucleotides occurring in one or the other strand of double-stranded DNA. The information is not apportioned in any way across the strands; information is not contained in a series of nucleotides alternating from one strand to the other. Genes themselves may be found on either strand of the double helix.
2. The discrete bits of information are in the form of triplets that are more commonly referred to as codons. One triplet, or **codon,** of three nucleotides specifies one amino acid.
3. The genetic code is unambiguous and redundant (or degenerate). It is unambiguous in that each codon specifies one amino acid only. At the same time, many amino acids have more than one codon, so the code is said to be redundant or degenerate.
4. There are three stop codons in the code. These triplets do not specify amino acids. They are "stop" signals, or positions contained in the message.
5. There are no punctuations within a message. A message is read from beginning to end without interruption. In cases in which an mRNA molecule carries more than one message, the mRNA is said to be polycistronic or polygenic. In such cases, individual messages are separated by special stop/start codons. Again, however, within a given message, there are no punctuations.
6. A message is "read" three nucleotides, or one codon, at a time. This means that once reading begins during protein synthesis the information in the message can be interpreted only three nucleotides at a time: that is, only between triplets.

   As we shall see in the chapter on mutations (Chapter 9), this strict reading of the message has serious consequences; deletions or additions of nucleotides scramble the reading frame and the message, resulting in the synthesis of an altered protein whose biological function may also be altered.
7. There is no overlapping among codons. In a given message, any nucleotide of a triplet is a part of only that triplet and not a part of any adjacent triplet.
8. The genetic code is nearly universal from viruses to human beings. With a very few exceptions, the codons specify the same amino acids in almost all organisms.

## THE 64 CODONS: DNA AND RNA

As we have seen, DNA is composed of four components, the four nucleotides A, T, G, and C. The information in genes is made up of these nucleotides arranged in unique sequences in each gene. This information concerns the type and location of amino acids to be found in a protein. (Our discussion, unless indicated otherwise, will concentrate on protein synthesis.)

The nucleotides are found in a series of groupings called codons, each of which is made up of three nucleotides

| AAA | Phe | AGA | Ser | ATA | Tyr | ACA | Cys |
|---|---|---|---|---|---|---|---|
| AAG | | AGG | | ATG | | ACG | |
| | | AGT | | | | | |
| AAT | Leu | AGC | | ATT | STOP | ACT | STOP |
| AAC | | | | ATC | | | |
| GAA | | GGA | Pro | | | ACC | Trp |
| GAG | | GGG | | GTA | His | | |
| GAT | | GGT | | GTG | | GCA | Arg |
| GAC | | GGC | | | | GCG | |
| | | | | GTT | Gln | GCT | |
| TAA | Ile | TGA | Thr | GTC | | GCC | |
| TAG | | TGG | | | | | |
| TAT | | TGT | | TTA | Asn | TCA | Ser |
| | | TGC | | TTG | | TCG | |
| TAC | Met | | | | | | |
| | | CGA | Ala | TTT | Lys | TCT | Arg |
| CAA | Val | CGG | | TTC | | TCC | |
| CAG | | CGT | | | | | |
| CAT | | CGC | | CTA | Asp | CCA | Gly |
| CAC | | | | CTG | | CCG | |
| | | | | | | CCT | |
| | | | | CTT | Glu | CCC | |
| | | | | CTC | | | |

FIGURE 6.1
DNA codons and the amino acids they specify. Anticoding or template strand of DNA is shown.

(triplets). Each codon specifies an amino acid, and the codon's position in the gene specifies that amino acid's position in the polypeptide.

The groundwork for deciphering the genetic code was established by Marshall Nirenberg and Johann Henrich Mathaei (1961). Their work was followed in rapid succession by the work of Nirenberg and Philip Leder (1964) and, independently, Har Gobind Khorana (1967) and Severo Ochoa. In total, the body of evidence they accumulated broke the code and established the relationship between the gene, the sequence of nucleotides, and the 20 commonly occurring amino acids. The work of these scientists was done using in vitro protein-synthesizing techniques in which synthetic mRNA molecules were used as the source of information for the sequence of amino acids. Nirenberg and Mathaei had established that a homopolymer of uracil (e.g., UUUUUUUUU) coded for a polypeptide consisting solely of the amino acid phenylalanine. It was then a matter of changing the sequences of the mRNA until all possible combinations were used and the respective polypeptides' amino acid sequences determined. Codons were then assigned to each amino acid.

The **genetic code** (more properly, the genetic dictionary) consists of all the possible triplets that can be made using four nucleotides taken three at a time ($4^3 = 64$). The fact that there are three nucleotides in each codon is not an accident. Three nucleotides per codon is the fewest number that would give enough codons to specify each of the 20 amino acids. If there were two nucleotides per codon, there would not be enough codons to cover all the amino acids; four amino acids would be without codons ($4^2 = 16$). And, if there were more than three nucleotides per codon, say four, there would be an enormous surplus of codons for the 20 amino acids ($4^4 = 256$).

Figure 6.1 lists the codons as they appear in DNA, and Figure 6.2 shows the codons as they appear in mRNA, which is the code commonly used. Note that of the 64 codons of mRNA (Figure 6.2), three codons are not as-

Second position

| First position (5′ end) | U | C | A | G | Third position (3′ end) |
|---|---|---|---|---|---|
| U | UUU Phe | UCU Ser | UAU Tyr | UGU Cys | U |
| U | UUC Phe | UCC Ser | UAC Tyr | UGC Cys | C |
| U | UUA Leu | UCA Ser | UAA STOP | UGA STOP | A |
| U | UUG Leu | UCG Ser | UAG STOP | UGG Trp | G |
| C | CUU Leu | CCU Pro | CAU His | CGU Arg | U |
| C | CUC Leu | CCC Pro | CAC His | CGC Arg | C |
| C | CUA Leu | CCA Pro | CAA Gln | CGA Arg | A |
| C | CUG Leu | CCG Pro | CAG Gln | CGG Arg | G |
| A | AUU Ile | ACU Thr | AAU Asn | AGU Ser | U |
| A | AUC Ile | ACC Thr | AAC Asn | AGC Ser | C |
| A | AUA Ile | ACA Thr | AAA Lys | AGA Arg | A |
| A | AUG Met | ACG Thr | AAG Lys | AGG Arg | G |
| G | GUU Val | GCU Ala | GAU Asp | GGU Gly | U |
| G | GUC Val | GCC Ala | GAC Asp | GGC Gly | C |
| G | GUA Val | GCA Ala | GAA Glu | GGA Gly | A |
| G | GUG Val | GCG Ala | GAG Glu | GGG Gly | G |

FIGURE 6.2
The mRNA genetic code. All 64 possible codons are listed with the amino acids they specify. The codons are read in the 5′ → 3′ direction. Note the single codon for methionine (MeT) and the three stop codons.

signed amino acids, they are termination sequences (UAA, UAG, UGA). These act as punctuation marks in mRNA, indicating the end of a message, the period at the end of a sentence. Also note that, with the exception of the amino acids methionine (Met) and tryptophan (Trp), all the amino acids have more than one codon (redundancy) but that no codon specifies more than one amino acid (unambiguity).

Methionine and tryptophan appear in proteins relatively rarely. The other amino acids appear more frequently. There is a correlation between frequency of appearance and number of codons. For example, serine (Ser) and leucine (Leu) are widespread in proteins and have the largest number of codons (6 codons each). Other amino acids that appear in proteins in moderate amounts have more codons (2–4 codons) than methionine and tryptophan but fewer than serine or leucine. Methionine also has a special role in protein synthesis, as we shall see.

## THE WOBBLE HYPOTHESIS

In 1955, Francis H. Crick suggested that the information in the genes, the nucleotide sequences, is translated into polypeptide amino acid sequences by way of an adaptor molecule. He proposed that the molecule carries a particular amino acid and "reads" the codons of mRNA. To facilitate this reading, the adaptor molecule, he said, should contain RNA to make base pairing between the adaptor and messenger molecules possible. Crick was correct. Shortly after his proposal was made, Mahlon B. Hoagland and Paul Zamecnik and their colleagues published a report (1958) identifying a soluble RNA molecule needed for protein synthesis. The molecule they discovered is now known to be transfer RNA (tRNA), and it has the characteristics Crick had predicted. For the moment, all we need to know is that tRNAs have a sequence of three nucleotides, called the **anticodons,** that base pair with codons of mRNA. And for each amino acid there is a specific tRNA to carry it to the site of protein synthesis where the message is read. (We shall deal with tRNA in more detail in Chapter 8.)

In 1966, Crick then proposed the **"wobble hypothesis."** In his publication, he said, "It is shown that such a wobble could explain the general nature of the degeneracy of the genetic code."

With few exceptions, the third base of a codon (at the 3′ end of the triplet) is not critical. Crick proposed that, via a wobble fit of the third base pairing, a single tRNA can read several codons for a particular amino acid. In this way, although there are 61 codons specifying amino acids, fewer than that number of tRNAs are required to read mRNA.

Table 6.1 shows the allowable base pairings of the wobble hypothesis. Notice the unusual pairings: C, U, or A pairing with I (inosine, a nucleotide intermediate in the synthesis of both adenine and guanine) and A-U, G-U.

*table 6.1*

WOBBLE PAIRINGS

| mRNA BASE AT 3′ END | tRNA BASE AT COMPLEMENTARY 5′ END |
|---|---|
| _ _ C<br>_ _ U<br>_ _ A | _ _ I (inosine) |
| _ _ C<br>_ _ U | _ _ G |
| _ _ A<br>_ _ G | _ _ U (modified nucleotides) |

## THE READING FRAME

As we have seen, codons are composed of three nucleotides. The sequence of codons (nucleotides) in an mRNA molecule between a start sequence and a stop sequence is defined as the reading frame.

table 6.2

DEVIATIONS FROM THE UNIVERSAL GENETIC CODE

| SOURCE | CODON | USUAL MEANING | NEW MEANING |
|---|---|---|---|
| Fruit fly mitochondria | UGA | STOP | Tryptophan |
| | AGA & AGG | Arginine | Serine |
| | AUA | Isoleucine | Methionine |
| Mammalian mitochondria | AGA & AGG | Arginine | STOP |
| | AUA, AUU & AUC | Isoleucine | Methionine |
| | UGA | STOP | Tryptophan |
| Yeast mitochondria | CUN* | Leucine | Threonine |
| | AUA | Isoleucine | Methionine |
| | UGA | STOP | Tryptophan |
| Protozoa nuclei | UAA & UAG | STOP | Glutamine |
| Higher plants | UGA | STOP | Tryptophan |
| | CGG | Arginine | Tryptophan |
| *Mycoplasma* | UGA | STOP | Tryptophan |

*Source:* From Robert F. Weaver and Philip W. Hedrick, *Genetics*, 2nd edition. 
*N = any base.

Once again, it was Crick who, in 1961, presented experimental evidence for the nature of the reading frame. His publication states, "The sequence of the bases is read from a fixed starting point. This determines how the long sequences of bases are to be correctly read off as triplets. There are no special 'commas' to show how to select the right triplets. If the starting point is displaced by one base, then the reading into triplets is displaced, and thus becomes incorrect" (see Chapter 9).

In short, there is a strict reading of the codons as triplets in mRNA. Once reading begins at the start sequence, the nucleotides are read three at a time, and there is no signal within the messenger molecule to indicate that a correct or incorrect triplet is being read. If a nucleotide is deleted from or inserted into the message, a frame shift occurs. The nucleotides in the mRNA are still read three at a time as triplets, but there is a displacement caused by the deletion or insertion. And, because the wrong sets of triplets would be read, an incorrect amino acid sequence would result. A polypeptide would still be made, but, in all probability, it would have little if any biological activity.

For example, if, beyond the starting point, the correct codon sequence is UUUGUCAACGGAUGG, the amino acid sequence will be phenylalanine (Phe)-valine (Val)-asparagine (Asn)-glycine (Gly)-tryptophan (Trp).

If now the first uracil (U) is deleted, we have a message that reads: UUGUCAACGGAUGG . . . This would give us an amino acid sequence of leucine (Leu)-serine (Ser)-threonine (Thr)-aspartate (Asp)-glycine (Gly).

If, instead of a deletion, an insertion occurred, there would be a similar scrambling of the message. The resulting change in the amino acid sequence might seriously distort the folding of the polypeptide or protein and thereby affect the molecule's biological activity (Chapter 8).

## THE ALMOST UNIVERSAL GENETIC CODE

Until 1979, the commonly held view was that all organisms, from viruses to human beings, used the same genetic dictionary. That is, in all organisms studied to that date, the same codons were found to specify the same amino acids. And, in fact, we know now that for virtually all organisms, whether their genomes are composed of DNA or RNA, the same coding dictionary is used.

The exceptions began to be reported in 1979. First, mitochondria from the fruit fly *Drosophila melanogaster* and from yeast were found to have a slightly different dictionary. Later, human mitochondria, some protozoa, bacteria, and plant chloroplasts were discovered with variations in their genetic codes. Table 6.2 lists all the known variations of the genetic dictionary. What effects these deviations might have on evolution or protein synthesis is still unclear.

So the genetic dictionary is virtually universal. And, although there appears to be no particular reason or advantage for a given codon to specify a particular amino acid, it does seem that once the code was established any major deviations would almost certainly be lethal for the organism. Many amino acids play crucial roles in establishing and maintaining particular protein conformations. These conformations are required for the biological activity of the protein; so a change

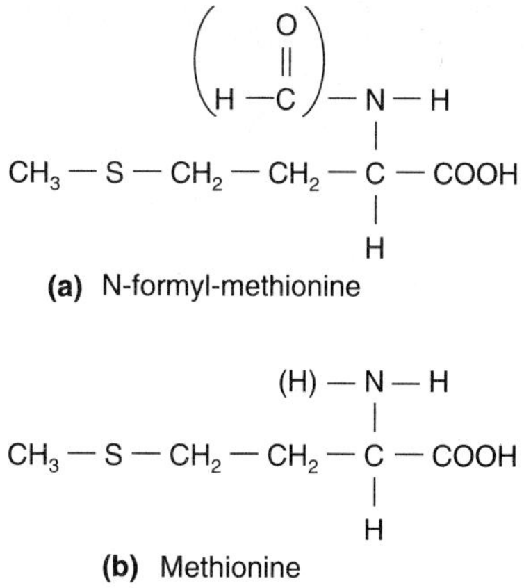

$$\begin{array}{c} \left(H-\overset{O}{\overset{\|}{C}}\right)-N-H \\ | \\ CH_3-S-CH_2-CH_2-C-COOH \\ | \\ H \end{array}$$

**(a)** N-formyl-methionine

$$\begin{array}{c} (H)-N-H \\ | \\ CH_3-S-CH_2-CH_2-C-COOH \\ | \\ H \end{array}$$

**(b)** Methionine

FIGURE 6.3
N-formyl methionine (a) and methionine (b). Note that one of the (H) of the nitrogen group in methionine is replaced by a formyl group $\left(H-\overset{O}{\overset{\|}{C}}-\right)$.

in conformation could lead to the production of a useless polypeptide. In the absence of the normal protein, the organism might not survive. Once a coding dictionary was established, therefore, it had to remain unchanged.

To alter the code would require not only changes in codons but also changes in tRNAs so that new codons specifying new amino acids could be read. In turn, the tRNA genes would have to be altered as would the enzymes that attach the proper amino acids to the proper tRNAs. It is not a simple thing to change the coding dictionary. And, even though the particular coding dictionary now in use might not have been inevitable, because of the dire consequences of massive changes, it is for all intents and purposes locked into living organisms.

## SPECIAL SEQUENCES: INITIATION AND TERMINATION

### Initiation

In both prokaryotes and eukaryotes, the process of reading the mRNA (called translation) begins at a unique codon, often referred to as **the initiation codon.** This is the triplet AUG, which specifies an unusual form of the amino acid methionine; the codon is located toward the 5′ end of mRNA.

In prokaryotes, when AUG is located in this position, as opposed to any other position within the reading frame, it is translated not as ordinary methionine (Figure 6.3b) but rather as **N-formyl-methionine (fMet)** (Figure 6.3a).

When found anywhere else in the message, AUG is read as ordinary methionine. A special tRNA ($tRNA^{fMet}$) translates AUG at the beginning of messages, whereas the common tRNAmet translates AUGs found within messages. Less frequently, the codon GUG, which codes for the amino acid valine, has also been found to act as the initiation codon.

The eukaryotes also use AUG as the starting or initiating codon and a special tRNA to translate this codon when it appears at the 5′ end of mRNA. However, unlike prokaryotes, higher organisms do not formylate the methionine before it is used as the initiating amino acid in protein synthesis. When AUG is found in the interior of a message, it is translated by ordinary tRNA.

Mammalian mitochondria are unusual in this regard. They use not only AUG as the initiating codon specifying N-formyl-methionine, but also AUA, AUU, and AUC. At the initiation position, all four of these codons specify N-formyl-methionine, and all are recognized by the special tRNA. These codons all code for methionine in the human mitochondrion when found within a reading frame of mRNA.

### Termination

Just as there is a special codon to indicate the start of a message or reading frame, so too there is a stop signal at the end of the message. In fact, all mRNAs studied to date have at least one stop codon, and, occasionally, tandem stop codons appear just after the last codon in a message specifying an amino acid.

The nearly universal **stop codons** or **termination sequences** are UAG, UGA, and UAA. In mammalian mitochondria, the codons AGA and AGG which normally code for the amino acid arginine, code instead as stop signals.

The three most commonly occurring stop signals have been given the names **Ochre** (UAA), **amber** (UAG), and **opal** (UGA). They are also frequently referred to collectively as **nonsense codons.** This misnomer was originally applied to the three termination sequences because no amino acids could be assigned to them. In fact, as early as 1966, Crick and his co-workers suggested the possibility that UGA is not recognized as an amino acid codon.

Termination of polypeptide synthesis occurs when, during the process of reading mRNA (translation), one of the stop codons arrives at the A site of the ribosome. This is where codon recognition by tRNA takes place. Since, however, there is no tRNA capable of reading the stop codons, no amino acid is added to the growing polypeptide chain.

At this point in the process, special **release factors** assist in several actions. They help remove the completed polypeptide from the tRNA holding it, and they help remove the tRNA from the ribosome. The special release factors then separate from each other, the mRNA and the ribosomal subunits. Precisely how these actions are brought about by the release factors is not clear at the moment.

**table 6.3**

TERMS USED IN REFERENCE TO THE GENETIC CODE

| TERM | MEANING |
|---|---|
| Anticodon | A sequence of three nucleotides in tRNA complementary to the codon of mRNA |
| Code letter | A nucleotide in a codon of DNA or mRNA |
| Codon | A sequence of three nucleotides specifying an amino acid or a stop signal |
| Commaless code | There are no punctuations between codons |
| Frameshift mutation | Changes in the reading frame resulting from additions or deletions other than multiples of three |
| Missense mutation | Changes in codons so that another amino acid is specified |
| Nonoverlapping code | A nucleotide is a part of one codon only |
| Nonsense mutation | Changes in codons producing stop sequences |
| Reading frame | The nucleotide sequence between the start and stop codons specifying a polypeptide |
| Redundant, or degenerate, code | An amino acid may have more than one codon |
| Unambiguous code | Each codon specifies one amino acid only |
| Universal code | The genetic code as used by all organisms |

The known prokaryotic release factors are **RF-1,** which recognizes stop codons UAA and UAG; **RF-2,** which recognizes either UAA or UGA; and **RF-3,** whose activity may be to facilitate the actions of RF-1 and RF-2. It is assumed that, since both RF-1 and RF-2 recognize UAA, the stop signal UAA is the most efficient terminator of polypeptide synthesis, although termination may be influenced by sequences flanking the stop codons.

Eukaryotes appear to have a single release factor, designated **eRF.** Unlike the prokaryotic system, eukaryotic termination is energy-dependent and requires the participation of the high-energy compound GTP.

## GENETIC TERMS

A great many words and phrases are used to describe the various characteristics and components of the genetic code. Table 6.3 lists some of the commonly used words and phrases.

## SUMMARY

The genetic information of all organisms, whether in DNA or RNA, is stored in the form of sequences of nucleotides (bases).

These sequences are triplets of nucleotides called codons. Each codon specifies an amino acid.

All the possible codons taken together make up the genetic code. There are 64 codons in the code, and all but three of them specify amino acids. The three codons not specifying amino acids are stop, or termination, signals.

The relationship between a gene and the polypeptide for which the gene codes is the genetic code. Genetic information is made useful to the cell via protein synthesis. The protein molecule is the physical manifestation of genetic information.

The genetic code is linear, contains discrete bits of information in the form of triplets, is redundant and unambiguous, has no punctuation marks, and is nonoverlapping.

Although each codon consists of three nucleotides, the first two nucleotides (beginning at the 5′ end) are critical for the codon's role as a store of information.

The third base is not vital and may form unusual base pairings during protein synthesis. This phenomenon is often referred to as the wobble fit.

The information of the codons in an mRNA molecule is read by another RNA molecule: transfer RNA, or tRNA. This species of RNA carries amino acids and reads, by way of base pairing, mRNA.

Each mRNA has a particular starting point toward its 5′ end, identified by the codon AUG, which is referred to as the initiation or start codon. The information of mRNA then ends at a stop, or termination, sequence located toward the 3′ end of the mRNA. The stop codons are triplets that are not read by any tRNA. Between the start and the stop signals is the open reading frame.

Virtually all organisms studied to date use the same genetic code. There are, however, some exceptions to be found among mammalian mitochondria, plant chloroplasts, and some protozoa and bacteria.

## Study Questions

1. List at least four of the general features of the genetic code and briefly explain the significance of each.
2. How many total codons are there in the genetic code?
3. How many of the codons of the genetic code specify amino acids, and what is the function of the remaining codons?
4. What is the importance of the wobble hypothesis?
5. What does the phrase *universal genetic code* mean, and why is the code said to be almost universal?

## Readings and References

Crick, F. H. C. 1966. Codon-anticodon pairing: The wobble hypothesis. *J. Mol. Biol.* 19:548–55.

Crick, F. H. C., L. Barnett, S. Brenner, and R. J. Watts-Tobin. 1962. General nature of the genetic code for proteins. *Nature* 192:1227–32.

Hoagland, M. B., M. L. Stephenson, J. F. Scott, L. I. Hecht, and P. C. Zamecnik. 1958. A soluble ribonucleic acid intermediate in protein synthesis. *J. Biol. Chem.* 231:241–57.

Hofstadter, D. R. 1982. Is the genetic code an arbitrary one, or would another code work as well? *Sci. Am.* 246 (March):18–29.

Khorana, H. G., H. Buchi, H. Gosh, N. Gupta, T. M. Jacob, H. Kossel, R. Morgan, S. A. Narang, E. Ohtsuka, and R. D. Wells. 1967. Polynucleotide synthesis and the genetic code. *Cold Spring Harbor Symp. Quant. Biol.* 31:39–49.

Nirenberg, M., and P. Leder. 1964. RNA codewords and protein synthesis. *Science* 145:1399–407.

Nirenberg, M., and J. H. Mathaei. 1961. The dependence of cell-free protein synthesis in *E. coli* upon naturally occurring or synthetic polyribonucleotides. *Proc. Natl. Acad. Sci. USA* 47: 1588–602.

Singer, M., and P. Berg. 1991. *Genes and genomes*. Mill Valley, Calif.: University Science Books.

Speyer, J. F., P. Lengyel, C. Basilio, and S. Ochoa. 1962. Synthetic polynucleotides and the amino acid code, IV. *Proc. Natl. Acad. Sci. USA*. 48:441–48.

CHAPTER SEVEN

# CONTROL OF GENE EXPRESSION IN PROKARYOTES

## OUTLINE



## CHAPTER OBJECTIVES

*This chapter will discuss:*

- The two ways by which cells adapt to changes in the internal and external environments
- The control of biosynthetic and biodegradative pathways
- Fine-tuning: how cells control enzyme activity
- Coarse control: how cells regulate gene expression
- What is an operon
- The operons: *lac*, *trp*, and *ara*; their components and mechanisms of control

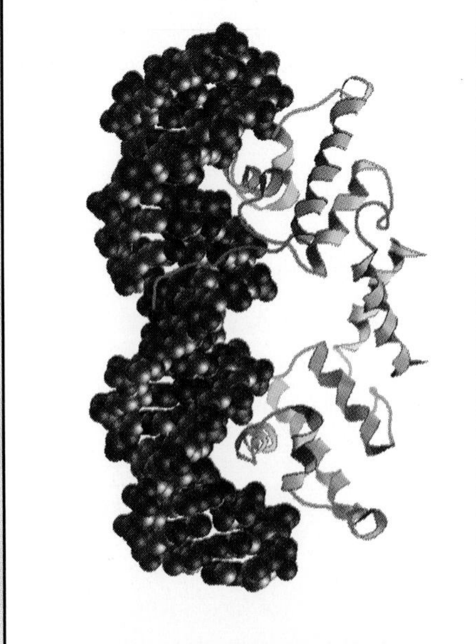

## INTRODUCTION

Having looked at genes and the genetic code we now move on to a discussion of how the information in genes (and the genetic code) is controlled and used by the cell. We are moving from structure to function.

Genes must first be activated, or expressed (i.e., turned on; the topic of this chapter), and then the information thus made available to the cell is used to synthesize proteins (the topic of Chapter 8). So, genes are expressed, and the information is used to produce enzymes, receptors, carriers, and so on (Chapter 4).

All cells, from the relatively simple bacteria to the complex, highly specialized cells of a multicellular organism such as the human being, must be able to respond to an ever-changing internal environment that, to a large extent, is at the mercy of the external environment. That is, as conditions outside the cell change, the cell must adapt. Not to do so risks the survival of the cell or of the organism.

The cell adapts by responding to signals generated either at the cell surface, the cytoplasmic membrane, or within the cytoplasm. If, for example, the type of carbohydrate available in the external environment changes or the internal concentration of a particular amino acid changes, the cell may respond in essentially two ways: One, it may adjust the level of enzyme activity. Two, it may stop or start the processes needed for enzyme synthesis. That is, the cell may increase or decrease enzyme activity, or it may activate or inactivate the genes encoding the enzyme, thereby changing the amount of enzyme available.

The first method, adjusting the level of enzyme activity, can be regarded as a second-to-second fine adjustment of the cell's metabolism. What is important is that the enzyme is already present; its *activity* is being regulated. The second method, starting or stopping enzyme synthesis, is coarse control; it is not the level of enzyme activity being modulated but rather the presence or absence of the enzyme.

An analogy to this type of control is the dimmer switch used to control lighting. First, the switch is used to turn the light on or off (coarse control); second, the switch can be used to control the level or intensity of the lighting (fine control). Just so, cells can control whether a gene is on or off and the level of activity of the gene's end product, the enzyme.

At this point, it might be helpful to explain why so much emphasis is placed on enzyme synthesis in discussing the various topics of molecular genetics rather than using or studying the control of the synthesis of other functional groups of proteins such as structural proteins, transport systems, or receptor proteins. The answer is that, of all the functional groups of proteins, the one with the easiest-to-measure biological activity is the enzyme group. We can and do study genes by studying their end products, the proteins in particular. And the easier it is to study the product, the easier it is to study the gene. So we study an easily identifiable **phenotype** (the observable characteristics that are determined by the genes) to learn about the genotype (the complement of genetic information that the cell contains).

## CONTROL OF METABOLISM

The metabolism (the sum total of all biological activity) of a cell can be divided into two broad categories: (1) biosynthesis, or **anabolism,** and (2) biodegradation, or **catabolism. Biosynthesis** involves the activities required to make all of the various cellular components—from the largest to the smallest, from membranes and organelles to DNA, proteins, and simple amino acids such as glycine. **Biodegradation** includes all those activities necessary to break down the nutrients such as sugars, proteins, and fats, that cells use to maintain and reproduce themselves.

As a rule of thumb, the biosynthetic pathways of prokaryotes are active and must be inactivated. Amino acids are in constant demand for the purpose of protein synthesis. The various pathways by which amino acids are produced must, therefore, be continually in operation and monitored. If there is a momentary abundance of any one amino acid, the rate of synthesis of that amino acid can be reduced by regulating the activity of a key enzyme. If, however, the concentration of the amino acid still is not reduced sufficiently, the genes responsible for the enzymes needed to synthesize the amino acid are turned off, so no more enzyme is produced. This type of control is called repression.

On the other hand, the biodegradative pathways, especially those used for the production of energy, are generally inactive, except for the main pathways, such as glycolysis, which are always active. Glycolysis is the pathway by which glucose, the major energy source of most organisms, is first degraded. The enzymes in this system are present whether or not the organism is using glucose as an energy source. These enzymes are said to be **constitutive;** they are present whether or not their substrate is present. In other words, the genes coding for these enzymes are generally active, or turned on.

In the event that glucose is not available to the cell, other genes are activated depending on what sugar is present. For example, if lactose (the milk sugar) replaces glucose, the genes coding for the enzymes needed to utilize lactose are activated. This type of control is called induction, and a modified form of **lactose** is the inducer. Figure 7.1 compares inducible and repressible systems.

In this chapter we shall look at three ways by which genes can be controlled at the level of mRNA synthesis.

| Pathway | Type | Regulatory mechanism | Repressor | Operon | Signal for gene expression |
|---|---|---|---|---|---|
| Anabolic (biosynthesis) | Uses energy | Enzyme repression (*trp* operon) | Inactive → activated and binds DNA | On → off | Absence of end product |
| Catabolic (biodegradation) | Releases energy | Enzyme induction (*lac* operon) | Active → inactivated and releases DNA | Off → on | Presence of nutrient |

FIGURE 7.1
Comparison of inducible and repressible operons.

The first example concerns the genes needed for lactose utilization and involves catabolic enzymes (i.e., enzymes used in biodegradative pathways). It shows how both negative and positive controls work. These genes are inactive and must be activated. This example also demonstrates how the process of catabolite repression operates. **Catabolite repression** is a means by which a cell is able to choose between glucose and other sugars as an energy source. The cell will preferentially utilize glucose even if the other sugars are available, thereby avoiding the necessity of having to synthesize new enzymes. The glucose-degrading enzymes are already present; they are constitutive enzymes. Glucose is the catabolite repressing the synthesis of other, inducible, enzymes.

The second example is the series of genes that code for the enzymes needed to synthesize the amino acid tryptophan. These are anabolic enzymes (i.e., enzymes needed in biosynthetic reactions). The genes are active and must be inactivated. In this case, the genes are regulated or controlled at two levels: repression and attenuation (i.e., a mechanism by which gene expression is controlled by terminating mRNA synthesis prematurely).

The final example of gene control at the mRNA level is the so-called *ara* cluster of genes. These genes code for the enzymes needed for the degradation of the sugar *arabinose*. We will look at this system because it demonstrates how both positive and negative control are possible using a single regulating protein molecule.

# FINE CONTROL OF CELLULAR ACTIVITY

Let us look briefly at some examples of how cells regulate enzyme activity so that we can gain some perspective on the regulation of gene expression, the subject of this chapter.

Enzyme activity is regulated by changes in the conformation of the enzyme (remember, all enzymes, except for ribozymes, are proteins; Chapter 4). The altering of conformation is also important in understanding how repressor proteins interact with the DNA of the genes to bring about regulation of gene expression.

## Regulation of Enzyme Activity

There are basically four methods of regulating enzyme activity. The conformation of the enzyme can be changed by an effector molecule, by attachment of a chemical group, by binding or releasing other polypeptides, or by proteolysis.

Some enzymes, called **allosteric enzymes,** are controlled by an **effector molecule,** a molecule other than the enzyme's substrate. The term *allosteric,* from Greek, means "another site." Allosteric enzymes, then, have two sites. One site is the **catalytic site.** This is the business part of the molecule where substrate molecules are attached and ultimately are converted to product. But for the substrate to fit the catalytic site, the site must be adjusted (remember the fit of oxygen in the hemoglobin molecule; (Chapter 4). The adjustment is brought about by the effector molecule, which attaches to the "other site" and causes a conformational change in the protein. The change allows the substrate to enter and bind to the catalytic site. So, whether the enzyme is active depends on the presence or absence of the effector molecule.

In the absence of the effector molecule, the enzyme is said to be in the T (tense) state; when, however, the effector molecule binds the allosteric site, the enzyme assumes the R (relaxed) state (Figure 7.2). This kind of control can be either positive (stimulatory) or negative (inhibitory).

An example of positive control is the effect of the metabolite β-D-fructose 2,6-*bis*phosphate on the rate of glycolysis. This effector molecule is a potent activator of the enzyme phosphofructokinase, the key control enzyme of the glycolytic pathway. The metabolite increases the affinity of phosphofructokinase for its substrate fructose and simultaneously decreases the inhibitory influence of ATP on the enzyme. The resulting increase in the rate of glycolysis, in turn, increases the rate of ATP synthesis.

An example of negative control is the action on the enzyme aspartate transcarbamolyase by cytidine triphosphate. This enzyme carries out the first reaction in pyrimidine (i.e., uracil, thymine, and cytosine) biosynthesis and is one of the best understood of the allosteric enzymes. High concentrations of the end product, cytidine triphosphate, of this anabolic pathway inhibit the enzyme.

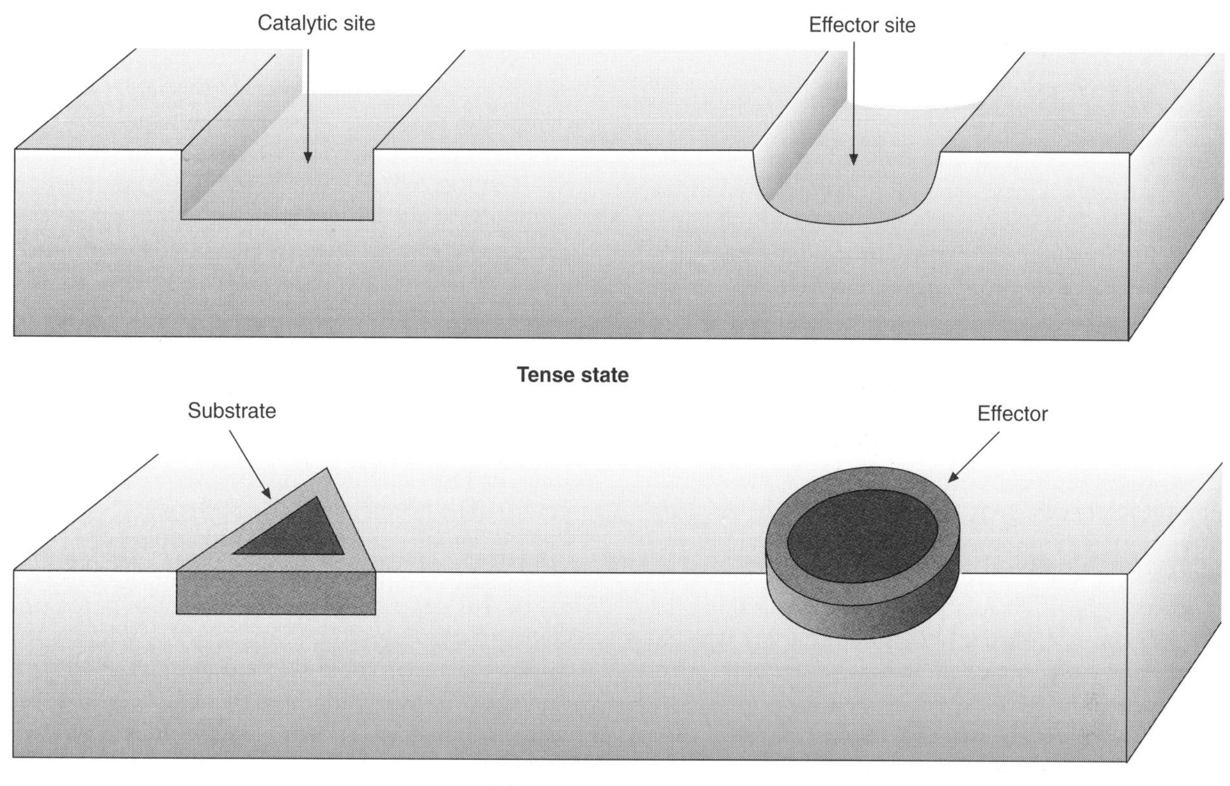


FIGURE 7.2
Regulation of allosteric enzyme activity.

Keep these examples of enzyme control via allosteric interaction in mind. They will go a long way in explaining the control or regulation of gene expression.

A second way by which enzymes can be controlled is via covalent modification of the protein (enzyme). Here, control is exerted by attachment of a chemical group, such as by addition of a phosphate, adenine, or methyl group to a specific amino acid of the protein. All these covalent additions are, of necessity, reversible.

A common method of control is to phosphorylate the enzyme at a serine residue. This method often involves the use of ATP as a phosphate donor and to provide the energy needed to couple the phosphate to the protein. An example of such control is the conversion of an inactive phosphorylase *b* to an active phosphorylase *a* in liver. The activation of this enzyme is critical to the breakdown of glycogen and the subsequent release of glucose into the bloodstream in response to the hormone epinephrine (adrenaline). This step is important in the so-called fight-or-flight response, by which individuals generate energy in order to either confront a danger or flee from it.

The third way of controlling enzyme activity is to bind or release other **polypeptides** to the protein to be controlled. The protein calmodulin is an example. Calmodulin is an intracellular calcium receptor that regulates several different enzymes, including many involved in the metabolism of cyclic nucleotides (e.g., cAMP) and glycogen. Calmodulin is sensitive to changes in the intracellular concentration of $Ca^{2+}$ ions. Increases in the $Ca^{2+}$ ion concentration activate calmodulin, which, in turn, binds to enzymes and activates them. Calmodulin may be the most common translator of one of the so-called second messages, changes in calcium ion concentrations. These second messages are triggered by such things as hormone interaction with membrane receptors—the first message. By this means, information from outside the cell can be relayed into the cell even though the informational molecule itself, the hormone, does not pass across the cell's plasma membrane. Many hormones such as adrenaline, insulin, and glucagon, because of their chemical makeup, cannot pass the membrane barrier, and the cell relies on second messages generated inside the cell to pass on the hormone's information.

Fourth, proteins in general and enzymes in particular can be influenced by proteolysis, a process by which proteins or polypeptides are digested. Unlike the preceding processes, proteolysis is irreversible and activates only inactive proteins; it does so by breaking peptide bonds and, in many cases, releases a small number of amino acids. The digestive enzyme chymotrypsin, for example, is converted from an inactive to an active form by this method, as are a number of enzymes needed for blood clotting.

## COARSE CONTROL OF CELLULAR ACTIVITY

Now let us look at the coarse control of metabolism, namely, regulation of gene expression. As indicated in the introduction to this chapter, an analogy to the dimmer light switch is useful in this regard. It should not be understood by the use of the word *coarse* that these systems are *inelegant*. The term *coarse* is used here to contrast this type of regulation, which is essentially an "all-or-nothing" affair, with the "fine-tuning" of enzyme activity just discussed. And it should also be kept in mind that, although fine-tuning is a second-to-second control, coarse control can also be a short-term control in prokaryotes, where the life span of mRNA is measured in minutes. Nonetheless, with this control we are concerned with the presence or absence of enzymes not with their activity; that is the important point. So the phrase "regulation of gene expression" implies the presence or absence of enzymes. Our task is to understand how the presence or absence of enzymes turns genes on and off.

Most of the models used to explain control of gene expression are based on work done with bacteria. The bacterial genome is relatively simple compared with that of the eukaryotes, and bacterial cells can easily be grown and manipulated.

Much more is known about the structure and function of the bacterial genome than is known about even the simplest eukaryotic genome, let alone the human genome. The genes of prokaryotes are tightly clustered, and there appears to be little DNA that does not encode information for protein or RNA synthesis. The bacterial DNA is a relatively large (compared with the cell), circular, double-stranded molecule.

The cell's external environment is constantly changing, and, for cells as small as the bacterium *Escherichia coli* (it would take something like 25,000 of these cells laid end to end to equal one inch), the changes can be rather drastic and sudden. Human cells, in contrast, are protected by virtue of both the physical presence and the activities of other cells.

Nutrients can and do appear and disappear with little, if any, warning to the cell. So the bacterium must be able to take advantage, very quickly, of a rapidly changing situation. Because of its small size, however, a bacterium cannot keep a full complement of its enzymes always at the ready. Such a concentration of protein would undoubtedly reduce the amount of free water available, and that reduction would surely affect adversely the cell's physiology. There is just not enough room in the cell.

But, perhaps more important, is the fact that it is an enormous waste of energy for the cell to continually synthesize what, at the moment, may be unnecessary proteins. As we shall see in this chapter and also in Chapter 8, each type of protein manufactured requires the production of a messenger RNA molecule. To make this molecule requires nucleotides in the triphosphate form, as with DNA synthesis, and these are produced at the expense of ATP.

It is to the cell's advantage to be able to control protein synthesis at the level of gene expression, where energy is used in the greatest quantities. Other steps in protein synthesis also require the expenditure of energy, so the entire process is energy-dependent.

### The Jacob and Monod Paradigm

One way by which cells control gene expression is by controlling the process of transcription, which is the synthesis of mRNA. The first, and still the best understood, example of the control of gene expression is that proposed by Francois Jacob and Jacques Monod in a paper published in 1961. In this paper, Jacob and Monod also proposed the existence of messenger RNA.

That organisms could respond to changes in the environment by synthesizing new enzymes, a process called induction, has been known at least since 1899, when E. Duclaux published his ideas on the subject. And, in 1900, F. Dienert noted that yeast cells grown on lactose or galactose as a carbon and energy source contained enzymes needed to metabolize those compounds. But when the cells were transferred to a glucose medium, the enzymes were lost. How the response to sugars other than glucose is brought about was explained by Jacob and Monod in their "operon model." They defined the operon as a cluster of genes coordinately controlled by an operator gene [*sic*].

The model they proposed, which has come to be called the *lac* operon model, deals with the transport and utilization of the disaccharide **lactose.** This sugar is composed of two smaller sugars, glucose and galactose (Figure 7.3). The *lac* operon model also explains how inactivation, or repression, of gene expression occurs.

Evidence for enzyme repression was first reported by Jacques Monod and Cohen-Bazire in 1953 in a paper describing the repression of the enzyme tryptophan synthase by the amino acid tryptophan. We shall return to this topic later in the chapter. For now, however, we can say that inducible operons, such as the one responsible for lactose utilization, are active only in the presence of the inducer; in this case, the inducer is a modified form of lactose. On the other hand, repressible operons, such as the one involving tryptophan synthesis, are active only in the absence of a **corepressor** a molecule that must bind the repressor protein in order for that repressor to function. A corepressor is often the gene end product; in the case of tryptophan synthesis, tryptophan is both the end product and the corepressor.

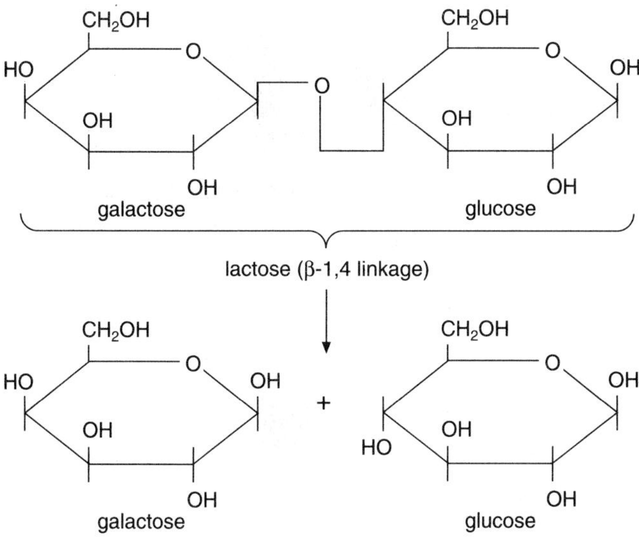


FIGURE 7.3
Lactose structure.

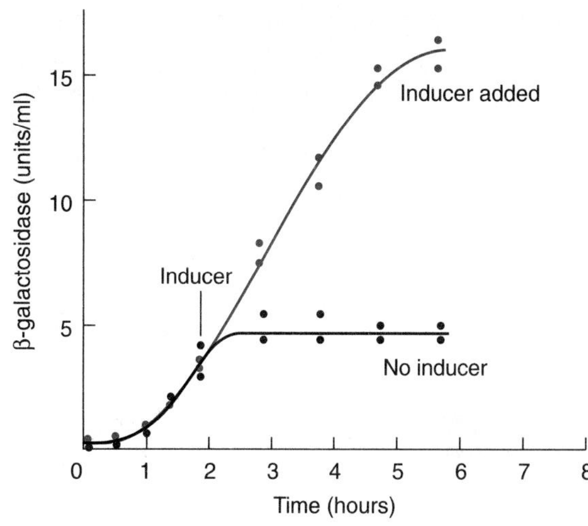


FIGURE 7.4
Effect of inducer on β-galactosidase synthesis.

# THE OPERON

Let us first define the concept and the term *operon* in somewhat greater detail. Then we shall describe which cellular activities are controlled by controlling gene expression.

The operon is a linear arrangement of genes whose activities are coordinately regulated. Prokaryotic genes, as noted, tend to be found in clusters, and rarely are individual genes transcribed. Rather, the gene cluster is controlled as a unit (coordinately regulated). The genes are activated or inactivated as a single unit. A cluster is regulated by a protein–DNA interaction. It is this ability to coordinately regulate gene clusters that allows prokaryotes to respond quickly to changing environmental conditions. In fact, what is being regulated is the synthesis of mRNA via regulation of the process of transcription.

The Jacob and Monod *lac* operon has a molecule of mRNA with a half-life of approximately 3 minutes. That is, at each 3-minute interval one-half of the remaining mRNA molecules are destroyed. Such a short half-life allows for a very rapid reversal of the effects of transcription, ultimately the synthesis of proteins. In other words, once the *lac* operon is activated and the proteins are made, the operon can be inactivated, and very quickly the synthesis of the proteins is terminated because of the loss of the mRNA. As a result, the cell saves energy and amino acids by not making the proteins.

The *lac* operon is an example of an inducible enzyme system. The term *inducible* describes those enzymes produced only when their substrate is present (in this case, lactose), as opposed to what are called the constitutive enzymes, which are always present (e.g., those needed to utilize glucose). At the beginning of this chapter, we saw that glycolysis is a pathway by which glucose is preferentially metabolized to yield energy in the form of ATP, so, it makes sense that the enzymes needed for glucose utilization are always at the ready. And, although the *lac* operon involves the transport of lactose into the cell and its subsequent separation into the monosaccharides glucose and galactose, these sugars are both, in fact, metabolized via glycolysis. So, essentially, lactose, like glucose, is metabolized by way of glycolysis.

# THE LACTOSE (*lac*) OPERON

In the Jacob and Monod paper of 1961, the essence of induction is pictured in a graph showing the effect of inducer on the synthesis of one of the enzymes of the **lac operon, β-galactosidase,** which splits lactose into glucose and galactose (Figure 7.4).

Although earlier we defined inducible systems as those operating only when the substrate (i.e., inducer) is present, that is not an entirely accurate statement. For example, *E. coli*, is able to bring into the cell and metabolize a small amount of lactose, the inducer, because it has available a small concentration of the required proteins (enzyme and transport proteins). This last point is important because it explains a seeming paradox. Lactose is taken into the cell by way of a special transport protein. But the protein appears to be synthesized only if the inducer, lactose, gets into the cell. If, however, no lactose can get into the cell because the transport protein has not been induced, how is the *lac* operon, including the transport protein, to be induced or activated?

The answer, as alluded to above, is that there is a basal level or minimal expression of the genes of the *lac* operon, and, therefore, some of the inducer, lactose, can be taken up by the cell. However, if lactose were the sole carbon and energy source and if for any reason, say because of mutation, the *lac* operon could not be activated, the cell would starve to death because of a lack of available energy. The minimal level of gene expression would not supply enough of the required proteins, transport protein and enzyme, to sustain the cell. In other words, not enough lactose would enter the cell and be metabolized to maintain the high level of energy production the cell requires. So, for a cell to survive on lactose as the sole carbon and energy source, it must take in enough lactose to derepress, or activate, the *lac* operon.

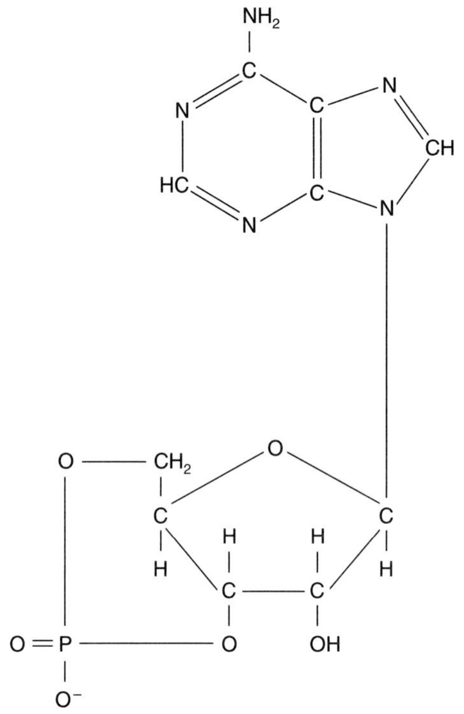

## Proteins Coded by the *lac* Operon

Once activation occurs (induction), three proteins are eventually synthesized. One is **galactoside permease,** more commonly referred to as just permease. This is the transport protein that carries lactose across the cell membrane. Another is **β-galactosidase,** the enzyme responsible for modifying lactose, thus producing the inducer proper, and for splitting lactose into glucose and galactose. The function of the third protein synthesized, transacetylase, is not fully known.

## Positive and Negative Control

The *lac* operon, as indeed most if not all inducible catabolic operons, is under dual control. For activation to occur requires the simultaneous presence of both cyclic adenosine monophosphate (cAMP; Figure 7.5) and the inducer, in this case lactose.

This requirement for cAMP is known as **positive control.** The cAMP is associated with the catabolite repression system and its concentration is, in fact, inversely related to the glucose concentration in the cell. The lower the glucose concentration, the higher the cAMP level and vice-versa. Later, we will see that the *ara* operon is controlled in a similar fashion.

The inactivation of the *lac* operon, which depends on the action of a repressor protein, is known as **negative control.**

## Two Kinds of Necessary Interactions

It should also be noted here that there are two kinds of actions occurring when the *lac* operon is activated or inactivated. The first is the protein–DNA interactions involving: (1) DNA-directed RNA polymerase, the transcription enzyme needed to synthesis mRNA; (2) the catabolite activator protein (CAP), the protein needed to prepare DNA for transcription; and (3) the repressor protein needed to inactivate the operon. The second kind of action involves proteins and molecules such as the inducer and cAMP, both of which cause conformational changes in proteins (see Figure 7.2 again). Both of these types of interactions (i.e., protein–DNA and protein–inducer/cAMP) are a common form of control of gene expression.

$NH_2$ N C C N CH HC N C N O CH2 O C C H H H C C H O=P O OH $O^-$

FIGURE 7.5
Cyclic AMP.

# COMPONENTS OF THE *lac* OPERON

It will be easier to understand the workings of the *lac* operon if we first list and briefly describe the various components of the model.

1. cAMP: Cyclic adenosine monophosphate, a signal molecule formed from ATP.
2. CAP/CRP: Catabolite activator protein/catabolite receptor protein (the terms are used interchangeably); positive regulator protein activated by cAMP.
3. inducer: Lactose (allolactose).
4. repressor: A protein encoded by the *lacI* gene.
5. DNA-directed RNA polymerase: The transcription enzyme.

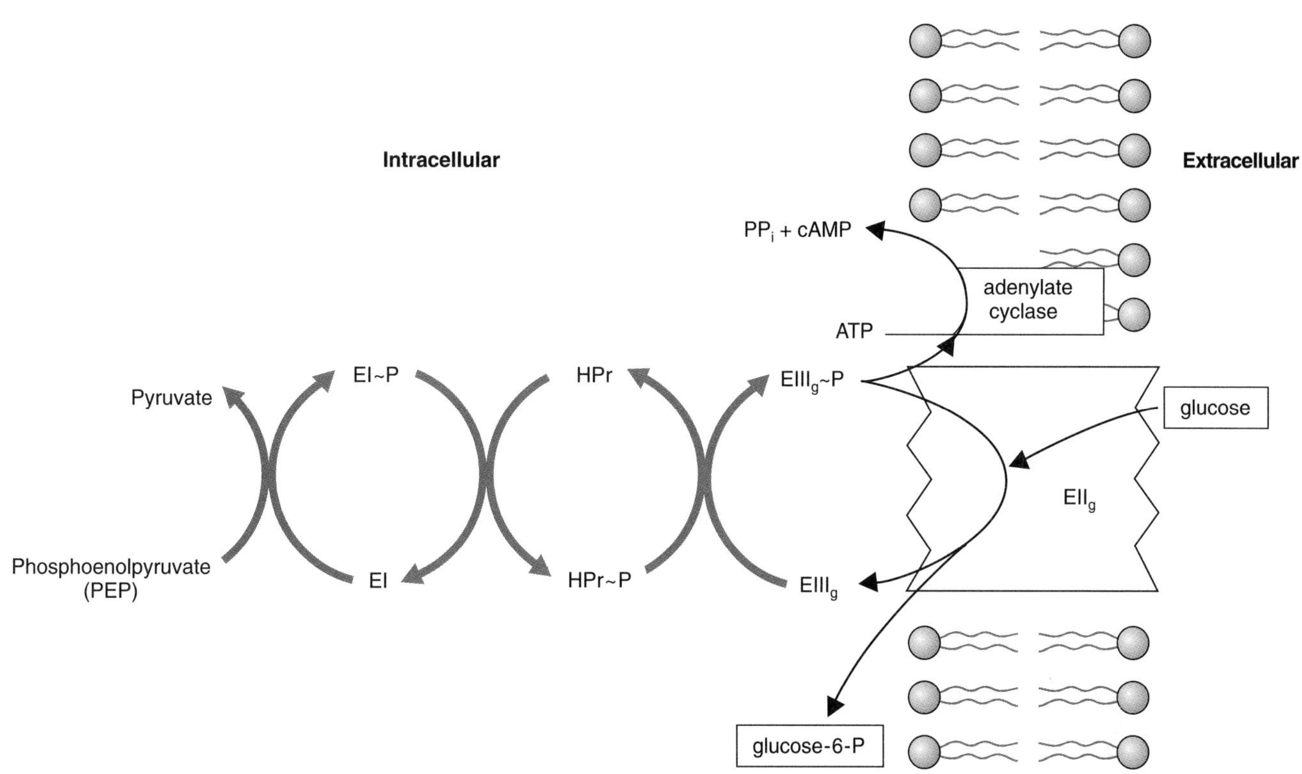


FIGURE 7.6
The phosphoenolpyruvate-dependent phosphotransferase system. PTS glucose transport/cAMP production.

6. regulatory sequences: The promoter and operator regions of the *lac* operon and sites of attachment of the RNA polymerase (5 above) and the repressor protein (4 above).
7. the genes:

   *lac*I: Codes for the repressor protein.
   *lac*Z: Codes for the enzyme β-galactosidase, which splits lactose into glucose and galactose.
   *lac*Y: Codes for galactoside permease, the transport protein.
   *lac*A: Codes for the enzyme transacetylase, whose role in lactose metabolism is poorly understood, so we shall not concern ourselves with it in any detail.

# cAMP

In *E. coli*, sugars, such as glucose, are transported into the cell via a group translocation mechanism, which is a variation of the active transport systems that are ATP-driven.

The transport of glucose is accompanied by the simultaneous phosphorylation of the sugar, which ensures that the sugar, now phosphorylated, will not leak from the cell (the cell membrane is impermeable to sugar phosphates owing to the charge on the phosphate). The energy needed to drive this transport is a compound called phosphoenolpyruvate (PEP), which, incidentally, is produced during glycolysis, the pathway that metabolizes glucose. This entire process is called the phosphoenolpyruvate-dependent phosphotransferase system, or PTS (Figure 7.6).

PTS is dependent on some enzymes and proteins that are common to the transport of all sugars and some other enzymes and proteins that are specific for particular sugars. The protein EIII$_g$, a transmembrane protein, is glucose-specific. In the transport of glucose, EIII$_g$ is phosphorylated by phosphoenolpyruvate and it then transfers the phosphate to glucose as the sugar is passing through the membrane.

If, however, glucose is depleted in the environment, the PTS then uses the phosphorylated EIII$_g$ protein to activate the enzyme adenylate cyclase, which, in turn, converts ATP into cAMP, thereby, increasing the cAMP concentration in the cell.

It is this variation in the concentration of cAMP in the cell—high when the glucose level is low and low when the glucose is abundant—that signals the cell that glucose is or is not available and that another sugar is or is not to be transported into the cell to be metabolized as a carbon and energy source.

As indicated earlier, glucose has a special role in energy metabolism and is, therefore, the sugar of choice to be transported into the cell; its degradative enzymes are constitutive. And, in its presence, by way of the low cAMP concentration, other operons, such as the *lac* operon, are said to be catabolite-repressed. So glucose, the catabolite, represses other sugar operons (catabolite repression). A high concentration of cAMP in the absence of glucose overcomes the catabolite repression by glucose.

## CAP

The **catabolite activator protein** (also called CRP, catabolite receptor protein) is an allosteric protein that is activated by a conformational change brought about by attachment of cAMP. The higher the cAMP concentration becomes, the greater the probability CAP will be activated.

CAP is a **dimer** (it is made up of two subunits) having two identical polypeptide chains, each of which is composed of 210 amino acids. When bound by cAMP, CAP undergoes a large conformational change that allows it to bind to DNA at a specific site or sequence of nucleotides. The CAP-cAMP complex is also active in operons other than the *lac* operon; it is a common way of activating genes. Each CAP molecule is activated by a single molecule of cAMP. The CAP-cAMP complex then binds to a site of about 22 bp located at the 5′ end of the *lac* operon promoter sequence. The CAP-cAMP attached at this site causes the DNA molecule to bend severely. This bending is thought to be necessary to form what is called the **open promoter complex.** The open promoter complex involves the breaking of hydrogen bonds across the double helix, making it possible for DNA-directed RNA polymerase to position itself along the nucleotide sequence of the promoter region just downstream, or to the 3′ end, of the promoter region next to the CAP-cAMP. This open promoter complex formation is believed to be a required step if transcription is to take place.

What is puzzling about the functioning of the activated CAP is that its binding sites in other operons differ relative to the starting point of transcription. The binding sites may be within, adjacent to, or at some distance upstream, or to the 5′ side of the promoter sequence.

## Inducer

**Inducers** are substances, generally small molecules, that are required to initiate the synthesis of proteins. As discussed earlier, the proteins of the *lac* operon are found in the cell at very low concentrations in the absence of the inducer, too low, in fact, to allow the cell to depend on them to meet its metabolic requirements. In the presence of the inducer, however, the rate of synthesis of the particular inducible proteins, such as β-galactosidase, may increase as much as 1000-fold and continue at that rate until the inducer is no longer available. At this point, the rate of synthesis of the proteins returns to its basal or minimal level.

The inducer of the *lac* operon is the disaccharide milk sugar lactose. That is, lactose is the apparent inducer. It is true that the sugar must be in the cell's medium. However, once inside the cell, even at low concentrations, the enzyme β-galactosidase, itself in low concentration, converts the lactose into its isomer, allolactose (Figure 7.7). The enzyme converts the β-1,4 galactoside linkage, joining the glucose and the galactose units into a β-1,6 galactoside linkage.

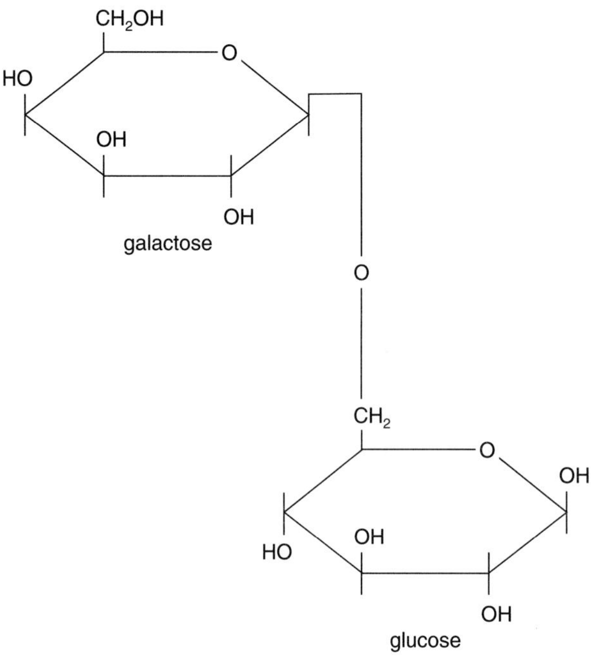


FIGURE 7.7
Allolactose (β-1, 6 linkage).

Note that the enzyme β-galactosidase performs two functions: (1) It converts the apparent inducer, lactose, into the real inducer, **allolactose;** and (2) it splits lactose into glucose and galactose.

## Repressor

The repressor molecule is a protein made up of four identical polypeptide subunits; it is a **tetramer.** Each polypeptide is composed of 360 amino acids. This protein alone accounts for something on the order of 0.002% of wild-type *E. coli* protein.

The tetramer contains four binding sites for allolactose—one site for each polypeptide chain (remember the binding of oxygen to hemoglobin; Chapter 4). Each of the polypeptides also contains a separate domain (a specific sequence of amino acids) needed to bind DNA.

The repressor protein is another allosteric molecule. Its conformation can be altered by the binding of the inducer, allolactose. The conformational change on binding allolactose reduces the repressor's affinity for DNA. In the absence of allolactose, the repressor binds to the operator sequence, a control element, of the *lac* operon, thereby physically preventing the DNA-directed RNA polymerase from reaching the transcription start sequence of the operon. In this way, transcription is stopped; no mRNA

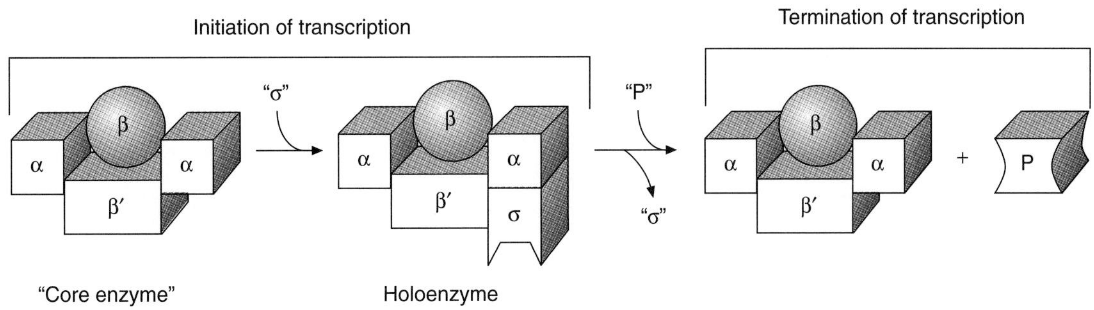


FIGURE 7.8
The core enzyme/holoenzyme (DNA-directed RNA polymerase).

synthesis can take place. The *lac* operon is said to be inactive; the genes are turned off.

When, on the other hand, the inducer, allolactose, is present, it binds directly to the repressor protein, which is itself bound to the operator sequence. The binding of allolactose results in the separation of the repressor from DNA, allowing the open promoter complex to form. Transcription of the *lac* genes can now occur. In short, activation of the *lac* operon requires the inactivation of the repressor protein.

## DNA-Directed RNA Polymerase

The triphosphate ribonucleotides ATP, CTP, GTP, and UTP (U = the pyrimidine uracil, which in RNA replaces thymine and base pairs with adenine) serve both as precursors to RNA and as energy sources for the biosynthetic reactions. In their presence, DNA-directed RNA polymerase synthesizes RNA using a segment of a strand of DNA (a gene) as the template (thus its name). This process of RNA synthesis, in particular mRNA synthesis, is called transcription.

It is the enzyme DNA-directed RNA polymerase that transfers the information stored in the sequence of nucleotides of the gene to the cell's protein-synthesizing machinery via the process of transcription. Unlike the eukaryotes, which have three different RNA polymerases, one for each type of RNA needed for protein synthesis (mRNA, tRNA, rRNA), prokaryotes make do with one polymerase to manufacture the three RNAs.

DNA-directed RNA polymerase (in this context the phrase *RNA polymerase* may be used, as we have been doing, but it should be understood that there are other RNA polymerases—for example, primase, the one needed in DNA replication) is made up of four polypeptide chains: two alpha chains, one beta chain, and one beta prime chain. Together, these chains make up what is called the **core enzyme.** When this core enzyme binds another polypeptide known as the **sigma factor, σ**, RNA synthesis can be initiated at the proper place along the DNA molecule. This five-chain enzyme is referred to as the complete enzyme, or **holoenzyme** (Figure 7.8).

Alone, in the absence of sigma, the core enzyme can bind DNA and synthesize RNA, but it cannot initiate RNA synthesis at the correct transcription start site or nucleotide sequence. For that, the sigma factor is required. The factor is a part of the open promoter complex along with the CAP-cAMP group.

A second factor needed by the core enzyme is one called **rho, ρ**. This polypeptide is needed to recognize **termination sequences** to bring transcription, mRNA synthesis, to a stop at the end of a gene.

The polypeptides of the core enzyme function as follows. The alpha subunits appear to be structural members needed to hold beta, beta prime, and sigma in the proper conformation and orientation to bind DNA at the appropriate site and to maintain this positioning so that the rho factor can attach when it is needed. The beta prime is essential for binding the enzyme to DNA. And beta is the polypeptide chain that actually polymerizes RNA while recognizing the template DNA nucleotide sequence. The sigma factor ensures that transcription is started at the proper transcription start site; sigma allows the enzyme to position itself along the DNA molecule so that the beta subunit is located correctly at the start site to begin transcription. And, last, rho brings transcription to a halt at the proper point by recognizing termination nucleotide sequences.

Once activated, RNA polymerase can "read" something like 42 nucleotides per second. When the new mRNA chain reaches a length of 8–9 or so bases, sigma is released, and the core enzyme continues on its own until joined by the rho polypeptide some time before the polymerase reaches a termination sequence where mRNA synthesis is to stop.

## Regulatory Sequences

The regulatory sequences are nucleotide sequences for which there are no transcriptional products such as RNA, in contrast to structural genes. The regulatory sequences of the *lac* operon, promoter and operator, are located at the 5′ end of the operon. The promoter region is followed by the operator region lying downstream. Together these two regions make up a sequence of some 82 bp of which 26 are the operator sequence. The promoter sequence is the binding site of the

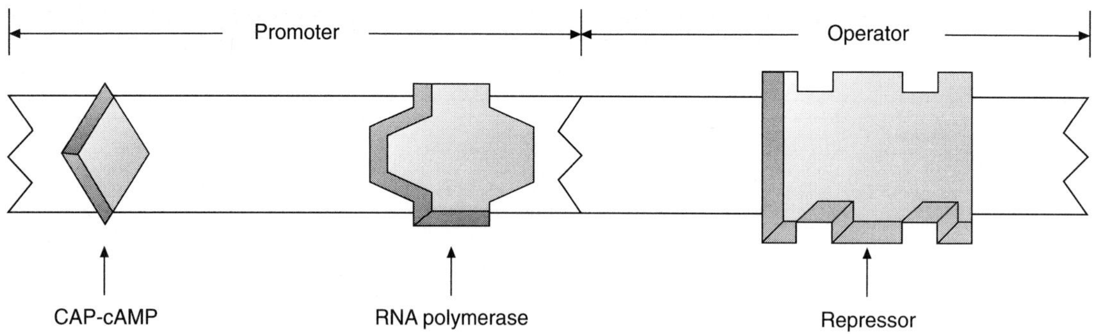


FIGURE 7.9
Promoter and operator binding sites of the *lac* operon.

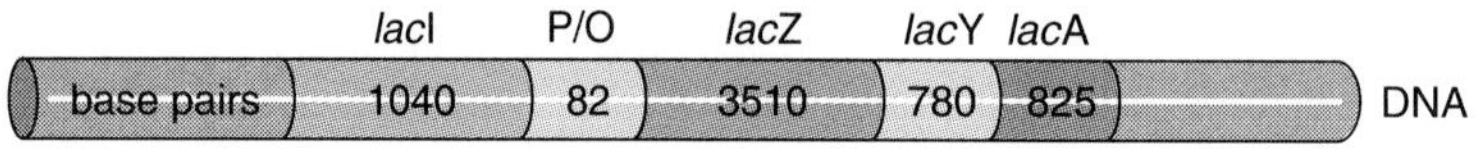

*lac*I P/O *lac*Z *lac*Y *lac*A
base pairs 1040 82 3510 780 825 DNA

FIGURE 7.10
The *lac* operon genes.

RNA polymerase, and the smaller operator region is the binding site of the repressor protein (Figure 7.9).

Together, the promoter and operator sequences lie between the regulator gene (*lac*I) upstream and the structural genes (*lac*Z, *lac*Y, and *lac*A) downstream.

The CAP-cAMP binding site is to the 5′ end of where the RNA polymerase attaches in the promoter region. So the promoter has two binding site sequences: one for CAP-cAMP, followed by the second, where the polymerase attaches (Figure 7.9).

As a note, these control or regulatory sequences are described as "*cis*-acting," meaning they control genes on the same DNA molecule and they do not code for a protein; in contrast, a "*trans*-acting" sequence, such as the *lac*I gene, codes for a protein that can influence other genes on the same or a different DNA molecule.

## The Genes

The *lac* operon contains a total of about 6200 bp of DNA (including the transacetylase gene). Of these, as we have noted above, 82 bp are the promoter and operator sequences. The remainder of the base pairs are divided into four structural genes (each codes for a specific protein) that constitute the *lac* operon:

**lacI:** 1040 bp (repressor protein)
**lacZ:** 3510 bp (β-galactosidase)
**lacY:** 780 bp (permease)
**lacA:** 825 bp (transacetylase)

The genes are arranged along the chromosomes as shown in Figure 7.10. Table 7.1 gives more information about these genes.

The transcription of the Z, Y, and A genes results in the synthesis of a **polycistronic,** or **polygenic, mRNA** molecule. That is, a single mRNA molecule is transcribed that contains more than one message. There are, in this mRNA, three messages with information for the synthesis of three proteins: one for β-galactosidase (Z), one for galactoside permease (Y), and one for transacetylase (A). Transcription is a coordinated, or simultaneous, expression of these genes.

The repressor protein, a tetramer, has a molecular weight of about 152,000 daltons. β-galactosidase is, likewise, a tetramer, but it is much larger, at approximately 500,000 daltons. Both of these proteins, upon synthesis, are found within the cell. Permease, on the other hand, is relatively small, at some 30,000 daltons. This protein is a membrane-bound transport protein required to take lactose into the cell. And, although the suffix *-ase* suggests that it is an enzyme, it is not. When first discovered, it was considered to be an enzyme, and the molecule was given the name permease. In its presence, the cell's membrane becomes "permeable" to lactose, thus, the term permease. The idea was that, since the protein is specific for a particular substrate, lactose, it was probably an enzyme. In fact, in their 1961 paper, Jacob and Monod referred to a "galactoside-permease" protein that "obeys . . . classical enzyme kinetics."

## ACTIVATION OF THE *lac* OPERON

We now want to look at how regulation of the *lac* operon occurs. But to understand the cell's response to environmental change—which is what the regulation of the *lac* operon, and indeed all operons, is all about—let us first see what *E. coli* does when the sugar in the culture medium is changed. We will look at what the organism does in the presence of glucose, lactose, and glucose plus lactose.

table 7.1

THE *lac* OPERON

| GENE | NUMBER OF AMINO ACIDS | MOLECULAR WEIGHT OF POLYPEPTIDE (DALTONS) | TYPE AND MOLECULAR WEIGHT (DALTONS) OF PROTEIN | GENE CODES FOR |
|---|---|---|---|---|
| I | 360 | 38,000 | Tetramer (152,000) | Repressor |
| Z | 1021 | 125,000 | Tetramer (500,000) | β=galactosidase |
| Y | 260 | 30,000 | Membrane (30,000) protein | Galactoside permease |
| A | 275 | 30,000 | Dimer (60,000) | Transacetylase |

*Note:* A dalton is equal to the mass of a hydrogen atoms, or $1.67 \times 10^{-24}$ g.

## Glucose

When *E. coli* is grown in a medium containing glucose as the sole carbon and energy source, glucose acts as a **catabolite repressor,** keeping the *lac* operon inactive or repressed. The uptake of glucose is an energy-driven process. It is what is called an active transport system. The energy, in this case phosphoenolpyruvate, is used to phosphorylate the $EIII_g$ protein, as we have seen. The phosphorylated EIIIg is then used to phosphorylate glucose as it passes into the cell.

The glucose is used immediately; the enzymes needed for its metabolism are constitutive. And, under these conditions, there is no CAP-cAMP available for the *lac* operon. Since $EIII_g$ phosphorylates glucose, adenylate cyclase is not activated, so no cAMP is produced. In the absence of CAP-cAMP, the open promoter complex is not formed, so DNA-directed RNA polymerase cannot efficiently bind to the *lac* operon promoter sequence. Nor could it begin transcription if it did bind properly because the operator region is bound by the repressor protein in the absence of allolactose.

## Lactose

### *Lactose Enters the Cell*

If, now, the cells are removed from the glucose medium, washed by centrifugation to remove any remaining glucose, and placed in a medium containing lactose as the sole carbon and energy source, the utilization of lactose is low for some minutes, it is at basal levels, because of the small amount of β-galactosidase and permease present in the cell.

At the membrane, the $EIII_g$ protein, not having glucose available to phosphorylate, instead phosphorylates the enzyme adenylate cyclase, which, in turn, converts ATP to cAMP.

### *Fate of Lactose in the Cell*

As lactose begins to enter the cell, more and more of it is converted into the inducer allolactose by the enzyme β-galactosidase. At this point, two things will happen: (1) as the concentration of allolactose increases, the probability that the repressor protein will be inactivated increases, and (2) as the level of cAMP increases, the probability that the CAP will be activated increases also.

The binding of four allolactose molecules to the repressor protein (a tetramer) results in a conformational change (remember, the repressor protein is an allosteric protein), causing the protein to disassociate from the operator sequence of the operon. The repressor has been inactivated.

### *The Role of cAMP*

The binding of the cAMP to CAP, on the other hand, has the opposite effect. Through a conformational change, CAP can now bind to DNA, initiating the open promoter complex.

### *mRNA Synthesis*

The release of the repressor protein from DNA frees the operator sequence, and the attachment of CAP-cAMP fosters the binding of DNA-directed RNA polymerase, the holoenzyme, to the promoter sequence. Now, in the absence of the repressor protein at the operator region, the polymerase can move across this sequence of nucleotides to position itself at the correct starting site of the first structural gene of the operon, *lacZ*, and begin to transcribe the genes into a polygenic mRNA molecule.

This mRNA molecule will then be translated into the proteins β-galactosidase, permease, and transacetylase. And, as a result of the presence of the proteins in the cell, the amount of lactose taken into the cell and converted to glucose and galactose will increase dramatically. The cell, now, is assured an adequate supply of carbon and energy.

Notice that lactose is converted to sugars that are readily metabolized by the cell's constitutive enzyme system, the glycolytic pathway. So the induction of the *lac* operon really supplies the cell with the substrates it was denied in the first place, glucose (or what is actually an intermediate in glucose metabolism, glucose-6-phosphate, produced by the phosphorylated $EIII_g$ as glucose is transported into the cell or via the conversion of galactose to glucose-6-phosphate).

What we see here, the induction of the *lac* operon by lactose (allolactose) is a system under both negative and positive control. Negative control is the repressor protein's binding the operator sequence, preventing RNA polymerase from functioning. Positive control is the necessary binding of the CAP-cAMP to the promoter sequence, which facilitates the binding of RNA polymerase via the open promoter complex.

### Glucose and Lactose

What will happen if *E. coli* is in a medium containing both glucose and lactose? In this case, $EIII_g$ will preferentially phosphorylate glucose to glucose-6-phosphate rather than phosphorylating adenylate cyclase and, thereby, activating the cyclase. The result is that the concentration of cAMP is kept below the concentration needed to activate CAP and assist in turning on the operon. So the synthesis of the polygenic mRNA molecule is prevented, and, consequently, β-galactosidase and permease are not produced. As a result, lactose is not transported into the cell until the glucose is depleted, and $EIII_g$ is free to phosphorylate adenylate cyclase, thereby initiating the chain of events leading to activation of the *lac* operon.

## INACTIVATION OF THE *lac* OPERON

The question now is: How is the *lac* operon inactivated, or turned off, once it has been turned on? The answer is that, as the concentration of lactose decreases in the medium, the environment, less and less lactose is transported into the cell and, therefore, fewer molecules of allolactose are made. The inducer molecules bound to the repressor protein are constantly bouncing in and out of the allosteric site of the repressor (remember, allosteric proteins are not covalently modified, so allolactose is reversibly bound). Once released from the repressor, the allolactose can be split by β-galactosidase into glucose and galactose and used as a carbon and energy source.

### The Need for Allolactose

Maintaining operon activity requires a continuous supply of allolactose. As the level of lactose entering the cell drops, the level of allolactose also falls. This decrease, in turn, frees some repressor proteins, which can once again bind the operator sequence and prevent transcription. Then, as the β-galactosidase, permease, and mRNA needed to synthesize those proteins reach the end of their "life spans" (their half-lives being predetermined, as all protein and RNA half-lives are), they are not replaced. The *lac* operon is shut down, though not completely, as the synthesis of the mRNA and the proteins is returned to the basal level.

## THE TRYPTOPHAN (*trp*) OPERON

As noted earlier, the *lac* operon is normally inactive; it must be activated. The proteins it codes for are involved in catabolic reactions—the metabolism of lactose.

The tryptophan **(*trp*) operon** of *E. coli*, on the other hand, is normally active, and control is exercised in order to inactive it. This operon's end products (enzymes) are involved in the anabolic reactions of amino acid (tryptophan) biosynthesis (Table 7.2).

Since protein synthesis is a continuous process and amino acids are the building blocks of proteins, it makes sense that the amino acid operons are normally active and must be inactivated or repressed only when the supply of the particular amino acid is high enough to meet the cell's immediate demands. There are a couple of reasons why the supply of the amino acid might be sufficient. There might be an exogenous, or outside, source of the amino acid. In that case, the cell need not produce it and save the energy required to synthesize the enzymes: e.g., mRNA need not be made. Or it might be that, for a time, the proteins in production require little of a particular amino acid, so the concentration of the amino acid is sufficient to meet the current demands; the operon is then inactivated.

This *trp* operon differs from the *lac* operon in that, instead of activating the operon, signal molecules, in this case tryptophan, inactivate the operon. In contrast, lactose activates the *lac* operon. But, like the *lac* operon protein components (CAP and repressor protein), the *trp* operon protein components are also subject to conformational changes that result in changes in activity. We will discuss these changes as the *trp* operon is described.

Another important aspect of the *trp* operon not found in the *lac* operon is that the trp operon is subject to two levels of control: repression and attenuation. Repression is a rather straightforward process, whereas attenuation is much more complex, involving conformational changes in mRNA rather than in proteins.

As it turns out, repression of the *trp* operon is relatively weak; that is, negative control is not complete. With a second level of control, namely, attenuation, control is enhanced. This second kind of control may be necessary because of the central importance of amino acid biosynthesis and because of the large amount of energy needed to synthesize the cell's required proteins. The amino acid biosynthetic pathways utilize numerous enzymes and need a large expenditure of energy for their manufacture. So the more efficiently the cell can respond to demand changes

table 7.2

THE *trp* OPERON

| GENE | PROTEIN | FUNCTION |
|---|---|---|
| L | Leader sequence (m-RNA) | Attenuation mechanism |
| E<br>D | Anthranilate synthase (two components) | Chorismate converted to N-(5′-phosphoribosyl)-anthranilate |
| C | N-(5′-phosphoribosyl)-anthranilate isomerase<br>Indole-3′-glycerol phosphate synthase | N-(5′-phosphoribosyl)-anthranilate converted to indole-3′-glycerol phosphate |
| B<br>A | Tryptophan synthase (two components) | Indole-3′-glycerol phosphate converted to L-tryptophan |

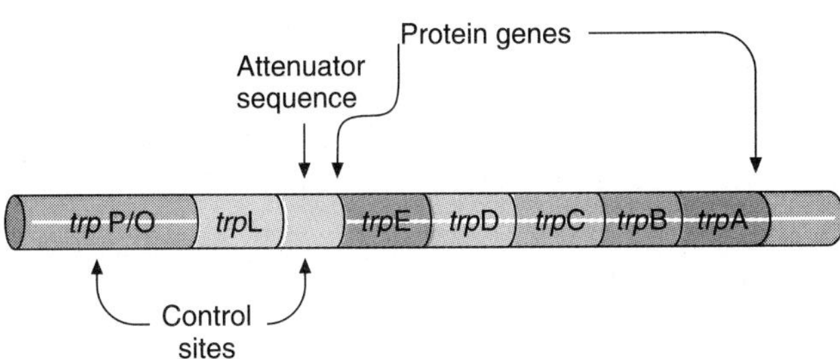


FIGURE 7.11
Organization of the *trp* operon genes.

via operon control the better. In fact, together, repression and attenuation control *trp* operon activity over a range of some 600-fold. Repression and de-repression change the level of *trp* operon activity over a range of about 70-fold. Attenuation adds another 8–10-fold range.

# COMPONENTS OF THE *trp* OPERON

As we did with the *lac* operon, we will first look at the parts of the *trp* operon (Figure 7.11) individually before describing how, in combination, they function to control transcription of *trp* genes.

1. repressor: A dimeric protein (two-polypeptide molecules) encoded in the *trp*R gene.
2. corepressor: Tryptophan, associates with the repressor protein and activates it.
3. control elements: Attenuator nucleotide sequence, promoter and operator sequences.
4. the genes: Five structural genes coding for three proteins:

   *trp*E and *trp*D (anthranilate synthase).
   *trp*C (anthranilate isomerase and indole-3′-glycerol phosphate synthase: one gene product catalyzes sequential reactions).
   *trp*B and *trp*A (tryptophan synthase).

## Repressor

The repressor protein is composed of two identical subunits, each containing 107 amino acids. When the level of tryptophan exceeds the required concentration, tryptophan, the end product, binds to the repressor, producing a complex whose protein component (the dimeric protein) undergoes a conformational change that allows the repressor to bind the *trp* operator sequence. The rate of transcription of *trp* genes is thereby decreased by a factor of about 70. Thus, the availability of the metabolic end product, tryptophan, controls its own synthesis. This effect is the first level of control, namely, repression.

## Corepressor

As indicated above, tryptophan must complex with the repressor protein in order to activate it and inactivate the operon. Since tryptophan is also necessary for repression to occur, it is called the corepressor.

## Control Elements

Located just downstream (in the 3′ direction) from the *trp* promoter/operator sequences and just ahead of the first structural gene of the operon (*trp*E) is a sequence of approximately 162 nucleotides that codes for a transcript (mRNA) called **the leader.** Within this sequence is a group of nucleotides referred to as the attenuator sequence. It is this attenuator sequence's transcript that controls the operon at the second level—attenuation. More precisely, as we shall see, it is the internal base pairing of the transcript that makes possible two mutually exclusive hairpin structures that control the operon (the single-stranded mRNA folds back on itself, positioning nucleotides so that hydrogen bonds between appropriate bases can occur; this folding is unusual for mRNA) (see Figure 7.12). It is the creation of either of these hairpins in the leader mRNA that determines whether the *trp* structural genes will be decoded by DNA-directed RNA polymerase and, ultimately, whether the enzymes needed for tryptophan synthesis will be produced.

The promoter/operator sequences serve the same functions in the *trp* operon as they do in the *lac* operon. That is, they are sites of attachment for DNA-directed RNA polymerase (promoter) and for the repressor protein (operator).

## CONTROL OF THE *trp* OPERON: REPRESSION AND ATTENUATION

### Repression

As indicated earlier, repression of the *trp* operon relies, like the control of the *lac* operon, on a conformational change in an allosteric protein. But, unlike the *lac* operon, which is an inducible system, the *trp* operon is a repressible system. That is, the genes are normally on, and control is exerted by turning the genes off—stopping the process of transcription.

We saw in the *lac* operon model that a signal molecule, the inducer allolactose, forms a complex with the repressor protein. The result is a conformational change in the protein. That change forces the repressor protein to disassociate from the operator region of the DNA, thereby allowing RNA polymerase to proceed beyond the operator sequence to the first structural gene where transcription is to begin.

Repression of the *trp* operon is, in essence, the opposite series of events brought about, however, by a common act. In this case, tryptophan, the corepressor, complexes with a protein, bringing about a conformational change in the allosteric protein. The end product of *trp* operon activity, tryptophan, is the signal molecule rather than the starting compound as in the *lac* operon model.

At this point, we have inactivation, instead of activation, of the genes. The tryptophan-protein complex (amino acid–protein or corepressor–repressor protein complex) binds to the operator of the *trp* operon and physically blocks RNA polymerase from proceeding beyond the promoter region. The operon is inactivated, though not 100%, and the rate of transcription of the structural genes is considerably reduced. Binding of the repressor-corepressor to the operator sequence is not terribly efficient, and, at any given moment, the operator may be freed of the complex and transcription may occur.

Both repression and attenuation of the *trp* operon are responses to the internal level of tryptophan. Repression, acts directly via the production of the tryptophan–repressor protein complex. While attenuation acts indirectly, as we shall see next, via the availability of tryptophan-charged tRNA.

When the level of tryptophan is high enough to meet the cell's demands, the operon is repressed. Should some of the RNA polymerase get beyond the operator region, transcription of the leader sequence would begin, but it would be terminated at the attenuator region so that the polymerase would never reach the operon's structural genes.

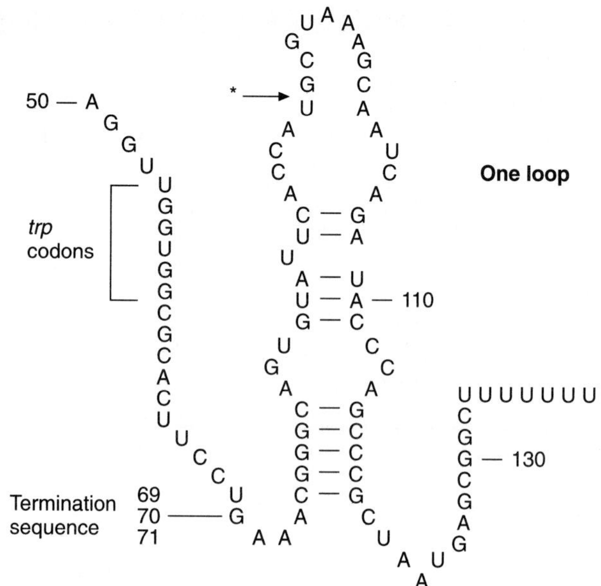


**(a)** Transcription continues

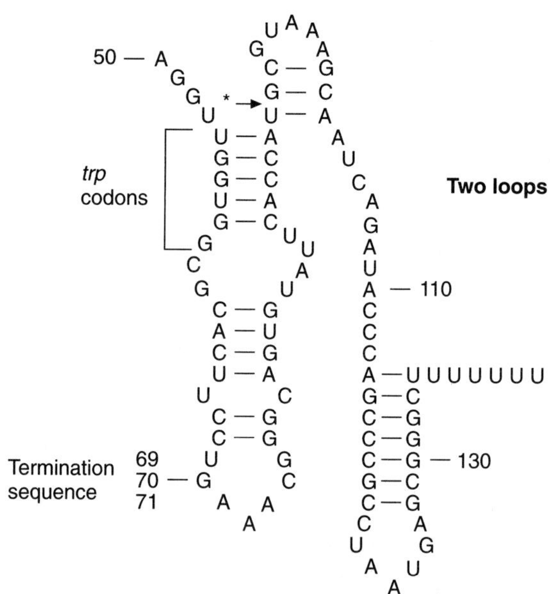


**(b)** Transcription terminated

FIGURE 7.12
mRNA hairpin structures.

### Attenuation

Before proceeding with a discussion of attenuation, let us take a closer look at the mRNA **hairpin** structures (Figure 7.12) briefly alluded to above (see Control Elements) to see under what conditions they are formed. The attenuator sequence is located within the leader sequence. When the attenuator region is reached by RNA polymerase, a series of nucleotides is produced to form the attenuator mRNA. This is the sequence of nucleotides that can form the two mutually exclusive hairpin structures.

This mRNA can be divided into four sections or sequences of nucleotides:

section 1: nucleotides 50–68
section 2: nucleotides 75–92
section 3: nucleotides 108–121
section 4: nucleotides 126–134

Section 4 is followed by a series of seven uracils ending at nucleotide 141.

Sections 1 and 2 are complementary and, thus, can base pair. Sections 3 and 4 are also capable of base pairing. These pairings constitute one structure made of two hairpin loops (sections 1 and 2 base pair and make up one hairpin loop, and sections 3 and 4 base pair to form the second hairpin loop; Figure 7.12).

But, and this is the crux of the matter, if section 1 *does not* base pair with section 2, then section 2 is free to base pair with section 3, leaving section 4 single-stranded, or unpaired. The resulting single loop made up of sections 2 and 3 is the second of the two possible hairpin structures. Refer to Figure 7.12 for these two possible structures. We shall see presently how this looping, or hairpin, formation controls the *trp* operon.

It is also important to note here that between sections 1 and 2 there is a stop codon (UGA) located at nucleotide positions 69, 70, and 71. Furthermore, and crucial to the operation of the control mechanism of the system, within section 1, nucleotides 54–59 make up two consecutive *trp* codons (UGGUGG).

Although the *trp* attenuation model is, perhaps, the best known and understood of this type of control mechanism, at least half a dozen other operons of amino acids are also known to be controlled by attenuator sequences.

The question now is: How is termination of transcription at the attenuator sequence accomplished? As transcription of the leader mRNA approaches the attenuator, one of two hairpin structures can form (Figure 7.12). One, the single hairpin loop allows transcription to continue into the *trp*E structural gene; two, the double hairpin loops terminate transcription, or mRNA synthesis.

Whether continuation or termination of mRNA synthesis occurs depends on whether ribosomes that attach to the mRNA molecule as soon as it begins to form remain attached to the mRNA or are released. Attachment of ribosomes to mRNA while the mRNA is still being synthesized is possible in bacteria because, in these cells, DNA is not separated from the cytoplasm by a nuclear membrane, as is the case in eukaryotes. (Ribosomes are the structures on which peptide bonds between amino acids are formed to produce a polypeptide chain; Chapter 8). In prokaryotes, the two processes of protein synthesis, transcription and translation, are coupled, making attenuation possible.

So protein synthesis begins (i.e., translation of the leader sequence), but, along the leader mRNA, the ribosome encounters the successive tryptophan codons (UGG), and this is the "decision" point.

Let's now look at the two possibilities. What happens when tryptophan is in short supply, and what happens when it is abundant (Figure 7.13)?

## *Low Levels of Tryptophan*

When the level of the amino acid falls below a critical concentration, the *trp* operon is de-repressed. That is, the tryptophan acting as the corepressor is released (it is not covalently bound), and it is utilized. The repressor protein changes conformation so that it can no longer bind DNA.

Now, as the RNA polymerase moves into the *trp* leader sequence, transcription begins and continues through about 90 nucleotides of this sequence. As this is happening, ribosomes begin to attach to the 5′ end of the newly transcribed mRNA molecule. DNA directed RNA polymerase pauses at about the ninetieth nucleotide, allowing translation of this first portion of the leader mRNA. This process involves the movement of the ribosome along the mRNA. To continue past this point, however, the RNA polymerase must be approached by the translation ribosome (see Figure 7.13).

The ribosome continues to move along the mRNA until it encounters the back-to-back tryptophan codons (UGGUGG) in section 1. These codons can be translated only if there is a sufficient supply of tryptophan to ensure the charging or attachment of the amino acid to its tRNA.

It is important that translation of the leader sequence occur while the downstream nucleotides of the leader RNA are being transcribed in order that one or the other of the hairpin structures can form. In other words, the ribosome must be moving along the mRNA molecule toward the tryptophan codons.

Now, since the level of tryptophan is low, the amino acid codons lying back to back cannot be translated, because there is no charged tRNA available. The ribosome stalls in section 1, where the codons are located. And that stalling does two things:

1. RNA polymerase moves forward as the ribosome approaches the tryptophan codons, and
2. the ribosome stalled in section 1 acts as a physical block.

The ribosome stalled here, unable to move forward for the lack of tryptophan-charged tRNA, prevents section 1 from base pairing with section 2. Section 2 is, therefore, free to base pair with section 3 as soon as section 3 is synthesized by the moving RNA polymerase. A single hairpin loop is formed.

Now that section 3 is paired, section 4, as it is made, remains single-stranded, and the polymerase continues transcription into the structural genes of the operon starting with *trp*E. The operon has been activated.

(I) Low tryptophan level: transcription continues

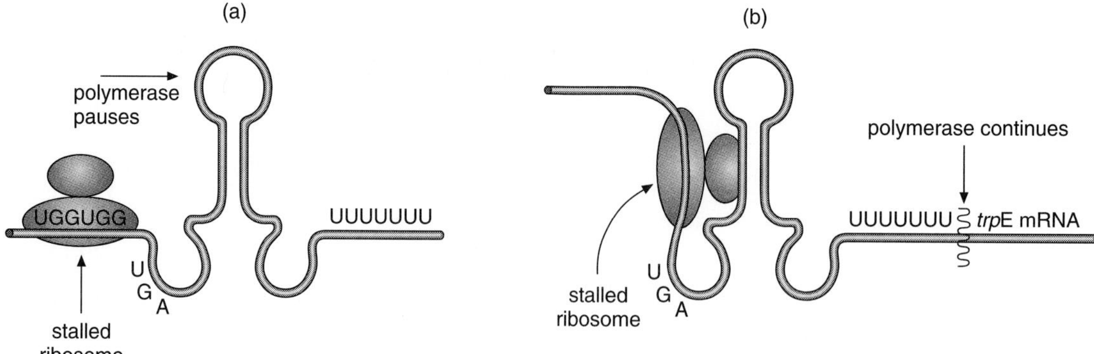


(II) High tryptophan level: transcription stops

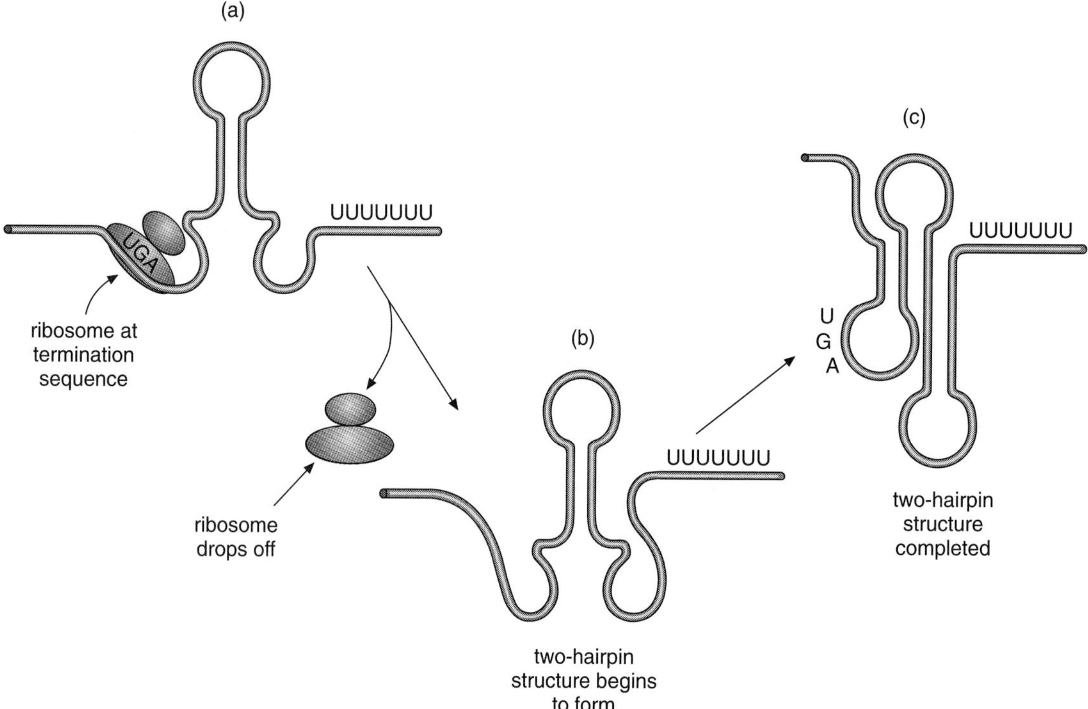


FIGURE 7.13
The tryptophan attenuation model.

## *High Levels of Tryptophan*

At the other extreme is the case in which the tryptophan level is sufficient for the cell's current requirements. As with low tryptophan concentrations, at high tryptophan levels, RNA polymerase begins transcription of the leader sequence and pauses at the usual place, at about the ninetieth nucleotide, allowing attachment of the ribosome to the newly synthesized mRNA.

Translation of the leader sequence begins. The ribosome moves to the successive tryptophan codons, "bumping" the polymerase forward. But the ribosome *does not stall* at this point. It doesn't stall because there is enough tryptophan present in the cell to ensure an adequate supply of tryptophan-charged tRNA, so the ribosome can move beyond the double *trp* codons. And, just ahead of these codons, the ribosome encounters the terminator sequence, UGA, and "drops off" the mRNA at that position. Meanwhile, RNA polymerase, having been bumped ahead by the approaching ribosome, has continued transcribing in front of the ribosome and is now into section 4 of the leader mRNA.

Once the ribosome is removed from section 1 of the mRNA, this section is now free to base pair with the already synthesized section 2 of the mRNA to form the first hairpin loop. As section 3 is made, it remains unpaired. Section 4 is then produced. These two sections (3 and 4) are free to base pair, forming the second hairpin loop. This second loop is held together tightly because of the

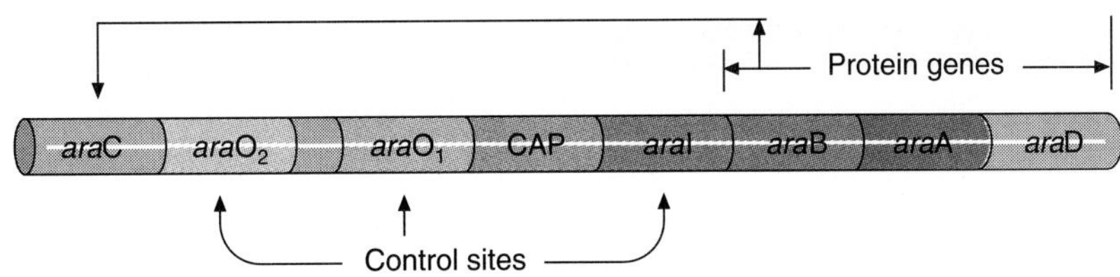


FIGURE 7.14
The *ara* operon.

numerous triple hydrogen bonds between the guanosines and cytosines in this loop.

It is the second hairpin structure that terminates transcription. Apparently, this tightly bonded structure exerts enough force on the relatively weak double hydrogen bonds between the DNA's adenines and the mRNA's uracils (remember the sequence of seven uracils) to cause the mRNA molecule to separate from the DNA and to separate the RNA polymerase from DNA. The uracil-adenine (RNA-DNA) double-bonded base pairings require less energy to break than do the triple-bonded guanine-cytosine pairings.

The separation of the polymerase at this point, of course, prevents the enzyme from ever reaching the *trp* structural genes. So the *trp* operon is inactivated, at least until the concentration of the amino acid drops to the level that those molecules binding the repressor are utilized. As the tryptophan concentration decreases, there is a smaller and smaller probability that there will be sufficient tryptophan-charged tRNA to prevent the ribosome from stalling at the *trp* double codons.

The combination of an inactivated repressor protein and an insufficient supply of tryptophan-charged tRNA will activate the operon once again.

## THE ARABINOSE (*ara*) OPERON

Thus far, we have seen an operon, (*lac*) controlled by two proteins, (CAP and repressor), exerting positive (CAP) and negative (repressor) control at one level. We have also seen an operon (*trp*) controlled on two levels, repression and attenuation. Now we turn to a third operon of *E. coli*. This one, the *ara* (arabinose) operon (Figure 7.14), is controlled by yet another method.

In this model, one protein exerts both positive and negative control. In this example, there are echoes of the *lac* operon, CAP and cAMP, and of catabolite repression by glucose.

Arabinose is a five-carbon sugar (a pentose) found in plant polysaccharides. And, although it is not utilized by human beings, bacteria, such as *E. coli,* found in the human intestinal tract can use arabinose as a carbon and energy source. By converting arabinose to xylulose-5-phosphate, the bacteria can then transform the xylulose-5-phosphate into the glycolytic intermediates fructose-6-phosphate and glyceraldehyde-3-phosphate, normal intermediates in the metabolism of glucose via glycolysis produced by constitutive enzymes.

The *ara* operon, then, is another inducible catabolic operon. It is normally inactive unless arabinose and cAMP are available. Like the *lac* operon, this one is under dual control.

Two major distinctive features of the *ara* operon are:

1. The C protein, serves as both a positive and a negative regulator, depending on its conformational state, which is, in turn, dependent on whether or not arabinose is present; it is an allosteric protein.
2. The presence of *three* control regions, identified as *ara*$O_1$, *ara*$O_2$, and *ara*I.

## COMPONENTS OF THE *ara* Operon

We will first look at the components (Table 7.3) of the system before describing how it functions.

1. cAMP: Cyclic adenosine monophosphate formed from adenosine triphosphate.
2. CAP: The catabolite activator protein, a regulator protein activated by cAMP.
3. C protein: A control, or regulator, protein having dual action, positive and negative regulation.
4. inducer: Arabinose.
5. DNA-directed RNA polymerase (RNA polymerase): The transcribing enzyme.
6. control regions (operators): *ara*$O_1$, *ara*$O_2$, *ara*I.
7. the genes:

   *ara*C: Codes for the C protein.
   *ara*B: Codes for ribulokinase.
   *ara*A: Codes for arabinose isomerase.
   *ara*D: Codes for ribulose-5-phosphate epimerase.

## cAMP

As in the *lac* operon model, the level of cAMP here is dependent on the availability of glucose; the two compounds are inversely related. As the concentration of glucose decreases, the concentration of cAMP increases.

table 7.3

THE *ara* OPERON

| GENE | PROTEIN* | FUNCTION |
|---|---|---|
| I | — | Control site: BAD genes |
| $O_1$ | — | Control site: C protein |
| $O_2$ | — | Control site: BAD genes |
| A | L-arabinose isomerase | Converts L-arabinose to L-ribulose |
| B | L-ribulose kinase | Converts L-ribulose to L-ribulose-5-phosphate using ATP |
| D | L-ribulose-5 phosphate epimerase | Converts L-ribulose-5-phosphate to D-xylulose-5-phosphate |

*—, DNA sequence within the *ara* promoter.

## CAP and the C Protein

The C protein is a regulator protein that is autoregulated. That is, when C protein and CAP-cAMP levels are low, more C protein is produced. Remember, the CAP-cAMP levels are low when glucose is abundant and the operon is, therefore, to be inactive. The C protein, besides regulating its own production, also regulates the transcription of the B, A, and D genes by binding to control sequences *ara*$O_2$ and *ara*I. The C protein is an allosteric protein.

## Inducer

The five-carbon sugar L-arabinose reacts with the C protein, causing a conformational change that results in release of the inactive *ara* operon and the transcription of the BAD genes. The transport of arabinose into the cell requires a permease system encoded in the *ara*E and *ara*F genes, which are found in two independent operons.

## DNA-Directed RNA Polymerase

As in the other operons, DNA-directed RNA polymerase is responsible for decoding the genes and producing mRNA in the process of transcription.

## Control Regions

Ara$O_1$ is the control sequence for the C protein, and *ara*$O_2$ and *ara*I together are the control sequences for the BAD genes.

# CONTROL OF THE *ara* OPERON

In control of the *ara* operon, the level of C protein is maintained at the appropriate concentration by an autoregulation mechanism; the C protein is a self-regulating protein. That is, as the C protein concentration falls below a critical level, RNA polymerase is stimulated to initiate transcription of the *ara*C gene. The enzyme decodes the *ara*C gene in a direction *away* from the BAD genes. Then, as the concentration of the C protein increases, the rate of *ara*C transcription decreases.

When the concentration of the C protein is high and the level of cAMP is low, and regardless of the availability of arabinose, no transcription of the BAD genes takes place (again, catabolite repression by glucose occurs). In this case, the C protein reacts with all three control sequences (*ara*$O_1$, *ara*$O_2$, and *ara*I). This reaction causes the DNA molecule to fold in such a way as to bring the *ara*$O_2$ and *ara*I sequences close together (Figure 7.15).

These two sequences are linked by a single molecule of the C protein. In this way, both *ara*C and the BAD genes are prevented from being transcribed. Should, however, the level of C protein fall, the result of normal half-life metabolism, unbending of the DNA occurs, making *ara*$O_1$ available once more to RNA polymerase—transcription of *ara*C can begin.

But to transcribe the *ara* genes requires not only unbending of the DNA but also a sufficient amount of cAMP to react with CAP. The CAP-cAMP complex binds to DNA at a site between *ara*$O_1$ and *ara*I, thereby facilitating the binding of RNA polymerase to the transcription initiation sequence of the structural B gene. Also required is the presence of arabinose.

As we saw in the *lac* operon model, the concentration of cAMP is dependent on the level of glucose available. If this, the sugar of choice, is available, the transport system phosphorylates glucose using phosphoenolpyruvate (PEP) as the phosphate donor as the glucose is entering the cell. If, on the other hand, the level of glucose falls, the PEP is, instead, available to the enzyme adenylate cyclase, which will convert ATP to cAMP.

And if, as the concentration of glucose is falling, arabinose is presented to the cell, two things happen. First, cAMP reacts with CAP and the CAP-cAMP complex associates with the CAP site located between *ara*$O_1$ and *ara*I. Second, arabinose reacts with the C protein, which is binding *ara*$O_2$ and *ara*I. Arabinose causes a conformational change in the C protein, which, as a result, releases the *ara*$O_2$ site and allows the DNA molecule to unbend.

With (1) the unbending of DNA, (2) the association of the CAP-cAMP with the CAP site, and (3) the linking of arabinose to the C protein at the *ara*I site, RNA poly-

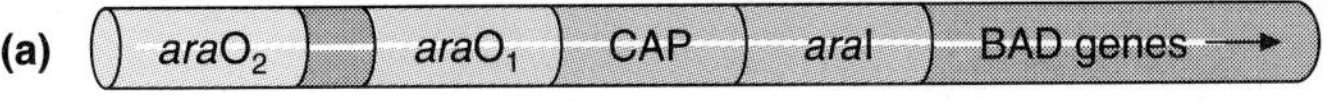


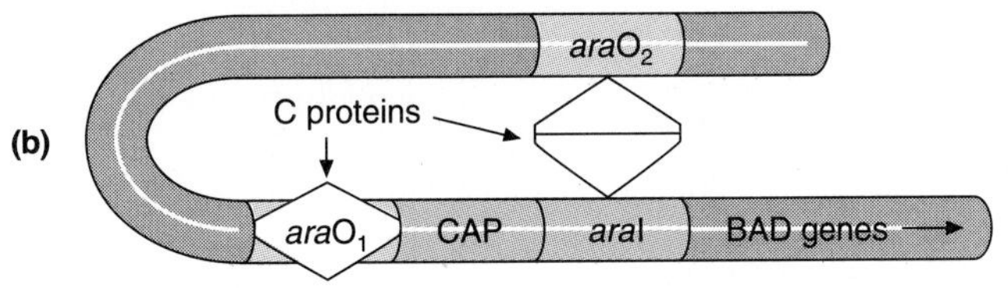


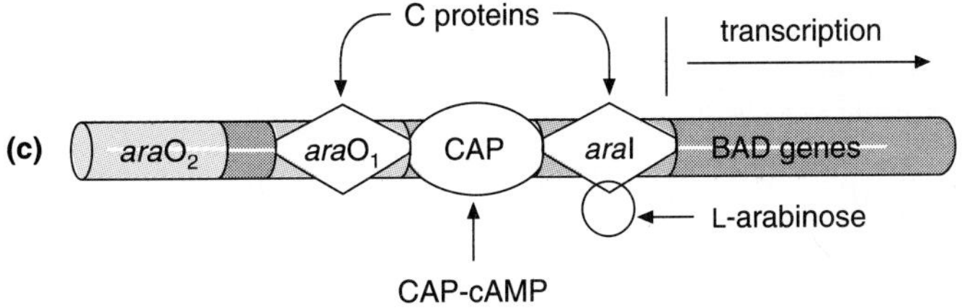


FIGURE 7.15
C protein binding and bending DNA; (a) the *ara* operon; (b) transcriptionally inactive *ara* operon; (c) transcriptionally active *ara* operon.

merase can bind to the initiation sequence of the B gene and begin transcription of the BAD structural genes.

Transcription of the genes can continue until the arabinose bound to the C protein at *ara*I is metabolized. When this happens, the C protein is released and reverts to its previous conformational state. Now, since *ara*I is bound by C protein without arabinose present, DNA is again bent, because *ara*$O_2$ and *ara*I are again cross-linked by the C protein, and transcription ceases. The *ara* operon is now inactivated.

## SUMMARY

In order to respond to changes in the internal and external environments, cells must have ways to regulate their metabolism. They must be able to control biosynthetic and biodegradative pathways, such as amino acid synthesis and carbohydrate breakdown.

Metabolism can be controlled either by control of enzyme activity or by control of enzyme synthesis.

Enzyme activity can be regulated by controlling enzyme conformation by means of allosteric enzymes (by attachment of chemical groups) or covalent modification (by binding or releasing other polypeptides to or from the enzyme). It can also be controlled by partially digesting the enzyme to release a small number of amino acids. The first three methods of control are reversible; the last one is not.

Control of enzyme synthesis involves the regulation of gene expression: that is, the turning on and off of the information-transfer process, called transcription. This method of control depends on the presence or absence of signal molecules: inducers or corepressors. How the control is exerted depends on whether the genes—clusters of which are called operons—encode information for the synthesis of enzymes needed for biosynthesis or for biodegradation. If the former, the presence of the signal molecule (e.g., amino acids) will work to inactivate the operon. If the operon encodes information for enzymes needed in degradative systems, the signal molecule (e.g., lactose or arabinose) will bring about the activation of the operon. In either case, what is being controlled is, in fact, the process of transcription, which results in the synthesis of messenger RNA, or mRNA.

The regulation of transcription, however, can be accomplished in a number of ways. The most straightforward is represented by the *lac* operon model, a biodegradative system, in which a form of lactose (allolactose) inactivates a repressor protein, thereby making the operon available to the transcription enzyme, DNA-directed RNA polymerase. However, whether the operon is activated also depends on the presence or absence of cAMP. The level of this compound is inversely related to the level of glucose available to the cell. So turning on the *lac* operon genes requires not only lactose but also cAMP. While lactose (allolactose) inactivates the repressor protein, cAMP activates CAP, another protein which, in turn, helps to establish the open promoter complex. This complex allows RNA polymerase to bind DNA at the proper site to begin transcription. This binding leads to the synthesis of the protein needed to transport lactose into the cell and the protein needed to begin the metabolism of the carbohydrate.

The control of the tryptophan (*trp*) operon, on the other hand, is more complicated. It relies on two levels of control—repression and attenuation. And, being an operon encoding information for enzymes needed in a biosynthetic pathway, the *trp* operon is normally activated and must be inactivated. Tryptophan, the end product of the enzymes encoded by the operon, acts as a regulator of its own synthesis. It functions as a corepressor.

When the level of the amino acid is high, it becomes associated with the repressor protein and together they turn off the operon. This repression, however, is not 100% efficient. Should transcription continue under these conditions, attenuation would begin. A leader mRNA is produced before the genes of the enzymes are reached and transcribed. A ribosome attaches to this newly synthesized leader mRNA and begins moving along as translation is occurring. The ribosome soon reaches a stop codon and "drops off" the mRNA. As a result, two hairpin loops of the RNA can form. These hairpin loops cause termination of transcription and the detachment of the leader RNA from the DNA template.

Should the level of tryptophan be low, neither repression nor attenuation would occur. There would not be a sufficient amount of the amino acid to activate the repressor. Nor would there be enough tryptophan to charge *trp* tRNA, and that is the key. Now, as the ribosome moves along the leader mRNA, it stalls at a position where two consecutive *trp* codons are located. The ribosome cannot proceed beyond this point because there is no charged tRNA to read these back-to-back *trp* codons. The stalled ribosome prevents the formation of the double hairpin loops. As a result, transcription continues into the *trp* genes and the mRNA for the enzymes is completed.

The *ara* operon is different in that a single protein, the C protein, controls its own synthesis and the activity of the operon. This operon is sensitive to the concentration of the carbohydrate arabinose.

To activate the *ara* operon requires a low level of glucose, thus a high level of cAMP, and arabinose. As in the *lac* operon, cAMP binds CAP and an open promoter complex forms. The arabinose binds the C protein, the repressor, causing a conformational change in the protein, which now can no longer bind the DNA and so drops off. Release of the C protein causes the DNA to unbend, exposing the BAD genes to RNA polymerase, and transcription can begin. Transcription leads to the synthesis of the enzymes needed to utilize arabinose as a carbon and energy source.

## Study Questions

1. By what two methods can cells respond to environmental changes?
2. What are the two broad categories of metabolism?
3. How do the categories of question 2 differ from each other?
4. Explain the four ways by which enzyme activity can be regulated.
5. Define the term *operon*.
6. List the components of the *lac*, *trp*, and *ara* operons.
7. Describe the role of each of the components listed in the answer to question 6.
8. How does the *trp* operon differ from both the *lac* and *ara* operons?
9. What is the difference between repression and attenuation?
10. How is control of the *trp* operon brought about by high and low concentrations of tryptophan?
11. How does the C protein control its own synthesis?
12. How does the C protein regulate the *ara* operon?
13. Explain how varying concentrations of arabinose affect the *ara* operon.

## Readings and References


Gilbert, W., and B. Muller-Hill. 1966. Isolation of the *lac* repressor. *Proc. Natl. Acad. Sci. USA* 56:1891–98.

Jacob, F., and J. Monod. 1961. Genetic regulatory mechanisms in the synthesis of proteins. *J. Mol. Biol.* 3:318–56.

King, R. C., and W. D. Stansfield. 1990. *A dictionary of genetics*. New York: Oxford University Press.

Kolter, R., and C. Yanofsky. 1982. Attenuation in amino acid biosynthetic operons. *Annu. Rev. Genet.* 16:113–34.

Lewin, B. 1990. *Genes IV*. New York: Oxford University Press.

Lobell, R. B., and R. F. Schleif. 1990. DNA looping and unlooping by araC protein. *Science* 250:528–32.

Miller, J. H., and W. S. Reznikoff, eds. 1978. *The operon*. Cold Spring Harbor, N.Y.: Cold Spring Harbor Laboratory.

Schultz, S. C., G. C. Shields, and T. A. Stietz. 1991. Crystal structure of CAP-DNA complex: DNA is bent 90°. *Science* 253:1001–07.

Stent, G. S., and R. Calendar, 1978. *Molecular genetics: An introductory narrative*. 2d ed. San Francisco: W. H. Freeman.

Stryer, L. 1988. *Biochemistry*. 3rd ed. San Francisco: W. H. Freeman.

Voet, D., and J. G. Voet. 1995. *Biochemistry*. 2d ed. New York: John Wiley & Sons.

Weaver, R. F., and P. W. Hedrick. 1997. *Genetics*. 3rd ed. Dubuque, Iowa: Wm. C. Brown.

Yanofsky, C. 1981. Attenuation in control of expression of bacterial operons. *Nature* 289:751–58.

Yanofsky, C. 1988. Transcription attenuation. *J. Biol. Chem.* 263:609–12.

CHAPTER EIGHT

# PROKARYOTIC PROTEIN SYNTHESIS

## OUTLINE



## CHAPTER OBJECTIVES

*This chapter will discuss:*

- The relationships between DNA, RNA, and proteins as a flow of information
- The distinction between polypeptides and proteins
- What and where the energy requirements of protein synthesis are
- How the components of protein synthesis function
- The mechanism by which the information in DNA is converted into proteins
- How proteins assume their biologically active forms

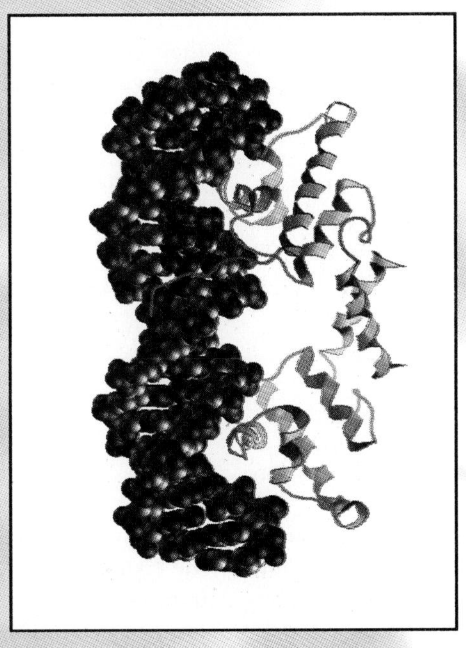

## INTRODUCTION

To this point in our study of molecular genetics, we have looked at the structure of DNA and how information is stored in genes. We have seen how genetic information is passed on to the next generation via replication of DNA, which is the process during which inheritance occurs.

Now we shall begin the study of how the information stored in the genetic material, DNA, is made available to the cell. The mechanism by which genetic information is made available to the cell is protein synthesis.

All cellular organisms, from single-celled bacteria to multicellular organisms such as human beings, are totally dependent on the reactions occurring within single cells. These reactions, whether catabolic or anabolic, collectively referred to as metabolism, rely almost completely on proteins. Recall the various categories of proteins discussed in Chapter 4.

Chapters 4 through 7 were a lead-in to the present chapter. First, we took a brief look at the various types of proteins needed to keep a cell or organism functioning. Then, in Chapter 5 we looked at the nature of the gene to see how information is stored. Chapter 6, The Genetic Code, introduced us to the relationship between the information in the gene and the structure of a polypeptide/protein. Last, Chapter 7 showed us how gene activity is controlled: how genes are switched on and off, thereby controlling when and what genetic information is available to the cell. In this chapter, we will see how the cell uses that available information to construct the hundreds upon hundreds of proteins the cell requires. This is done by way of protein synthesis.

Protein synthesis is a process requiring a number of steps, which are normally grouped into two major categories: transcription and translation. Transcription involves the decoding of genes and the concomitant synthesis of mRNA, the intermediate between DNA (genes) and polypeptides (proteins). This step in protein synthesis concerns the switching on of genes, (e.g., the *lac* operon model).

Translation follows transcription. It is during **translation** that mRNA is read. The codons of mRNA specify what amino acid is to be incorporated and what position that amino acid is to occupy in the polypeptide. Translation requires the participation of tRNA molecules both to read mRNA and to carry amino acids to the site of protein synthesis, the ribosomes.

Both transcription and translation are three-stage processes involving initiation of synthesis, elongation (synthesis) of the mRNA or polypeptide, and termination of synthesis.

Once polypeptide synthesis is completed, because of the unique sequence of amino acids making up each polypeptide, the molecule will fold into a particular conformation that gives it biological activity. This folding may or may not require factors that assist the polypeptide to acquire its proper conformation.

The information in the gene specifying the linear sequence of amino acids is ultimately the information determining the protein's final structure and, therefore, its biological activity. So the primary structure of the protein determines the protein's function.

Protein synthesis represents the flow of information in a cell from gene to polypeptide. Linear nucleotide sequences are converted to linear amino acid sequences, and the relationship between these sequences has been given the name **colinearity.** Figure 8.1 is a schematic overview of protein synthesis and the flow of information in a cell.

## THE CENTRAL DOGMA OF MOLECULAR BIOLOGY

By the late 1950s the idea that information flow during protein synthesis was a one-way transfer from DNA to protein had been accepted for several years, but it remained for Francis Crick to give the concept a name. This he did in 1958 and expounded further on the idea in a 1970 paper. He termed the idea the central dogma of molecular biology. He meant by this not a doctrine that must be accepted simply because it is stated to be true by some unquestionable authority but rather that it was an idea "for which there was no reasonable evidence." The notion that information in a cell is passed either from one DNA molecule to another during replication (Chapter 2) or from DNA to proteins via mRNA is, in fact, true for all cells.

Some viruses, however, do not follow this path. Retroviruses (e.g., HIV) pass information from RNA to DNA during replication (Chapter 3). Other single-stranded RNA viruses either use their genomic RNA directly as mRNA (e.g., RNA phage and poliovirus) or use their genetic RNA as the negative strand to synthesize a positive strand that acts as mRNA (e.g., rabies and influenza viruses). The double-stranded RNA viruses (e.g., reoviruses) transcribe one of the RNA strands into an mRNA molecule for protein synthesis.

## COLINEARITY

During protein synthesis, the information stored in DNA giving the type and location of amino acids in proteins flows in one direction (Figure 8.1). This flow is really about the transmission of sequences of chemical species. The sequence of deoxyribonucleotides in one strand of DNA, called the coding strand, is used to construct an mRNA molecule of nucleotides whose sequence matches that of the coding strand. The process of transferring information from the gene (the coding strand) with the concomitant synthesis of mRNA is called transcription. In the next

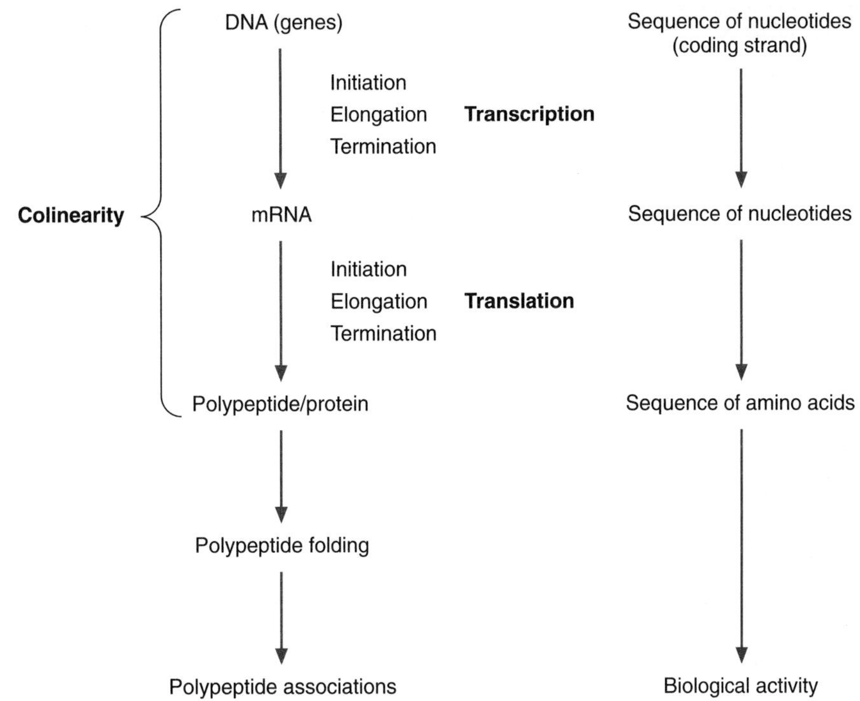


FIGURE 8.1
Flow of information in the cell. If one sequence is known plus the genetic code, all other sequences can be gotten. But, because of redundancy, going from an amino acid sequence to a gene sequence may not provide the exact nucleotide sequence of the gene.

process, called translation, the nucleotide sequence of mRNA is converted into a sequence of amino acids via three steps: initiation, elongation, and termination. Each unique sequence of amino acids is then able to fold into a unique conformation that has biological activity.

This is what is meant when we say the information flow is from DNA to protein; that is, one sequence, in DNA, determines another sequence, in a protein. A nucleotide sequence determines an amino acid sequence. This concept of linear sequences coding for other linear sequences is called colinearity. In other words, a string of one chemical species, nucleotides, determines a string of another chemical species, amino acids, to produce a polypeptide.

The sequence of codons of a gene specifying amino acids also orders these amino acids in a polypeptide, thereby achieving a biologically active molecule. In effect, then, DNA stores information for biologically active molecules by storing information concerning the building blocks of the protein and by storing information about how those building blocks are to be used (types and positions of amino acids).

The experimental evidence for this colinear relationship between nucleic acid and protein was supplied by A. Sarabhai and, independently, by Charles Yanofsky and their colleagues in the mid-1960s. The former established the relationship between a gene and a head protein for the T4 phage of *Escherichia coli*, and the latter established the colinear relationship between the A protein of *E. coli* tryptophan synthetase and its gene.

An example of colinearity would be the following (refer to Figure 6.2 for codons and the amino acids they specify).

The amino acids and their codons:

| | |
|---|---|
| alanine | GCU |
| glycine | GGA |
| leucine | UUA |
| threonine | ACU |

DNA: TTAACTGCTGGA: coding strand
AATTGACGACCT: anticoding (template) strand
mRNA: UUAACUGCUGGA
Protein: leucine-threonine-alanine-glycine

Note that the coding strand of DNA and the mRNA have identical sequences (remember that uracil replaces thymine in RNA).

The one-to-one correspondence of codons and amino acids that is at the heart of colinearity is possible in prokaryotes because there are no noncoding sequences in prokaryotic genes (Chapter 5). The nature of the eukaryotic gene (Chapter 10), however, is different. These genes do contain noncoding sequences. That is, there are sections of eukaryotic genes that do not specify amino acids and that are not regulatory sequences as far as is known today. In fact, these genes are referred to as split genes. So, the one-to-one relationship between codons and amino acids does not hold in eukaryotes. We shall deal with this problem in greater detail in Chapter 10.

table 8.1

POLYPEPTIDE COMPOSITION OF VARIOUS PROTEINS

| PROTEIN | MOLECULAR WEIGHT (DALTONS)* | NUMBER OF CHAINS |
|---|---|---|
| Bovine ribonuclease I | 12,600 | 1 |
| Egg lysozyme | 13,900 | 1 |
| Horse myoglobin | 16,900 | 1 |
| Bovine insulin | 5,700 | 2 |
| Yeast hexokinase | 102,000 | 2 |
| Human hemoglobin | 64,500 | 4 |
| *Escherichia coli* tryptophan synthetase | 159,000 | 4 |
| Bovine glycogen phosphorylase | 370,000 | 4 |
| *E. coli* aspartate transcarbamylase | 310,000 | 12 |
| *E. coli* glutamine synthetase | 592,000 | 12 |
| Bovine pyruvate dehydrogenase complex | 7,000,000 | 160 |
| Tobacco mosaic virus coat | 40,000,000 | 2130 |

*A dalton is equal to the mass of a hydrogen atom, or $1.67 \times 10^{-24}$ g.

# PROTEINS/POLYPEPTIDES AND BIOLOGICAL ACTIVITY

You now know that genes code for polypeptides. The original idea of Garrod and Beadle and Tatum (Chapter 5) that one gene encodes information for one enzyme (protein) had to be modified when it was discovered that most proteins, though not all, consist of more than one polypeptide. Table 8.1 shows some examples of proteins with various numbers of polypeptide chains.

Chemically, there is no difference between the basic structure of a polypeptide and that of a protein. Both are composed of amino acids linked together by peptide bonds. It should be noted, however, that there are simple proteins, composed solely of amino acids, and conjugated proteins, which have nonamino acid groups associated with them. So, for example, insulin is a simple protein, and hemoglobin, a four-polypeptide protein, is a conjugated protein having four heme groups, each consisting of an iron atom and four pyrrole rings as integral parts of its structure. In both cases, however, the polypeptides are composed of amino acids joined by peptide bonds.

Although the terms *polypeptide* and *protein* are often used interchangeably, *polypeptide* is generally used to denote a single chain of amino acids. The term **protein,** on the other hand, is used for molecules that may consist of more than one polypeptide chain.

Though there are some biologically active polypeptides, such as insulin, ribonuclease, and serum albumin, the polypeptide components of multichain proteins are not biologically active. For example, any number of protein molecules can be treated to release the component polypeptides, and none of these chains will display the biological activity of the native protein. The individual polypeptide chains must be associated with one another for there to be biological activity.

## Levels of Structural Organization

Polypeptides, whether biologically active or not, can have three levels of structural organization, referred to as the primary, secondary, and tertiary structures. Proteins have an additional level of organization, the quaternary structure, which is an association of polypeptides.

The first three structures together are responsible for the three-dimensional arrangement of the polypeptide. For any protein to function properly, a correct three-dimensional structure of its polypeptide chains is absolutely essential. The correct three-dimensional arrangement must be maintained in order that the protein's reactive centers are properly formed and displayed. If the form or placement of the reactive sites is wrong, the protein will lose its biological activity. How much biological activity is lost will depend on how deformed or misplaced the reactive sites are. The four structures of a protein can be described as follows.

### *Primary*

The primary structure is the sequence of covalently linked amino acids coded for by the gene (Figure 8.2).

### *Secondary*

The **secondary structure** is the spatial arrangement of the molecule's backbone atoms. This arrangement can produce an alpha helix (a right-handed helix), a beta-pleated sheet,

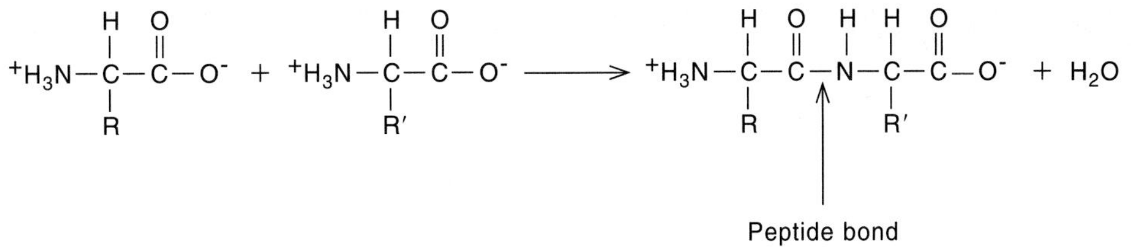

$$^{+}H_3N-\underset{\large R}{\overset{\large H}{C}}-\overset{\large O}{\overset{\|}{C}}-O^{-} + {}^{+}H_3N-\underset{\large R'}{\overset{\large H}{C}}-\overset{\large O}{\overset{\|}{C}}-O^{-} \longrightarrow {}^{+}H_3N-\underset{\large R}{\overset{\large H}{C}}-\overset{\large O}{\overset{\|}{C}}-\overset{\large H}{N}-\underset{\large R'}{\overset{\large H}{C}}-\overset{\large O}{\overset{\|}{C}}-O^{-} + H_2O$$

Peptide bond

FIGURE 8.2
The primary structure of a polypeptide/protein consists of many amino acids joined to one another by peptide bonds. The carboxyl group of one amino acid reacts with the amine group of the next amino acid to form the peptide bond.

a collagen triple helix, or turns in the molecule. Three bonds in the secondary structure (Figure 8.3) are critical to the shaping of the tertiary structure: the peptide bond, the bond between the alpha carbon and the amino nitrogen, and the bond between the alpha carbon and the carboxyl carbon. The peptide bond itself (C—N) is said to be planar; there is no significant rotation around this bond. In contrast, the bonds between the alpha carbon and its carboxyl and nitrogen groups can rotate freely. This freedom of rotation plays an important role in the molecule's ability to fold, turn, and twist.

## *Tertiary*

The **tertiary structure** (Figure 8.4) of a polypeptide is dependent on the amino acids present in the primary structure and the bonds discussed in the preceding paragraph.

The side chains of the amino acids of the polypeptide and the arrangement of alpha-helical chains and beta-pleated sheets with respect to one another all contribute to the folding of the polypeptide. Any prosthetic groups, such as the pyrrole rings of cytochromes, hemoglobin, and myoglobin, also affect the spatial arrangement of the tertiary structure. In very large polypeptides, sections of the molecule may fold independently of one another to create what are called domains. There may be helical and sheet domains within the same polypeptide.

## *Quaternary*

Finally, identical or different polypeptides may associate with one another to form much larger units (Figure 8.5), constituting the **quaternary structure.**

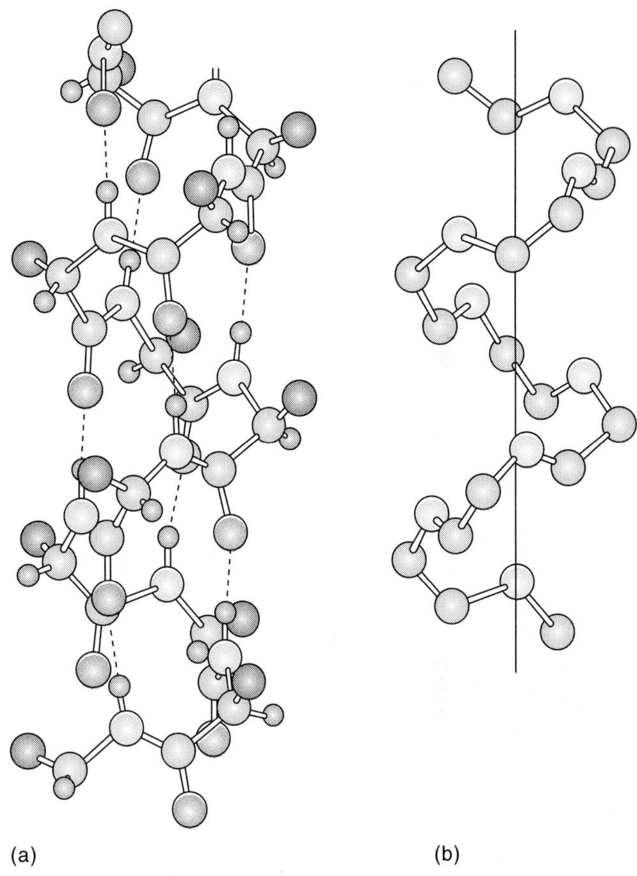


N C O R

FIGURE 8.3
The secondary structure of a polypeptide/protein: The α helix. (a) all the atoms of the right-handed helix are shown. (b) only the atoms of the helical backbone are shown. Note the direction of twist from bottom to the top—a right-handed helix is described.

# Protein/Polypeptide Complexion

The polypeptides or subunits are linked together by noncovalent bonds such as hydrogen bonds involving the polypeptide backbones and side chains, hydrophobic interactions of nonpolar groups, and interchain disulfide bonds (sulfur-to-sulfur bridges). Such large proteins may consist of two, three, four, or more polypeptides and are called dimers, trimers, tetramers, and so on.

An example of a tetramer is hemoglobin, which is made up of two identical alpha and two identical beta polypeptides. (Incidentally, because of this makeup, only two genes are required for hemoglobin, not four. One gene is needed for the alpha chains and another for the beta chains.)

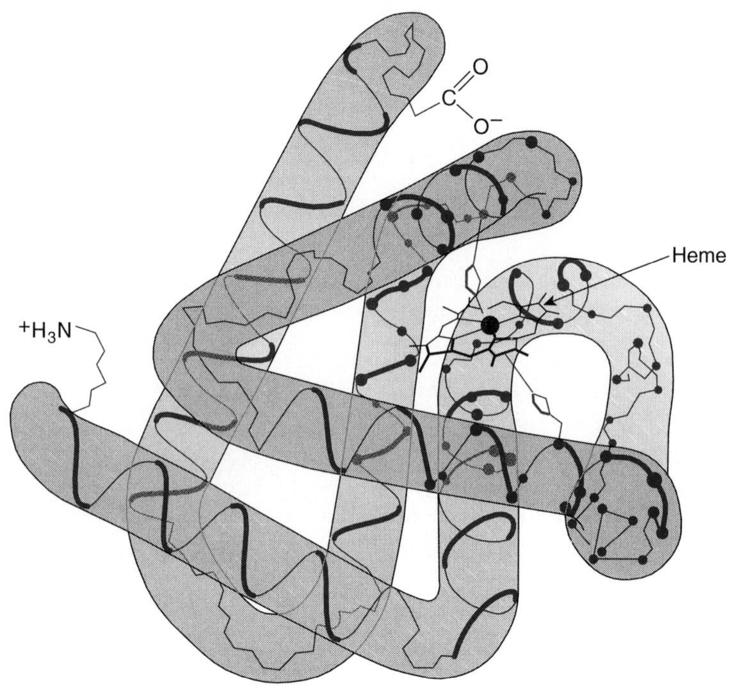


FIGURE 8.4
The tertiary structure of a polypeptide/protein. This is a representation of myoglobin, the oxygen-carrying protein of muscle. Myoglobin is a monomer made up of 152 amino acids in human muscle. The arrow points to the heme group. Also notice the α-helical spirals. The molecule is roughly spherical.

Individual biologically active proteins can also associate to form extremely large complexes. One of the largest such complexes known is the pyruvate dehydrogenase complex of *E. coli* needed for energy metabolism. This multienzyme system consists of three enzymes (pyruvate dehydrogenase, dihydrolipoyl transacetylase, and dihydrolipoyl dehydrogenase). Each complex has 24 copies each of the first two enzymes and 12 copies of the third enzyme. The complex has a mass of about 4600 kD (kilodaltons). The ribosome to be discussed later in this chapter is another example of a large complex. In this case, the complex is made up of both proteins and RNA; its mass is about 4220 kD.

The quaternary structure of proteins is subject to subtle changes in individual subunits. These changes are then propagated throughout the molecule. In the hemoglobin molecule, there is cooperative binding of oxygen; the binding of each molecule of oxygen stimulates the binding of the next oxygen molecule. A single hemoglobin molecule binds four oxygens in this way. In fact, the position of each polypeptide subunit of hemoglobin relative to the other subunits changes as hemoglobin is alternately oxygenated and deoxygenated.

Each of these structures (primary, secondary, tertiary, and quaternary) contributes to a protein's function. It is, however, the primary structure, the sequence of amino acids, that ultimately allows the formation of each of the subsequent structures. And the primary structure is determined by the sequence of nucleotides of the gene.

## THE ENERGY NEEDS OF PROTEIN SYNTHESIS

The synthesis of proteins is a very expensive proposition for the cell. For this reason alone, the control of protein synthesis is of paramount importance to the cell. Energy is expended in both transcription and translation and in binding amino acids to the proper tRNAs. We shall look at the energy requirements of each of these processes separately.

During transcription, mRNA synthesis requires a steady and large supply of energy in the form of triphosphate nucleotides. Remember, the synthesis of DNA and RNA requires energy to link the nucleotides via a phosphodiester bond.

As we saw in Chapter 2, each triphosphate nucleotide that is produced costs two ATPs. This production is a two-step reaction that begins with the 5′-monophosphate nucleotides produced by the purine and pyrimidine pathways. First, a specific nucleoside monophosphate kinase converts the monophosphate nucleotide (e.g., GMP) into a diphos-

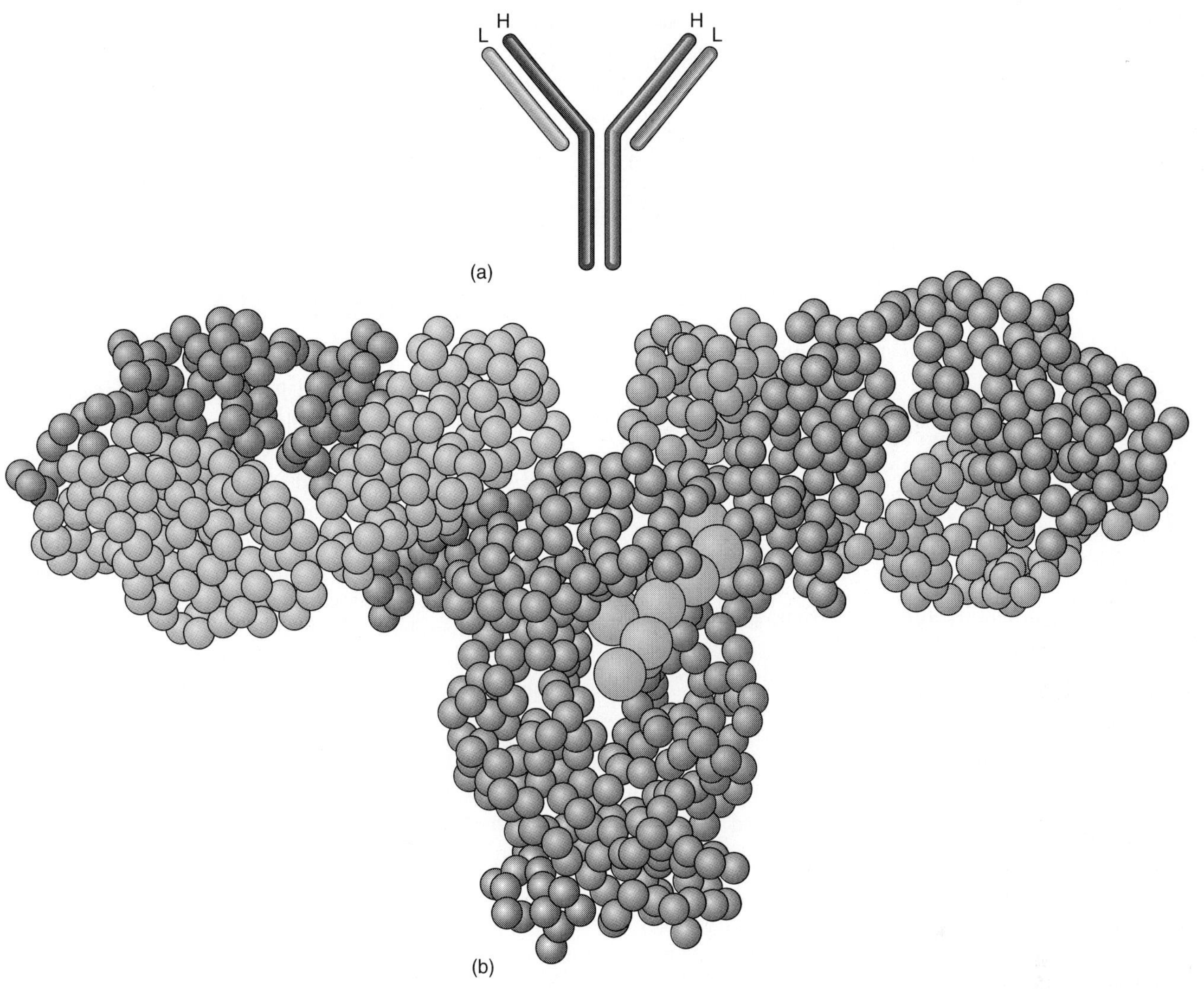


FIGURE 8.5
The quaternary structure of a polypeptide/protein. This is a representation of a typical antibody (immunoglobulin) molecule. It is composed of four polypeptides, two identical light chains (L), and two identical heavy chains (H).

phate nucleotide at the expense of one ATP: GMP + ATP → GDP + ADP. Then, a second enzyme, a nucleoside diphosphate kinase, phosphorylates the diphosphate nucleotide to the triphosphate form using another molecule of ATP: GDP + ATP → GTP + ADP.

If we make the not unreasonable assumption that a particular mRNA is 1000 nucleotides in length, the energy cost of synthesizing this molecule is 2000 ATPs. Because each different protein the cell produces requires its own mRNA in many copies, and hundreds, if not thousands, of proteins are being made at any one time, the energy cost is high. Add to this the short half-life of prokaryotic mRNAs requiring the resynthesis of the mRNAs over small time spans and we have a very high energy cost.

The synthesis of the other RNAs needed for protein synthesis, namely tRNA and rRNA, also require the expenditure of ATPs. Again, each nucleotide to be incorporated into tRNA and rRNA must be in the triphosphate form, which is produced at the expense of ATP.

During the process of translation, in which mRNA is read and polypeptides are synthesized, the high-energy compound guanosine triphosphate (GTP) is needed. GTP is required to form the initiation complex, which consists of mRNA, a special methionine-carrying tRNA, and the 30S ribosomal subunit (see Ribosomes, below, and Figure 8.9).

More GTPs are then needed as peptide bonds form. The attachment of each new amino acid–carrying tRNA (called charged tRNA) to the ribosome uses one GTP molecule, and the transposition of the growing polypeptide chain from one site on the ribosome to another needs another GTP. So for each amino acid added to a polypeptide, two GTPs are used.

From an accounting point of view, that is the equivalent of two ATPs. It may be that the high-energy GTPs are needed to produce conformational changes in the ribosome to promote the attachment of charged tRNAs and the necessary movement of the growing polypeptide chain from one site to another site on the ribosome as codons are read and amino acids are added to the polypeptide chain. The hydrolysis of GTP is rapid and irreversible, thus ensuring that whatever reactions GTP participates in will also be rapid and irreversible.

Furthermore, one of the functions of tRNA in protein synthesis is to carry amino acids to the ribosomes. Each amino acid is bound to its tRNA at the cost of one ATP. The amino acid must be "activated" by reaction with ATP before it can be attached to its tRNA (more about this reaction later in the chapter). This process is referred to as the **charging** of tRNA.

So, in the form of the nucleoside triphosphates, a great deal of energy goes into the production of a single type of protein. When one considers the thousands of proteins continuously being made by the cell, it is easy to understand the large energy cost of such an operation. These energy needs are met by the oxidation of sugars and fatty acids; this oxidation is coupled to the synthesis of ATP. So, sugar and fatty acid oxidations are linked to protein synthesis via ATP.

# COMPONENTS OF PROTEIN SYNTHESIS

An assortment of molecules is required for each polypeptide the cell produces. These molecules include three species of ribonucleic acid and an array of proteins that function as enzymes and as factors necessary in the two steps of protein synthesis, transcription and translation. What follows are the major components of protein synthesis found in *E. coli*.

1. DNA-directed RNA polymerase: The enzyme of transcription needed to decode the gene and produce mRNA.
2. sigma and rho: Two protein factors required for the proper initiation and termination of transcription.
3. ribonucleic acids: Include mRNA, tRNA, and rRNA, which are involved in the synthesis of polypeptides during translation.
4. ribosomes: Structures found in all cells; the sites of polypeptide synthesis.
5. translation initiation factors: Proteins needed to separate the ribosomal subunits and then to form the initiation complex.
6. translation elongation factors: At least three different proteins are required to elongate a polypeptide chain.
7. releasing factors: Another group of proteins to recognize the stop codons of mRNA.
8. peptidyl transferase: This enzyme forms the peptide bond between the newest amino acid to be incorporated into the polypeptide chain and the last amino acid of the chain.

## DNA-Directed RNA Polymerase

The enzyme DNA-directed RNA polymerase (not to be confused with the RNA polymerase and primase *E. coli* uses to begin replication of DNA) begins the synthesis of mRNA and adds ribonucleotides to the 3′ end of mRNA. The DNA molecule's **template strand** (hence, "DNA-directed") determines the base sequence of the mRNA. The ribonucleotide incorporation reaction requires energy provided by triphosphates of the four nucleotides. This reaction is driven by the hydrolysis of the pyrophosphate released upon incorporation of each nucleotide monophosphate into the growing mRNA chain. The reactions are as follows:

1. $(RNA)_n + NTP \leftrightarrow (RNA)_{n+1} + PP_i$
2. $PP_i \leftrightarrow P_i + P_i$

where NTP is any triphosphate nucleotide, *i* means inorganic, and *n* is the number of nucleotides in the mRNA chain.

The second reaction requires the enzyme inorganic pyrophosphatase to split the pyrophosphate, releasing the energy that drives the incorporation reaction.

This synthesis of mRNA is the process called transcription. As transcription occurs, two things are happening simultaneously. First, DNA is being decoded. That is, the polymerase enzyme "reads" the deoxynucleotides of the anticoding, or template, strand of DNA and base pairs those nucleotides with the appropriate ribonucleotides. Second, a phosphodiester bond is produced between the 3′ position of the last ribonucleotide of the mRNA and the 5′ position of the new ribonucleotide to be incorporated. So we have decoding of the gene and mRNA synthesis occurring at the same time.

The RNA polymerase itself, often referred to as the core enzyme, is composed of five polypeptide chains designated alpha (there are two alpha chains), beta, beta prime, and omega. This enzyme carries out the actual polymerization of mRNA.

Little is known of the actual functions of each of the polypeptides, though the beta subunit is known to provide the catalytic function of the enzyme, and beta prime appears to be necessary for binding DNA.

The prokaryotic RNA polymerase requires two protein factors to function properly: that is, to position itself exactly at the gene's start sequence and to terminate mRNA synthesis correctly. The first function needs sigma factor, and the second function needs the rho factor (more about those factors shortly).

When the core enzyme associates with sigma to begin transcription, the enzyme is known as the holoenzyme. This structure has a mass of approximately 480 kD (kilodaltons) and is among the largest of the known soluble enzymes, at about 100 Å in diameter.

The large size of the enzyme might be dictated by the number of functions it carries out in the synthesis of mRNA: (1) binding template DNA at a specific site, (2) initiation of mRNA synthesis, (3) mRNA elongation, and (4) chain termination and release from the template.

The other species of RNA, tRNA and rRNA, also require an RNA polymerase for their synthesis. *E. coli* makes do with a single enzyme to produce all three RNAs. Eukaryotes, on the other hand, have a different RNA polymerase for each species of RNA (Chapter 10).

## Sigma and Rho

### *Sigma Factor*

The sigma protein is essential for accurate binding of RNA polymerase to DNA and there to initiate mRNA synthesis. It is this factor that permits RNA polymerase to recognize promoter sites of genes.

Alone, RNA polymerase can tightly bind DNA, but it does so nonspecifically. It does not recognize promoter sequences, so attachments to promoters are hit-or-miss affairs.

The holoenzyme, however, binds loosely and nonspecifically to DNA. Via a process of binding and dissociation it moves along the DNA molecule until a promoter sequence is reached, at which point, sigma causes tight binding.

Once RNA synthesis has begun and perhaps 5–10 ribonucleotides have been joined, sigma dissociates from the polymerase, which continues on its own to transcribe the gene. The sigma subunit can then bind to another core enzyme and repeat its function.

### *Rho Factor*

RNA elongation continues, and at some point rho may attach. RNA synthesis proceeds until a termination sequence (UGA, UAG, UAA) is reached. At this point, RNA synthesis terminates in one of two ways: rho-independent or rho-dependent. (We shall discuss rho-independent termination later in the chapter; see Transcription.)

Rho-dependent termination requires the rho protein. How exactly transcription is terminated is still the subject of much speculation. However, a working hypothesis is as follows. After the release of sigma, rho binds strongly to the nascent mRNA chain and moves along the RNA molecule. The energy to do this is supplied by the hydrolysis of ATP.

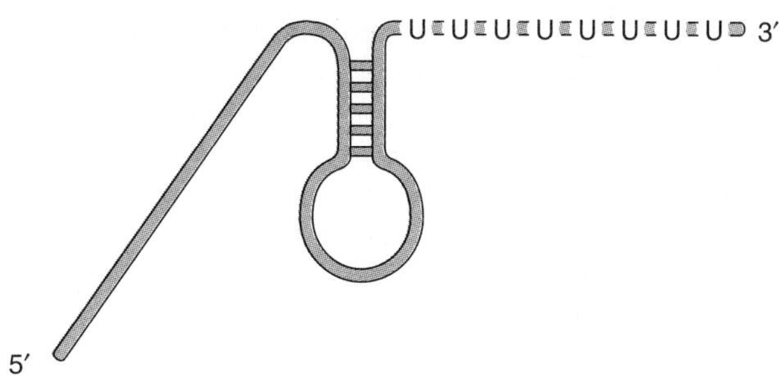


FIGURE 8.6
A stem-loop of RNA. The sequence of uracils at the 3′ end would base pair with a series of adenines in DNA. The relatively weak A=U bonds are all that hold the RNA to its DNA template.

When the RNA polymerase transcribes a terminator sequence, a so-called stem-loop of RNA is formed (Figure 8.6). Rho catches up to the polymerase at the stem-loop and causes the polymerase to pause. This rho-induced pause may itself terminate RNA synthesis and the release of RNA polymerase and the mRNA molecule from the DNA template. Alternatively, the rho factor may be the necessary agent to halt transcription and cause release of the enzyme and of mRNA.

## Ribonucleic Acids

### *Messenger RNA*

The function of messenger RNA as an intermediate between the gene (DNA) and the protein-synthesizing machinery of the cell was first proposed by Jacob and Monod in 1961, and evidence for its existence and function was published in the same year by S. Brenner, Jacob, and M. Meselson.

This species of RNA is a single-stranded molecule whose nucleotide content and length are dependent on the gene from which it is transcribed. On average, however, prokaryotic mRNAs contain between 900 and 1500 nucleotides (see Chapter 5).

Commonly, prokaryotic mRNA has a triphosphate adenosine at its 5′ end; though a triphosphate guanosine may occur in this position. And, where normally a thymine would occur in any of the DNA, another pyrimidine, uracil, is found in all RNAs. The nucleotide sequence of mRNA is complementary to the anticoding strand of DNA.

Like all other nucleic acids, mRNA is synthesized in a 5′ to 3′ direction. In prokaryotes such as *E. coli,* the molecule may contain more than one message and is then said to be polycistronic or polygenic to indicate that more than one gene has been transcribed. In other words, a single molecule of mRNA may contain information for the synthesis of different proteins/polypeptides involved in a particular pathway. For example, the information for the production of the enzymes needed in tryptophan synthesis or of those needed for the degradation of lactose is contained in one mRNA molecule (see Chapter 7). The rapid synthesis of mRNA allows for a quick response to changes in the environment (see Chapter 7).

Prokaryotic mRNA has a half-life of from 1 to 3 minutes, and it is enzymatically degraded (exactly how is still a matter of speculation). In contrast to prokaryotic mRNA, eukaryotic mRNA is stable for periods of time measured in hours and days.

## *Transfer RNA*

Crick first proposed the existence of what he called an adaptor molecule in 1958. The molecule was seen as the connection between the gene and the polypeptide: that is, between a sequence of nucleotides and a sequence of amino acids. His proposal turned out to be correct. Crick made a number of predictions concerning the adaptor molecule: First, amino acids are attached to the adaptor molecule; second, it is the adaptor molecule that "reads" the information carried by mRNA; and third, the adaptor molecule alone determines the position of the amino acids in the polypeptide during translation. Subsequent work using in vitro protein-synthesis techniques verified Crick's postulates. The adaptor molecule is now referred to as transfer RNA, or simply tRNA, in recognition of its role in transferring amino acids from the amino acid pools to the ribosomes, the sites of polypeptide synthesis.

More than 300 tRNA molecules from many different organisms have been sequenced to date. Though they vary in the number of nucleotides they contain, from 60 to 90, most are composed of about 76 nucleotides and have a number of features in common (Figures 8.7 and 8.8).

Transfer RNA is a single-stranded molecule that folds back on itself to form stabilizing internal base pairs. Beginning at the 5′ end of the molecule, it has the following characteristics:

1. a monophosphate at the 5′ end
2. a **D loop** containing the modified base dihydrouridine
3. an anticodon loop containing a sequence of three unpaired nucleotides, the anticodon, which reads the mRNA codon
4. a **variable loop** in which the greatest variation of nucleotide sequences occurs. This loop may contain from 3 to 21 nucleotides in different tRNAs.
5. a **T×C loop** in which a modified base, pseudouridine ($\psi$), appears
6. an **acceptor stem** of 7 bp, which may include unusual base pairings such as a guanine-uracil pair
7. a three-base sequence, CCA, at the 3′ end found on all tRNAs; the adenine has a free 3′ OH group to which the amino acid is enzymatically attached

All tRNAs have various combinations of so-called invariant, semivariant, and correlated invariants. These refer to positions in the molecule at which: the base is always the same (invariant); the base may vary but is always either a purine or a pyrimidine (semivariant); bases are outside loops that are base paired in all tRNAs (correlated invariants).

The modified bases of tRNAs are produced enzymatically after the molecules are synthesized, and they may account for up to 20% of the bases in the molecule. These modifications do not appear to be necessary for the tRNA's function—i.e., its binding to ribosomes, its translation of mRNA, or its ability to bind amino acids.

In *E. coli*, there are something like 60 tRNA genes located individually or in clusters along the chromosome, and a few are even interspersed within the operons of *E. coli*'s rRNA genes. Some clusters of these genes may contain multiple copies of a single tRNA gene, whereas other clusters may be made up of unrelated tRNA genes.

Each cluster, however, is transcribed as a unit from which individual tRNAs are cleaved, followed by base modification and the addition of a 3′-CCA end where needed. Although we might expect there to be some 61 different tRNAs since there are 61 codons specifying amino acids, this appears not to be the case. It seems that many of the tRNAs can translate more than one codon. That is, a single tRNA can be used to translate any of the codons that specify a given amino acid. So, for example, both codons for tyrosine are translated by one tRNA, and both lysine codons are likewise translated by a single tRNA. This is possible because of the wobble mechanism (Chapter 6), which allows unusual third base pairings.

To function properly during protein synthesis, tRNAs must carry the correct amino acids. That is, for a biologically active protein to be made, tRNAs must supply the correct amino acids and position those amino acids correctly in the polypeptide chain.

The tRNAs are matched with their amino acids by a group of enzymes collectively called **amino acyl tRNA synthetases.** There are at least 20 of these enzymes, one for each of the commonly occurring amino acids and their corresponding tRNAs. Some investigators have suggested that there may be as many as 30 different tRNAs, but there are only 20 amino acyl tRNA synthetases.

Although all of the enzymes catalyze the same reaction, structurally the enzymes are different. There is little amino acid sequence similarity between enzymes specific for different amino acids and the corresponding tRNAs. However, synthetases from different organisms specific for the same amino acid do show quite a bit of sequence similarity.

All amino acyl tRNA synthetases catalyze two sequential reactions involving three substrates. The substrates are

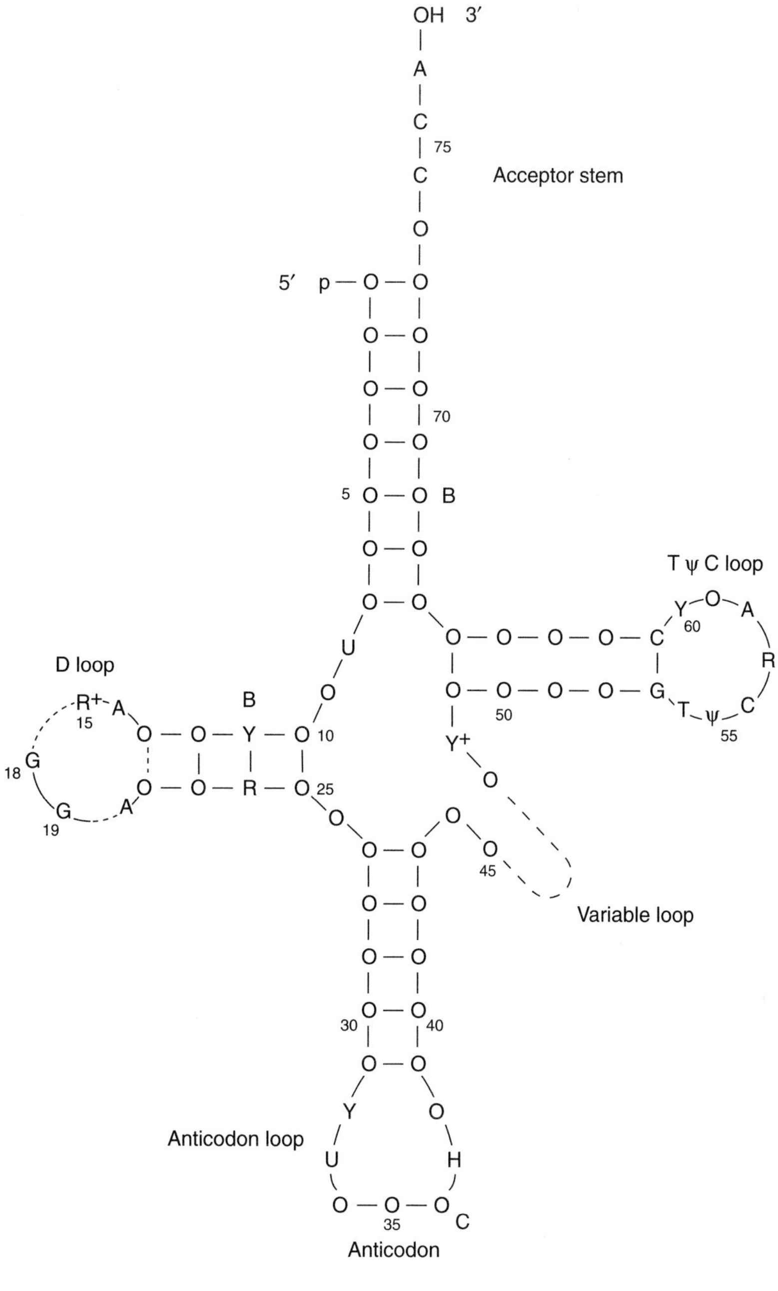

FIGURE 8.7
The cloverleaf structure of tRNA. The amino acid would attach at the 3′-OH position of the adenine of the acceptor stem. The anticodon is the series of 3 unpaired nucleotides (34-35-36) in the anticodon loop that reads the codon.

ATP, an amino acid, and a tRNA molecule. The reactions are as follows:

1. amino acid + ATP ↔ amino acyl~AMP + $PP_i$
2. amino acyl~AMP + tRNA ↔ amino acyl~tRNA + AMP

(The symbol ~ indicates a high-energy bond.) The overall reaction is:

amino acid + ATP + tRNA ↔ amino acyl~tRNA + AMP + $PP_i$

The reaction is driven by the hydrolysis of the inorganic pyrophosphate ($PP_i$).

In the first reaction, the amino acid is activated, and in the second reaction tRNA is charged; the same synthetase catalyzes both reactions. The amino acyl~AMP remains attached to the enzyme, which then accepts the tRNA as a substrate.

There is also a proofreading function associated with each synthetase. Occasionally, a synthetase may activate the

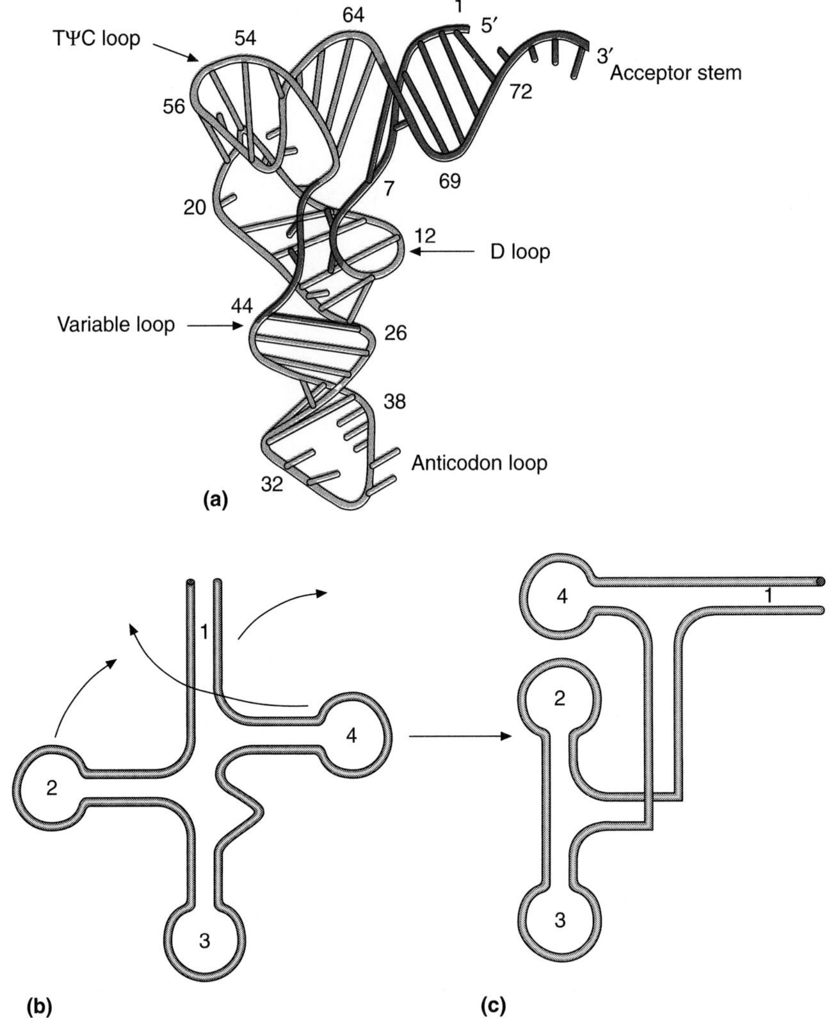


FIGURE 8.8
The three-dimensional structure of tRNA. The cloverleaf folds and twists (b,c) to produce the structure shown in (a).

wrong amino acid because of a structural similarity. For example, the isoleucine synthetase may on occasion activate the structurally similar amino acid valine. The activated amino acid, valyl~AMP, remains attached to the synthetase, but when the enzyme reacts with its correct tRNA (isoleucine tRNA), this reaction triggers an editing process. The valyl~AMP is hydrolyzed by the synthetase itself. Another example of this proofreading function is the formation of threonyl~AMP by meththionyl-tRNA synthetase and its subsequent hydrolysis. The efficiency of the correction system is such that the estimated misacylation results in only 0.17% of the total proteins produced being defective.

## *Ribosomal RNA*

Ribosomal RNA is found in **ribosomes,** the particles on which polypeptides are produced. The three classes of rRNA molecules are distinguished and referred to by their sedimentation coefficient(s). In *E. coli,* the molecules are 16S, 23S, and 5S. These rRNA molecules consist of approximately 1540, 2900, and 120 nucleotides, respectively.

*E. coli* has seven dispersed copies of a common genome sequence, each of which codes for a single transcript of about 5000 nucleotides. This RNA is then enzymatically cleaved to yield the 16S, 23S, and 5S rRNAs.

## Ribosomes

The ribosomes are the necessary particles on which polypeptides are synthesized (peptide bond formation). It is here that mRNA, containing the information for the amino acid sequence of the polypeptide, is brought together with the tRNAs carrying the amino acids. And it is here on the ribosome that the tRNAs read the mRNA.

Each ribosome is made up of a large and a small subunit with sedimentation coefficients of 50S and 30S, respectively (Figure 8.9). Each subunit contains its own distinct classes of RNAs and proteins (Table 8.2). Note in Table 8.2 that the complete ribosome has a sedimentation coefficient of 70S—not 80S, the sum of the large and small subunit coefficients. This is because sedimentation coefficients are not additive. They are derived on the basis of a particle's behavior in a gradient centrifuged at ultrahigh speeds. And, besides the particle's mass, other factors such as the particle shape affect its rate of sedimentation.

The tRNAs associate with the large subunit, while the mRNAs associate with the small subunit for polypeptide synthesis.

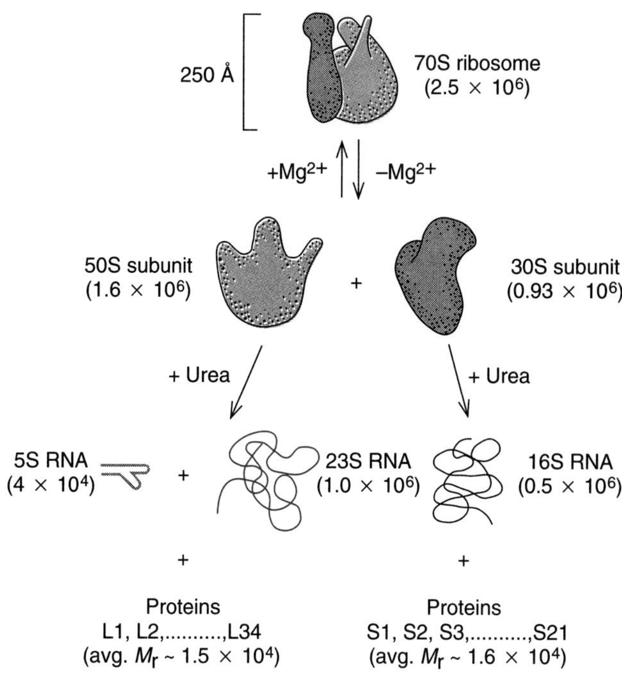


FIGURE 8.9
The *Escherichia coli* ribosome. From top to bottom—successive stages in the dissociation of the ribosome.

## Translation Initiation Factors

To begin the translation process, three proteins, acting at different times and sites, are required. These proteins are collectively called **initiation factors** and are designated **IF-1, IF-2,** and **IF-3.** Together these proteins assist in the establishment of an initiation complex and are then released and take no further part in polypeptide synthesis.

*table 8.2*

COMPONENTS OF RIBOSOMES

| COMPONENT | COMPLETE RIBOSOME | SMALL SUBUNIT | LARGE SUBUNIT* |
|---|---|---|---|
| *ESCHERICHIA COLI* | | | |
| Sedimentation coefficient | 70S | 30S | 50S |
| RNA sedimentation coefficient | | 16S | 23S + 5S |
| Nucleotides | | | |
| number of | | 1542 | 2904 + 120 |
| % of mass | 66 | 60 | 70 |
| Polyeptide chains | | | |
| number of | | 21 | 34 |
| % of mass | 34 | 40 | 30 |
| EUKARYOTE (RAT LIVER) | | | |
| Sedimentation coefficient | 80S | 40S | 60S |
| RNA sedimentation coefficient | | 18S | 28S + 5.8S + 5S** |
| Nucleotides | | | |
| number of | | 1874 | 4718 + 160 + 120 |
| % of mass | 60 | 50 | 65 |
| Polypeptide chains | | | |
| number of | | 33 | 49 |
| % of mass | 40 | 50 | 35 |

*Two values pertain to major and minor RNAs.
**Three values pertain to one major and two minor RNAs.
Source: Data from B. Lewin, *Genes IV*, 1990, page 157. Cell Press, Cambridge, MA.

IF-1 is required to help IF-3 separate the 30S subunit from the 50S subunit of an inactive ribosome and to join an mRNA, at its 5′ end, to the freed 30S subunit.

IF-2 then helps to bind a special charged initiator tRNA and GTP to mRNA and the 30S subunit. This initiator tRNA recognizes the initiation codon, AUG, at the 5′ end of mRNA. Once this complex is formed, a 50S subunit is added to complete a functional ribosome. For each ribosome that attaches to the mRNA, the same steps are followed.

If the mRNA is a polycistronic molecule, IF-2 will assist the complete 70S ribosome to reform an initiation complex at the next initiation codon (AUG) following a stop sequence in the mRNA. However, the greater the distance between a stop codon and the next initiation codon, the less likely it is that the ribosome will make it to the second or third initiation codon before separating into subunits. Some, of course, do make it, and, in fact, this may be the mechanism for maintaining a given ratio of different polypeptides.

## Translation Elongation Factors

In order to bind each successive amino acid to the polypeptide chain, three more protein factors are needed. These are called **elongation factors** and are designated **EF-Tu, EF-Ts,** and **EF-G.**

EF-Tu binds charged tRNA and GTP to the ribosome and mRNA. EF-Ts releases GDP, a GTP hydrolysis product, from EF-Tu. EF-Tu is needed to regenerate EF-Tu–GTP. Last, EF-G, using GTP as an energy source, promotes movement of the ribosome along the mRNA as each codon of mRNA is read and as a new amino acid is incorporated into the growing polypeptide chain.

## Releasing Factors

Another set of three proteins, called releasing factors (RF), is needed for the separation of the chain from the ribosome once polypeptide synthesis is completed. RF-1 recognizes the stop codons UAA and UAG. RF-2 recognizes stop codons UAA and UGA. RF-3 appears to stimulate the binding of RF-1 and RF-2 to the ribosome in the presence of GTP.

The releasing factors acting together or in cooperation with the enzyme peptidyl transferase then stimulate breakage of the bond holding the polypeptide to the last tRNA, thereby separating the polypeptide from the ribosome.

## Peptidyl Transferase

The enzyme **peptidyl transferase** is an integral part of the large (50S) ribosomal subunit and catalyzes a reaction in which the polypeptide chain is translocated to the next amino acid to be added to the chain. So, in effect, rather than an amino acid's being added to the polypeptide chain, the chain is added to the newest amino acid. The two are then joined via a peptide bond synthesized by the transferase. The nature of the components providing transferase activity is not well known.

# TRANSCRIPTION

Transcription is the process by which stored information is taken from the genetic material and ultimately made available to the cell in the form of protein or RNA. In protein synthesis, transcription is the process of mRNA synthesis during which the nucleotide sequence of DNA (the gene) is reproduced in the ribonucleotide sequence of mRNA. (Keep in mind that in all RNA synthesis the pyrimidine uracil replaces the pyrimidine thymine.)

As indicated in Chapter 5, cellular forms of life and even some of the larger viruses store genetic information in genes that may be located on either strand of the DNA molecule.

To avoid confusion concerning which strand is the "sense" strand, we shall always refer to the strand containing the information in the gene's nucleotide sequence as the **coding strand;** its complementary strand will be referred to as the **anticoding strand,** or the template strand (or simply the template), for example:

5′ ATGCCGTTACGC 3′: coding strand<br>
3′ TACGGCAATGCG 5′: anticoding or template strand

In other words, the transcribed strand is the template, the anticoding strand. Remember, because of the base-pairing rule, we cannot directly copy the coding strand. Rather, to get the sequence of nucleotides of the coding strand, the coding strand's complement, the anticoding strand, is transcribed. And because of the base-pairing rule, the transcription product, mRNA, has the base sequence of the coding strand.

Transcribing the template in the above example yields the sequence 5′ AUGCCGUUACGC 3′. (Remember U replaces T.) At a glance, it can be seen that this sequence is identical to the sequence of the coding strand. With regard to the relative positions of the promoter/operator sequences and the structural gene sequences in the coding, template, and mRNA strands, a representation might be as follows:

(promoter/operator/structural gene)<br>
Coding strand:<br>
5′ GCGCCG/TGAAG/ATGGTTGGAGTA . . . 3′<br>
Template strand:<br>
3′ CGCGGC/ACTTC/TACCAACCTCAT . . . 5′<br>
mRNA:<br>
5′ GCGCCG/UGAAG/AUGGUUGGAGUA . . . 3′

Also note that if, in the template's structural gene sequence, uracil replaces thymine, we have another RNA sequence (UACCAACCUCAU). And if this sequence is divided into triplets *beginning at the 3′ end,* we have a series of triplets that will base pair with the codons of mRNA *beginning at its 5′ end.* These new triplets—UAC, CAA, CCU, CAU—are the same triplets to be found in tRNAs and are referred to as the anticodons. These anticodons are what "read" mRNA during translation.

So DNA, the gene, via the coding strand, supplies the sequence of codons specifying the amino acids and their positions in the polypeptide. The anticoding strand gives the anticodons, or tRNAs carrying these anticodons, to be used to read the mRNA. Transcription involves three stages: initiation, elongation, and termination.

## Initiation

Transcription begins when the DNA-directed RNA polymerase associates with the sigma factor to produce the holoenzyme. The sigma factor allows the polymerase to bind specifically at the gene's promoter sequence.

In particular, the binding is at a six-nucleotide sequence called the Pribnow box (Chapter 5). The sequence is commonly TATAAT and is centered at about –10 nucleotides upstream of (before) the transcription start site. Further upstream, at about –35 nucleotides upstream from the transcription start site, sigma binds a second sequence, TTGACA.

These sequences are sometimes called **consensus sequences** because of the high probability of their being found in promoter regions. To date, more than 100 promoters have been found to have these, or very similar, sequences.

In the presence of sigma, the RNA polymerase binds tightly to the DNA in such a fashion that it is situated over the promoter sequence and may extend as far as the start sequence for transcription—i.e., into the first gene of the operon to be transcribed.

Assuming now that other required factors are present, CAP~cAMP for example, and no repressor proteins are attached to the downstream operator region (see The Lactose (*lac*) Operon, Chapter 7) transcription can begin.

The first stage is the formation of an open promoter complex (Figure 8.10). This complex requires the separation of the DNA strands; sigma assists in this separation. The separation of the strands is needed because any one nucleotide can base pair with only one other nucleotide. So, for mRNA to be made, the polymerase must base pair ribonucleotides with the deoxyribonucleotides of the template strand in accordance with Chargaff's rule (complementarity, Chapter 1). The polymerase alone will continue to separate the DNA strands and synthesize mRNA. Once transcription of a segment of the template has taken place and the polymerase moves on, the DNA behind the enzyme closes as the hydrogen bonds of the DNA base pairs reform.

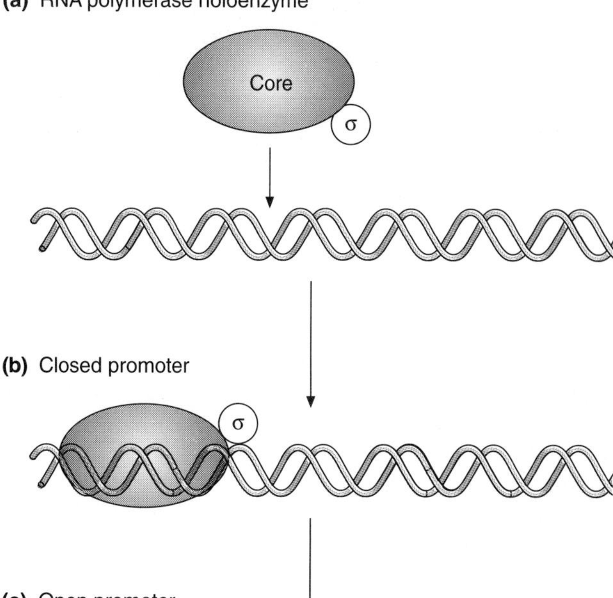


FIGURE 8.10
The binding of DNA-directed RNA polymerase plus sigma to DNA. The hydrogen bonds across the double-stranded DNA are broken and an open promoter complex is formed in preparation for mRNA synthesis (transcription).

The enzyme reads the template strand in a 3′ to 5′ direction as it synthesizes mRNA in a 5′ to 3′ direction. Note that the coding strand lies in a 5′ to 3′ orientation, so the gene and the mRNA have the same orientation.

In prokaryotic RNA synthesis, the first ribonucleotide is almost always a purine, and most often it is an adenine. In fact, the adenine (or occasionally a guanine) is a 5′-triphosphate nucleoside.

## Elongation

In the second stage of transcription, RNA polymerase continues reading the template strand and joining ribonucleotides by addition to the 3′ end of the growing chain. Synthesis of mRNA is therefore in the 5′ to 3′ direction as the template is decoded in a 3′ to 5′ direction. The strands of double-stranded nucleic acids, even in temporary hybrids such as DNA-RNA molecules, must be antiparallel if hydrogen bonding across the strands is to take place.

The energy required for synthesis is provided by the triphosphate ribonucleosides. These ribonucleosides are

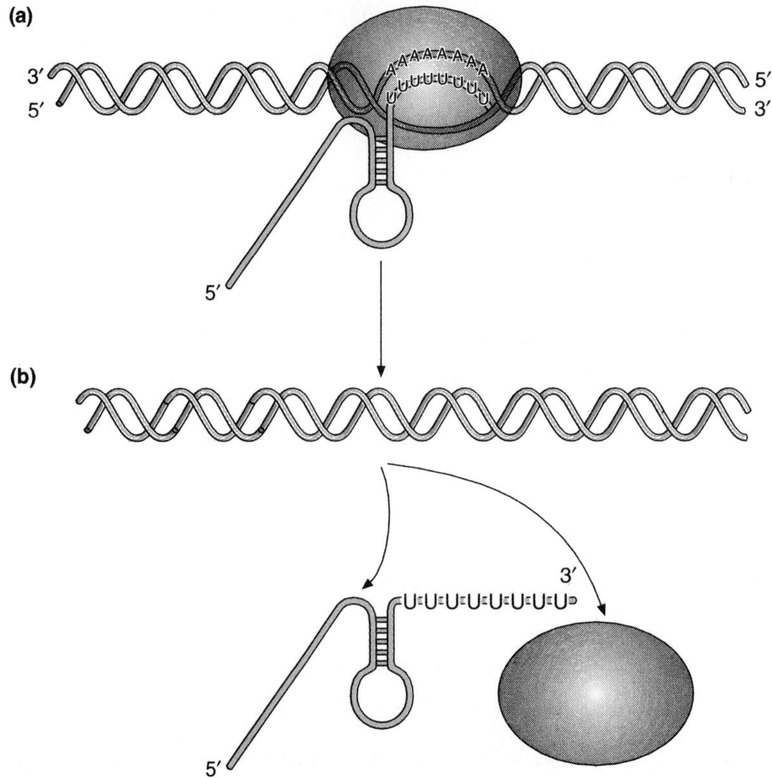


FIGURE 8.11
ρ-independent termination of mRNA synthesis/termination of transcription. (a) RNA polymerase synthesizes a poly-U 3′ end. (b) The mRNA and enzyme pull away from the DNA template.

the energy sources, the building blocks, and the information components for mRNA synthesis. (Recall the roles played by the triphosphate deoxyribonucleotide in DNA synthesis; Chapter 2.)

Synthesis of mRNA, then, looks like this:

Coding strand: 5′ AATGCCGTTACGCCC 3′
Template: 3′ TTACGGCAATGCGGG 5′
⟶
(reading direction)
mRNA: 5′ ppp-AAUGCCGUUACGCCC 3′

Using the incorporation of radioactive ribonucleosides into *E. coli* mRNA as a measure, researchers have estimated the rate of mRNA synthesis to be between 20 and 50 nucleotides per second, with an error frequency of about 1 nucleotide for every $10^4$ nucleotides incorporated. Though this error frequency is relatively large compared with that found in DNA replication, the mRNA error frequency has been explained as tolerable because: (1) most genes are repeatedly transcribed, so the likelihood that the same error would reappear is small; (2) the redundancy of the genetic code allows for some errors, and (3) often, amino acid substitutions in polypeptides do not appreciably alter the protein's biological activity.

# Termination

The last stage in mRNA synthesis is chain-growth termination. Synthesis of mRNA is ended when one of two DNA sequences is reached.

The DNA sequences, often referred to as transcription terminators, are either rho-dependent or rho-independent. In either case, a so-called **stem-loop,** or hairpin structure, is formed (recall attenuation; Chapter 7). RNA synthesis terminates shortly after this structure is formed (Figure 8.11).

The stem-loop forms at the 3′ end of the mRNA, because at the 5′ end of the template DNA an unusual sequence of nucleotides occurs. This sequence is known as an inverted repeat. That is, read in a 5′ to 3′ direction, the DNA nucleotide sequences of the two strands are identical. For example:

5′ TACGAAGTTCGTA 3′
*
3′ ATGCTTCAAGCAT 5′

At the asterisk, the G-C pairing, is the point of symmetry. When mRNA is transcribed from the template strand (3′→5′), the resulting sequence is:

*
5′ UACGAAGUUCGUA 3′

The molecule is self-complementary around the point of symmetry, $\overset{*}{G}$. So a hairpin, or stem-loop, can form:

```
5′ U–A 3′
   A–U
   C–G
   G–C
   A–U
   A U
    G
    *
```

Because of the turn at the bottom, the last A-U pairing does not form, but a loop is produced.

In rho-dependent termination, the template inverted repeat of DNA is followed by a series of adenines. This series produces a run of perhaps half a dozen uracils in the mRNA.

So we have:

```
5′ U–AUUUUUU 3′
   A–U
   C–G
   G–C
   A–U
   A U
    G
```

At the point where the poly-U sequence is attached to the DNA sequence, the hybrid DNA-RNA is unusually weak (A—U bonds are weak), and it requires very little energy to break the hydrogen bonds holding the two strands together. When separation occurs, mRNA synthesis, transcription, stops. This type of termination is rho-independent; no termination factor is required.

Rho-dependent termination (Figure 8.12) also uses a hairpin mRNA formation, but dissociation of the DNA-RNA hybrid needs the assistance of the protein rho and no poly U follows the hairpin.

Rho appears to attach itself to the mRNA while it is still under construction but after sigma is released. Rho moves along the RNA behind the RNA polymerase. The formation of the stem-loop apparently stalls the polymerase at the loop or shortly after the loop forms. The stalling allows rho to approach the enzyme. Rho then brings about release of the polymerase and mRNA from the DNA template. Rho may be needed because of the lack of the poly-U sequence.

The precise mechanisms by which rho-dependent and rho-independent termination are achieved still have to be worked out. Rho's activity, however, does require ATP hydrolysis, which is thought to provide the energy to move the protein factor along the mRNA.

It also appears that termination is not absolutely rho-dependent or rho-independent. Rather, rho-independent termination can utilize rho, and rho-dependent termination can proceed in the absence of the protein.

**(a)** ρ pursues polymerase

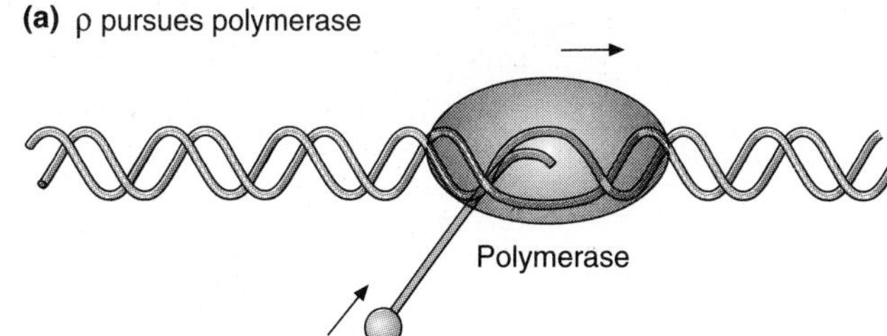


**(b)** Stem-loop forms; polymerase pauses; ρ catches up

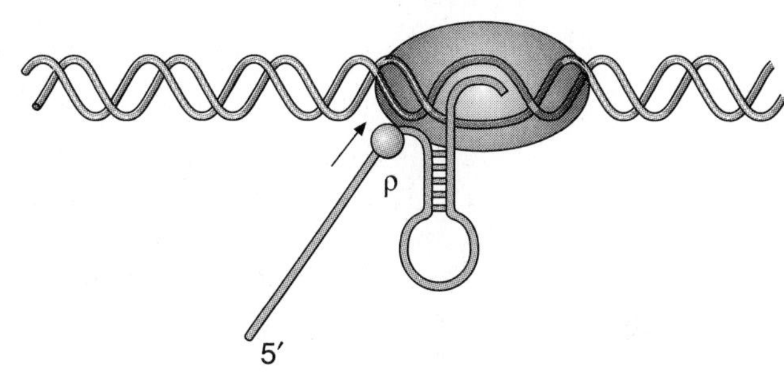


**(c)** ρ causes termination

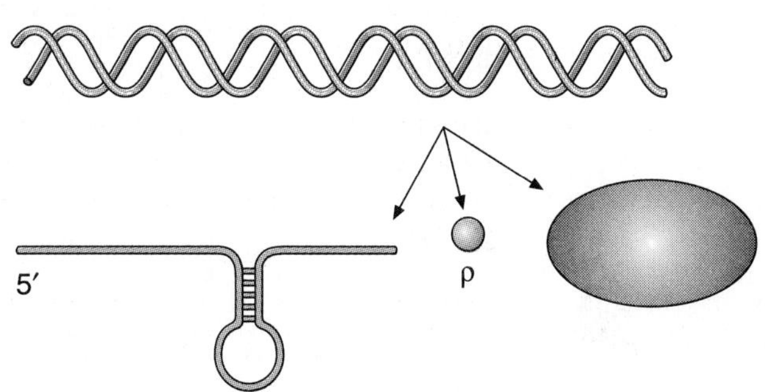


FIGURE 8.12
ρ-dependent termination of mRNA synthesis/termination of transcription. (a) Rho attaches to mRNA as it is being synthesized. (b) RNA polymerase pauses after stem-loop is synthesized, allowing ρ to catch up. (c) By some unknown mechanism, ρ causes the release of the mRNA and enzyme.

To confuse matters even more, at least in in vitro experiments, some rho-dependent terminations require more rho protein than do others. And, although rho seems to be an essential protein for *E. coli*, there are few rho-dependent terminator sequences in its genes.

## TRANSLATION

Translation is a series of events that converts the language of the genes, sequences of nucleotides, into the language of polypeptides, sequences of amino acids.

As does transcription, translation involves three stages, which are also called initiation, elongation, and termination. Together these stages result in the synthesis of a polypeptide.

In prokaryotes, the absence of a nuclear membrane makes possible the coupling of transcription and translation. Even before the completion of mRNA synthesis, the synthesis of a polypeptide can begin. Furthermore, as soon as 100–200 nucleotides of the mRNA have been translated (30–40 amino acids have been joined), another ribosome can attach to the mRNA, and the production of another polypeptide, identical to the previous one, can start. In this way, successive ribosomes can join the mRNA, and, on each ribosome, polypeptide synthesis can take place. The number of ribosomes that can attach to a particular mRNA, depends on the length of the mRNA molecule. Ribosomes apparently, however, do not touch one another and are spaced several codons apart.

These multiribosome-mRNA complexes are referred to as **polyribosomes** or simply as **polysomes.** Each ribosome moves the entire length of the mRNA, so at the completion of the journey each ribosome has attached to it a complete polypeptide molecule. It should be underscored that the polypeptide chains do not move from one ribosome to another as translation of mRNA occurs. The polypeptide chains are bound to the individual ribosome to which the first amino acid was attached.

## Initiation

Initiation, the first stage in translation, begins with the attachment of a 30S ribosomal subunit to the 5′ end of the mRNA molecule. This binding process requires the following (Figure 8.13):

1. mRNA
2. 30S ribosomal subunit
3. IF-1, IF-2, IF-3
4. GTP
5. special charged tRNA: N-formyl-methionyl~tRNA, abbreviated fMet-tRNA.

At the 5′ end of *E. coli*'s mRNA is a purine-rich sequence of from 5 to 8 nucleotides known as the **Shine-Dalgarno (SD) sequence.** This sequence binds the pyrimidine-rich 3′ end of the 16S rRNA of the 30S subunit in the presence of IF-1 and IF-3.

This is then followed by the binding of IF-2, GTP, and the special N-formyl-methionine–charged tRNA. The special tRNA binds only at the 5′ ends of mRNAs, where the initiation codon, AUG, is located. The AUG codon specifies the special amino acid, N-formyl-methionine (see Figure 6.3).

If AUG appears anywhere else in a message, it is read simply as the common amino acid methionine. In fact, it is the function of IF-2 to distinguish between the special tRNA-carrying N-formyl-methionine and the ordinary tRNA-carrying methionine. The distinction is made on the basis of the presence or absence of the N-formyl group. As we shall see, the N-formyl group may serve another function in protein synthesis: to help ensure polypeptide elongation in the right direction.

Once the initiating tRNA has been bound, the 50S ribosomal subunit binds to the 30S subunit to form a fully functional initiation complex. At this stage, the protein initiation factors are all released.

The completed ribosome makes available two sites for tRNA attachment. The first, toward the 5′ end of mRNA, is the **P site (peptidyl site),** and the second, toward the 3′ end of mRNA, is the **A site (amino acyl site,** or **acceptor site).** The P site is initially occupied by the special tRNA-carrying N-formyl-methionine.

Reading of mRNA is in the 5′→3′ direction and, in fact, takes place at the A site, where the latest charged tRNA associates with the ribosome. The D loop and the TψC loop of tRNA are thought to bind the 50S subunit while the tRNA's anticodon base pairs with the mRNA's codon. We, therefore, have three points of attachment to stabilize each charged tRNA molecule in the A site. This binding to the A site requires the participation of EF-Tu and GTP.

In fact, EF-Tu and GTP bind the charged tRNA and thus help to align the charged tRNA in the A site. Base pairing between the tRNA anticodon and mRNA codon occurs, GTP is hydrolyzed, and the complex EF-Tu-GDP is released from the ribosome. At this point both the P and A sites are occupied (Figure 8.14).

The enzyme peptidyl transferase then transfers the N-formyl-methionine to the amino acid at the A site. A peptide bond is formed between the carboxyl carbon of the N-formyl-methionine and the nitrogen of the amine group of the amino acid in the A site (Figure 8.15). We have a dipeptide joined to the tRNA occupying the A site.

## Elongation

The factor EF-G now binds GTP to the ribosome and promotes translocation of the dipeptidyl~tRNA from the A site to the P site. When this occurs, the ribosome moves one codon in the 3′ direction of mRNA, thereby placing a new codon in the A site to be read by the next charged tRNA. And the process is repeated.

So, in a three-stage cycle, amino acids are joined to produce a polypeptide:

1. Amino acyl~tRNA occupies the A site.
2. The growing polypeptide chain is transferred to the newest amino acid.
3. The chain is translocated from the A site to the P site, a one-codon advancement of the ribosome, and back to stage 1.

   Approximately 40 amino acids are added per second.

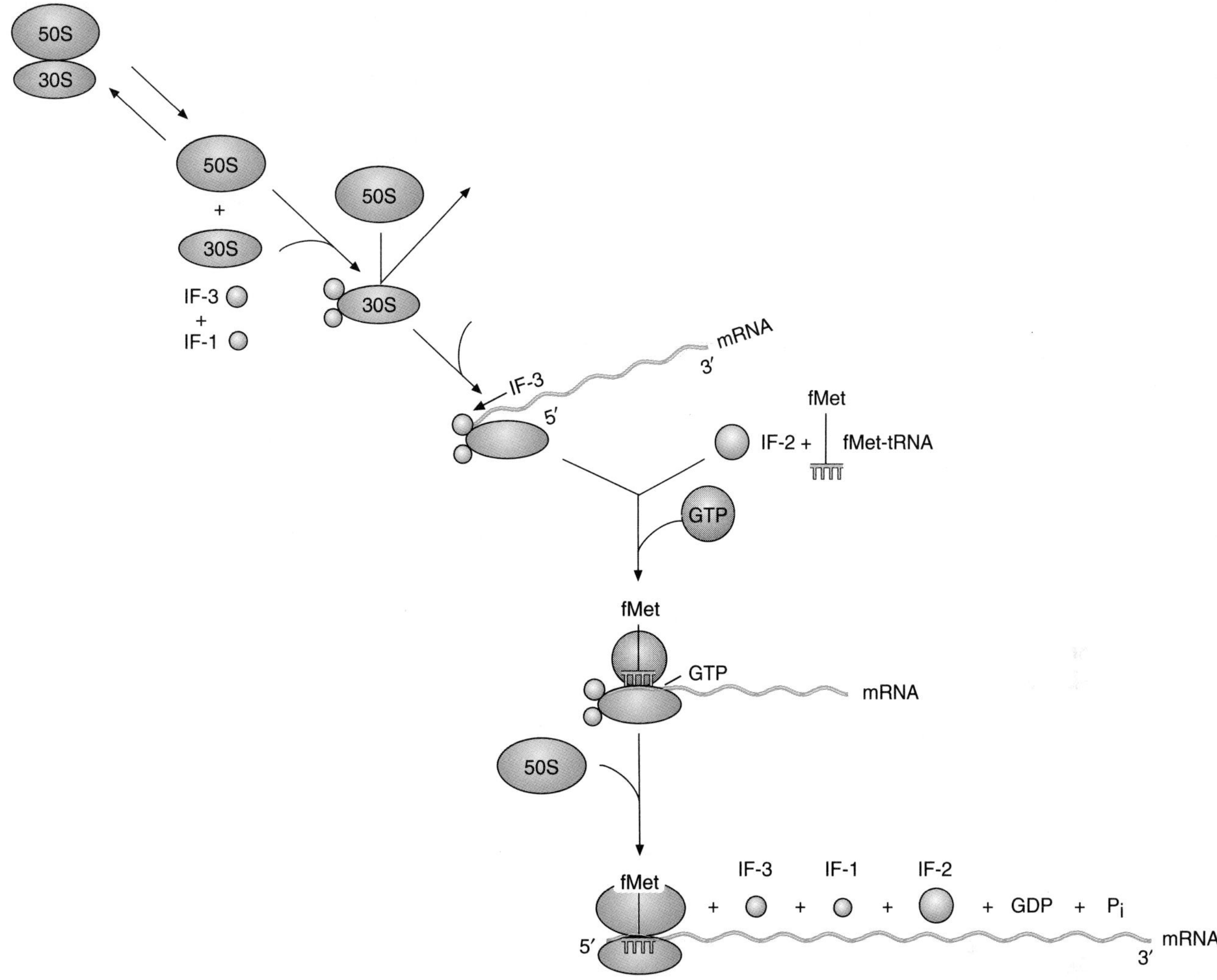


FIGURE 8.13
The initiation complex of *Escherichia coli*. The complete 70S ribosome dissociates, releasing the 50S and 30S subunits. Initiation factors IF-1 and IF-3 then participate in the binding of mRNA to the 30S subunit. Initiation factor 2 (IF-2), plus GTP, and the initiation fMet-tRNA bind the 30S-mRNA complex. This is followed by the binding of 50S subunit to complete the ribosome and the initiation complex. Elongation can now begin.

Regardless of the number of ribosomes associated with the mRNA, the same sequence of events occurs on each ribosome. As noted earlier, the growing polypeptide chains remain attached to their initiating ribosomes. On each ribosome, in a series along the mRNA, a polypeptide of different length is attached. The closer the ribosome is to the 3′ end of the mRNA, the longer is the polypeptide chain attached to it because more of the message has been read (Figure 8.16).

## Termination

When a stop codon is reached (UAA, UAG, UGA), a lack of tRNAs to translate these codons signals termination of protein synthesis. With the assistance of release factors RF-1, RF-2, and RF-3, the polypeptide is hydrolyzed from the tRNA. Precisely how this is accomplished is still a mystery.

If the mRNA is polycistronic—that is, carries more than one message—and the distance between the stop codon and the next initiation codon AUG in the next message is not too great, the 70S ribosome will not be released from the mRNA. Instead, the ribosome will continue on to the AUG codon, where another initiation complex forms and another round of polypeptide synthesis begins.

If you were now to look at the completed polypeptide, you would find that it was synthesized beginning at its N-terminus. That is, at one end, the first amino acid incorporated, N-formyl-methionine, has a nitrogen that is not part

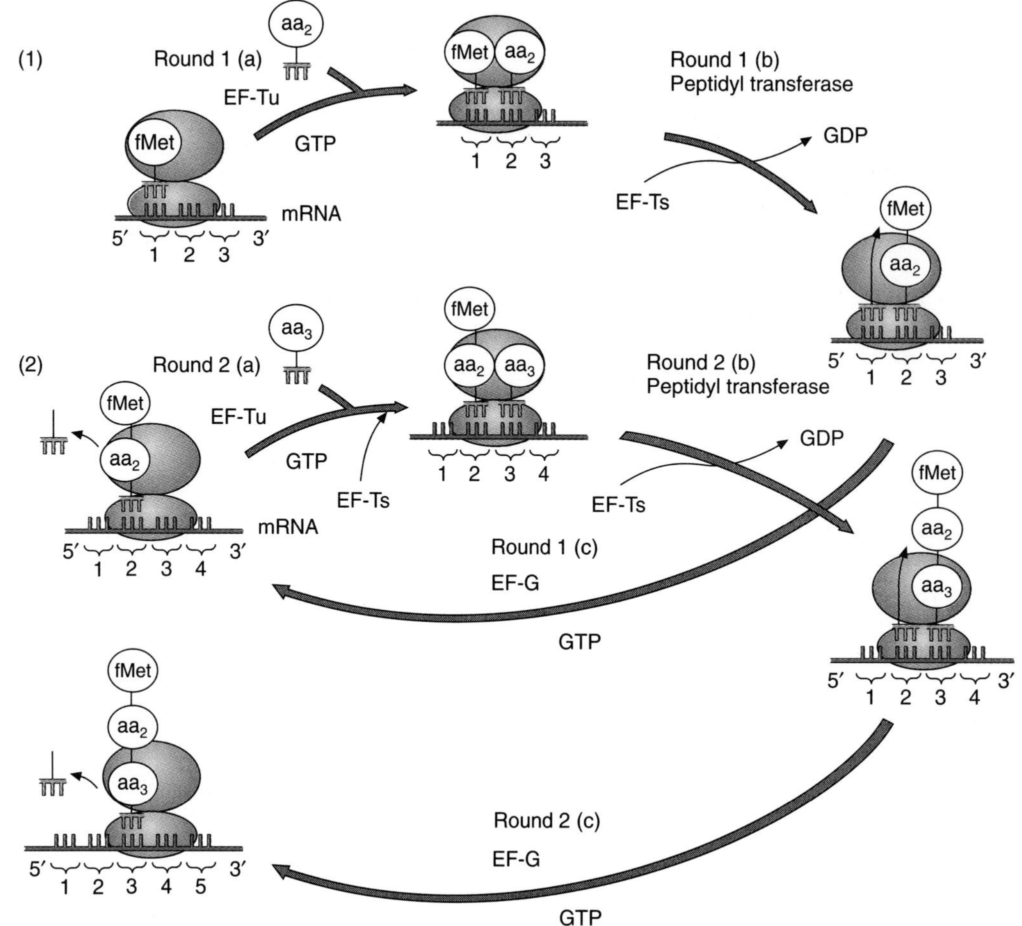


FIGURE 8.14
Protein synthesis. The process of elongation consists of repeating steps: (1) amino acids are brought to the A site (2) the chain in the P site is transferred to the amino acid in the A site by peptidyl transferase which joins them via a peptide bond (3) the mRNA is move one codon (5′ → 3′) and the process is repeated. As note, the process requires energy (GTP) and the factors EF-Tu, EF-Ts, and EF-G.

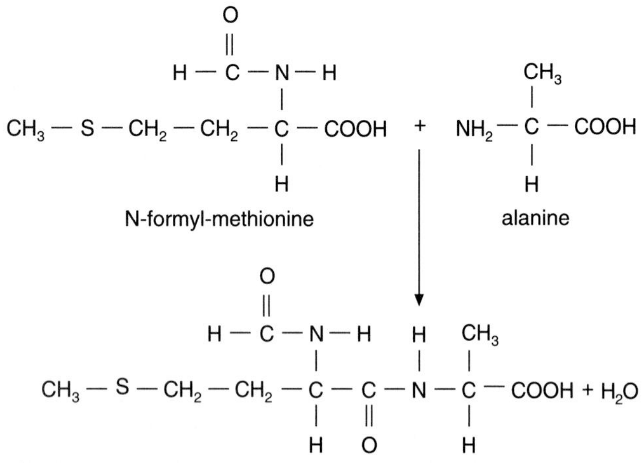

$$CH_3-S-CH_2-CH_2-CH(NH-CHO)-COOH \;+\; NH_2-CH(CH_3)-COOH$$

N-formyl-methionine + alanine

$$\downarrow$$

$$CH_3-S-CH_2-CH_2-CH(NH-CHO)-CO-NH-CH(CH_3)-COOH + H_2O$$

FIGURE 8.15
A peptide bond. N-formyl-methionine, via its carboxyl carbon, is joined to alanine, at its nitrogen group, to form a peptide bond. Note that the nitrogen of N-formyl-methionine cannot form a bond with alanine's carboxyl carbon because of the formyl group.

of a peptide bond and the last amino acid has a carboxyl which is not part of a peptide bond; this is the polypeptide's C-terminus.

Upon completion of the polypeptide, the N-formylmethionine is removed, so the second amino acid incorporated becomes the first amino acid of the new polypeptide. The chain now has a free nitrogen group at one end and a free carboxyl group at the other (neither is a part of a peptide bond). These are the N and C termini. The N-terminus corresponds to the 5′ end of the mRNA, and the C-terminus corresponds to the mRNA's 3′ end.

## PROTEIN FOLDING

Under the appropriate conditions—i.e., physiological conditions—many proteins spontaneously assume their correct secondary and tertiary structures. The key is that

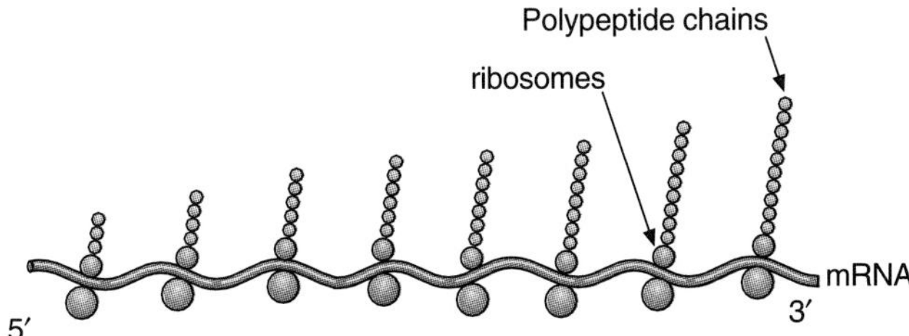


FIGURE 8.16
Elongation of a polypeptide chain. As the ribosome moves toward the 3′ end of mRNA, more codons are read. The chain, therefore, increases in length. The elongating chain remains attached to its original ribosome. Each ribosome attaches at the 5′ end of the mRNA molecule.

the primary structure, the sequence of amino acids, be correct. Given this condition, a polypeptide may begin to fold even before it is completed.

The folding may involve a complex set of sequential steps in which the formation of one section facilitates the formation of the next section. The folding is rapid, occurring within a few seconds. What is achieved is a molecule in the most stable thermodynamic conformation.

An example of this spontaneous folding is the refolding of the enzyme ribonuclease. The molecule can be denatured in vitro either chemically or by heating, at which point it loses its biological activity. A reversal of the conditions that caused the denaturation will restore the active conformation without any further information or substances added. All that is needed is that the sequence of amino acids, its primary structure, remain intact; that is all the information needed to reestablish the enzyme's biological activity.

The process may, however, require several hours. This time can be reduced to minutes if the enzyme protein disulfide isomerase is added to the ribonuclease solution. Apparently, the isomerase catalyzes random cleavage and reformation of disulfide bonds and continues to do so until the substrate protein achieves its most thermodynamically stable conformation.

For some proteins—those, for example, having cofactors such as metal ions (e.g., hemoglobin)—the cofactor must be present in order for the protein to attain its proper conformation. Still other proteins, multimeric proteins, may require the presence of at least one subunit for the proper assembly and stable conformational state to be reached.

In other cases, proteins may require other proteins, sometimes called molecular chaperones, to achieve the correct conformation. The chaperones have been found in *E. coli* and in plants. The molecular chaperones may function to prevent the formation of incorrect, biologically inactive states from which the protein cannot escape.

Although the primary structure of a protein determines the higher-order structures of that protein, it is possible for two proteins to arrive at very similar tertiary structures with very different amino acid sequences. For example, the globin proteins of different species vary considerably in primary structure but not in tertiary structure. And, often, proteins with similar biological functions have very similar structural features. This characteristic can sometimes be used to predict the protein's cellular activity.

The importance of protein folding to biological activity cannot be overemphasized. A case in point is the neurodegenerative disease amyotrophic lateral sclerosis (ALS), more commonly called Lou Gehrig's disease, after the famous New York Yankee first baseman. The gene responsible for one hereditary form of the disease has been identified. It codes for an enzyme called superoxide dismutase, an enzyme that helps cells safely dispose of oxygen free radicals, an extremely toxic species of oxygen released by a number of normal cellular activities.

In the mutated gene, any of 12 different amino acids can be replaced by other amino acids. One replacement is especially common: An alanine at position 4 is replaced by a valine. None of the mutations is anywhere close to the enzyme's catalytic or active site. Rather, they are all found in regions of the protein critical in maintaining the molecule's three-dimensional structure. For example, the alanine replacement occurs at the junction where two polypeptides must join to form a dimer. But because of this and the other mutations, the dimer either does not form or is not a stable conformation. (A still unanswered question is, Why is the disease often manifested relatively late in life?)

## SUMMARY

Protein synthesis is the process by which the information stored in the genes in the sequences of nucleotides or of codons is made available to the cell. DNA nucleotide sequences are converted to amino acid sequences to form a polypeptide/protein. DNA nucleotide sequences are colinear with mRNA nucleotide sequences and via that intermediate with amino acid sequences of polypeptides/proteins.

So the flow of information in the cell is from DNA to mRNA to polypeptide/protein. This flow is known as the central dogma of molecular biology.

The flow of information takes place in a two-stage process. Together the two stages constitute protein synthesis. The first stage is called transcription, during which the codon sequence of DNA is transcribed into codon sequences of mRNA. Then this sequence is read in the second stage, called translation. The reading is done by tRNA and sequences of amino acids that are joined by peptide bonds to form a unique polypeptide/protein. The tRNA molecules both read, via their anticodons, and deliver amino acids to the sites of protein synthesis, the ribosomes. Ribosomes are particles composed of RNA and proteins.

Each tRNA molecule is charged with a particular amino acid in a two-step reaction catalyzed by one of 20 enzymes collectively called amino acyl tRNA synthetases. These enzymes ensure that each tRNA is carrying its proper amino acid.

In general, polypeptides are not themselves biologically active. These molecules associate with other polypeptides to form the active protein molecules the cell needs.

Both transcription and translation consume large quantities of energy in the form of the nucleotide triphosphates used to construct mRNA and the polypeptide/protein itself.

Both transcription and translation require a number of enzymes and protein factors. The enzymes include DNA-directed RNA polymerase, responsible for the synthesis of mRNA, and peptidyl transferase, needed to form the peptide bond joining amino acids.

Among the protein factors required are sigma and rho, whose functions are proper initiation and termination of mRNA synthesis. Other proteins are needed to initiate protein synthesis, to elongate the polypeptide chain, and to terminate synthesis.

Upon completion of synthesis, and because of its unique amino acid sequence, its primary structure, each polypeptide either alone or in association with cofactors, or so-called molecular chaperones, will fold into a stable conformation state that will include secondary and tertiary structural motifs. Folding of the polypeptide/protein is critical for obtaining the proper three-dimensional structure of the molecule necessary for biological activity. The polypeptide itself may be biologically active, or it can associate with other similar or different polypeptides to produce a biologically active protein.

## Study Questions

1. Briefly explain the concept of the central dogma of molecular biology.
2. What does the phenomenon of colinearity tell us about protein synthesis?
3. What two things concerning amino acids in polypetides do codons of DNA specify?
4. What, if any, are the chemical and biological differences between polypeptides and proteins?
5. What are the various structures of polypeptides/proteins, and how are they related to one another?
6. What are the sources of energy for protein synthesis, and where during the process is energy needed?
7. What are the three types of RNA used during protein synthesis, and what is the role of each during the synthesis?
8. What is the difference between the core enzyme and the holoenzyme, and why is the difference crucial to protein synthesis?
9. List three types of protein factors needed for protein synthesis, and briefly explain the role of each.
10. What is needed to charge tRNA? Give the necessary reactions, in order, for the charging process.
11. Briefly explain what goes on during transcription and translation.
12. By what means are polypeptide molecules folded into stable conformations?

## Readings and References


Bohen, S. P., A. Kralli, and K. R. Yamamoto. 1995. Hold'em and fold'em: Chaperones and signal transduction. *Science* 268:1303–04.

Brenner, S., F. Jacob, and M. Meselson. 1961. An unstable intermediate carrying information from genes to ribosomes for protein synthesis. *Nature* 190:576–81.

Campbell, M. K. 1991. *Biochemistry*. Philadelphia: Saunders College Publishing.

Crick, F. H. C. 1958. On protein synthesis. *Symp. Soc. Exp. Biol.* 12:138–63.

Crick, F. H. C. 1970. The central dogma of molecular biology. *Nature* 227:561–63.

Deng, H.-X., et al. 1993. Amyotrophic lateral sclerosis and structural defects in Cu,Zn superoxide dismutase. *Science* 261:1047–51.

Echols, H., and S. M. van der Vies. 1991. Molecular chaperones. *Annu. Rev. Biochem.* 60:321–47.

Hoagland, M. B., M. L. Stephenson, J. F. Scott, L. I. Hecht, and P. C. Zamecnik. 1958. A soluble ribonucleic acid intermediate in protein synthesis. *J. Biol. Chem.* 231:241–57.

Jacob, F., and J. Monod. 1961. Genetic regulatory mechanisms in the synthesis of proteins. *J. Mol. Biol.* 3:318–56.

Lewin, B. 1997. *Genes VI*. New York: Oxford University Press.

Marx, J. 1993. Role of gene defect in hereditary ALS clarified. *Science* 261:986.

Pabo, C. O., and R. T. Sauer. 1992. Transcription factors: Structural families and principles of DNA recognition. *Ann. Rev. Biochem.* 61:1053–95.

Sarabhai, A., A. Stretton, S. Brenner, and A. Bolle. 1964. Colinearity of the gene with the polypeptide chain. *Nature* 201:13–17.

Singer, M., and P. Berg. 1991. *Genes and genomes*. Mill Valley, Calif.: University Science Books.

Voet, D., and J. G. Voet. 1995. *Biochemistry*. 2d. ed. New York: John Wiley & Sons.

Weaver, R. F., and P. W. Hedrick. 1997. *Genetics*. 3rd ed. Dubuque, Iowa: Wm. C. Brown.

Yanofsky, C., G. Drapeau, J. Guest, and B. Carlton. 1967. The complete amino acid sequence of the tryptophan synthetase: A protein and its colinear relationship with the genetic map of the A gene. *Proc. Natl. Acad. Sci. USA* 57:296–98.

CHAPTER NINE

# MUTATIONS AND REPAIR OF DNA

## OUTLINE



## CHAPTER OBJECTIVES

*This chapter will discuss:*

- The definition of mutation
- The various mechanisms by which point mutations are brought about
- The rates at which mutations occur among different organisms and within a single organism
- Techniques used to isolate mutant organisms
- Why some types of mutations can go undetected
- How various biological, chemical, and physical agents act to cause mutations
- How some mutations, including cancer, inborn errors of metabolism, and chromosomal rearrangements, affect the organism
- How mutations can be corrected by a number of different enzyme systems
- How mutations can be used to detect both carcinogens and mutagens

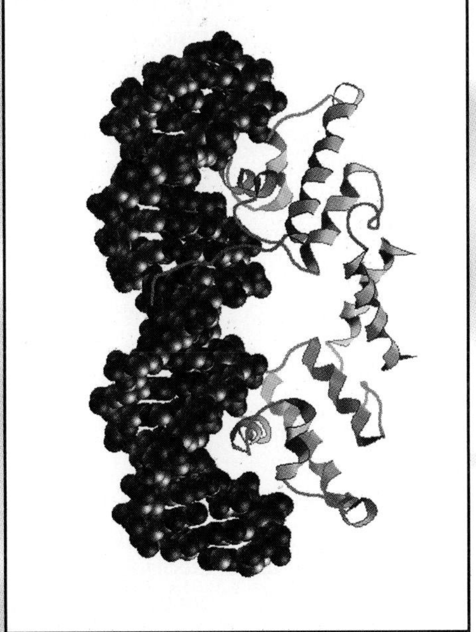

## INTRODUCTION

In Chapter 1, we listed a number of characteristics or abilities a molecule must have if it is to function as genetic material. One of these characteristics is that the molecule must be stable enough so that changes in the information content, the sequences of nucleotides, are rare occurrences. Yet, paradoxically, there must be some instability built into the molecule to allow for just such changes in information content to occur.

So the genetic material must at one and the same time be stable enough to preserve necessary information and unstable enough to allow for changes that are necessary to correct mistakes made during replication and to allow for changes that add new information to the genome. These changes in the genome are referred to as mutations.

In 1901, the Dutch botanist, and one of the rediscoverers of Mendel's laws, Hugo de Vries suggested that the process by which different alleles (copies) of the same gene come into existence be given the name *mutation*. He also proposed that the process was sudden and spontaneous and could be drastic. The mutations, he foresaw, were a source of genetic diversity.

Changes in the genetic material may involve as small a change as a single nucleotide or be as large as those involving entire arms of chromosomes containing millions of nucleotides. The effects of these changes on the organism may range from the acquisition of new beneficial information, to inconsequential effects, to lethal effects depending on the size of the disturbance and its location within a particular gene. Relatively large changes may produce nearly harmless effects; relatively small changes may produce severe effects. At least for the smallest mutations, there are repair mechanisms that function to correct or minimize the damage.

Much of the work involving mutations has centered on the deleterious effects of change, primarily because they are relatively easy to detect and, of course, because of their consequences.

Once again, microorganisms have been and, in many cases, still are the organisms being used in the study of the mechanisms of mutation and repair. More-complex organisms are by choice and necessity being used more and more often, but it still is much easier to use haploid microorganisms in mutation studies. In general, the ease with which a particular organism can be manipulated and mutations detected in that organism has been a decisive factor in the use of particular organisms in genetic studies.

The hope is that, ultimately, via the study of mutation and repair, it will be possible to diagnose and correct nonlethal troublesome mutations and lethal mutations in human beings. Tremendous progress has been made as far as diagnoses are concerned, and steady and considerable progress is being made with regard to repair.

## MUTATION DEFINED

By definition, a mutation is any change in the nucleotide (base) sequence of the genetic material, whether DNA or RNA. Although mutations directly affect an organism's genome, they are most easily seen and studied as expressions or manifestations of the effects. The nucleotide sequence of a gene is changed, and, as a result, a protein or RNA molecule is also changed. (Remember, genes code for polypeptides or RNA.) We can say, then, that a change in the genotype is expressed in the phenotype. The genotype is the organism's genetic constitution, whereas the phenotype is the observable or measurable expression of the genotype. So a change in the nucleotide sequence of a gene shows up as a change in the sequence of amino acids of a polypeptide (colinearity) or the nucleotide sequence of an rRNA or tRNA molecule.

An organism might have genes for certain enzymes, pigment, size, biochemical pathways, and so on. Those genes are a part of its genetic makeup, its genotype. The genes that are expressed then give us an organism of a particular phenotype; the organism is of a certain size and color and it is able to carry out certain biochemical transformations.

A mutation is set in DNA after a second round of replication. First, a change in the parental DNA occurs. For example, a mutagen chemically modifies a base; then, when the strand containing the modified base is used as the template for a round of DNA replication, the new base pairs with a "wrong" nucleotide. In the second round of replication, the wrong base pairs with its correct mate, resulting in the second base pair's replacing the first. Shortly, we shall see in more detail how this occurs.

A mutation can be detected by comparing the mutated organism, or mutant, with another nonmutated organism of the same type. The nonmutated organism is called the **wild type.** The wild type would be an individual organism or population from which the mutant was derived. For example, bacterial mutants are compared with the population of cells from which the mutants were gotten.

Occasionally, a **back mutation** occurs in which the wild type gene nucleotide sequence is restored. This phenomenon is referred to as reversion. Reversions may occur by the same mechanisms that produce other kinds of mutations.

## MECHANISMS OF MUTATION

A convenient way to deal with mutations is to look at the mechanisms by which mutations, changes in genomic nucleotide sequences, are brought about and the consequences to the organism of those changes. In other words,

**table 9.1**

**Some Types of Mutations and Their Phenotypic Expression**

| Mutation | Phenotypic expression |
|---|---|
| Behavioral | Altered mating behavior |
| Morphological | Altered physical traits |
| Nonsense | Amino acid codon altered to stop codon |
| Nutritional | Inability to synthesize or utilize nutrients |
| Regulatory | Gene is permanently activated or inactivated |
| Temperature-sensitive | Gene is inactivated at a particular temperature |

how are changes in the nucleotide sequences of genes brought about (changes in the genotype), and how are those changes manifested in the organism (changes in the phenotype)? Remember that mutations are any changes in gene nucleotide sequences that primarily affect the amino acid sequence of a protein. Changes in the amino acid sequence (the primary structure of the polypeptide; Chapter 8) can affect the other structures of a protein—secondary, tertiary, and quaternary. And these alterations in proteins (e.g., enzymes) can affect the behavior, morphology, and physiology of the organism (Table 9.1).

Mutations can affect any section of a chromosome and involve any number of nucleotides, ranging from a single base to millions of bases. Mutations involving very large numbers of nucleotides can affect chromosomal structures as well as function. These types of mutations can be grouped into four categories:

1. deletions: Large sections of chromosome are missing.
2. duplication: Segments of nucleotide sequences are repeated.
3. inversion: Nucleotide sequences of portions of a chromosome are reversed.
4. translocation: Parts of a chromosome are moved to another chromosome.

## POINT MUTATIONS

A complete description of all types of mutations is well beyond the scope of this book. So we shall concentrate on those types of mutations involving single-base changes, which are known as **point mutations.** These types of mutations are among the most intensely studied and, therefore, among the best-understood mutations. Point mutations may occur as single events in a gene or as multiple events involving bases at more than one location. Although we use the expression "change of a single base or nucleotide", keep in mind that bases or nucleotides occur in pairs in double-stranded DNA. So a change of a single base necessarily involves a base pair. But our concern is the base change in the gene or coding strand of DNA and that change's effect on polypeptide/RNA molecules and, therefore, on the cell's ability to function properly.

In eukaryotic organisms, mutations can affect the germ cells (i.e., egg and sperm), in which case the next generation can be affected. The mutation can affect nongerm cells, somatic cells, in which case the individual's phenotype may be altered, but the mutation is not passed on to the offspring.

Point mutations themselves can be classified according to the type of change in the nucleotide sequence. There are five principal ways by which base sequence changes can occur:

1. base deletions
2. base insertions
3. base substitutions: tautomeric shifts
4. proofreading errors
5. chemical changes in bases

Let's take a brief look, now, at a nucleotide sequence change brought about by the first four methods. Chemical changes are discussed in greater detail later in the chapter (see Some Mutagenic Agents).

### Deletions and Insertions

Though the precise mechanism of how nucleotides might be inserted or removed from a sequence is not completely understood, a workable model is as follows. Occasionally, during DNA replication, in which a particular base is repeated several times, a looping out of the strand of DNA may occur. Depending on whether the parental strand or the progeny strand loops out, a **deletion** or an **insertion** will occur at the second round of replication (Figure 9.1).

The result of these deletions or insertions is that, from the point of change, the nucleotide sequence of the gene is garbled. The reading frame of the gene has changed, so these mutations are also known as **frame-shift mutations.** (Remember, codons are read one at a time. Nucleotides are read in groups of three.) Figure 9.2 illustrates the insertion of an extra nucleotide into a gene's base sequence and how that affects a polypeptide's amino acid sequence and, therefore, the polypeptide's biological activity.

Insertions or deletions can be corrected and the normal reading frame reestablished if a second mutation of the

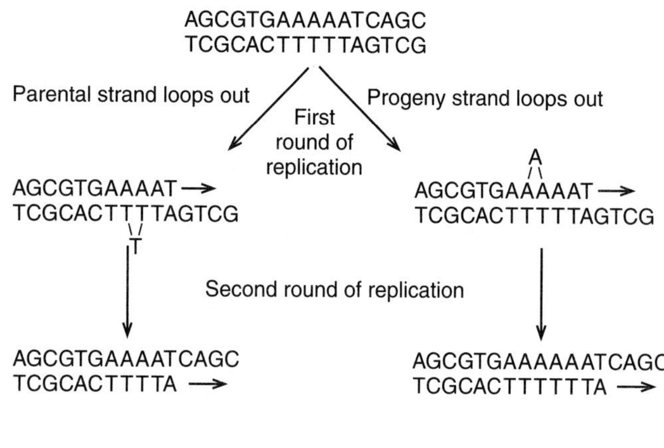


FIGURE 9.1
Frame-shift mutation: a possible mechanism. The mutation occurs in the string of adenines or thymines. If a thymine of the parent strand loops out during replication, it does not base pair with an adenine and a deletion will occur in the subsequent round of replication. However, if a progeny strand acquires an additional adenine during synthesis and it loops out, then all of the thymines will be paired, but at the next round of replication the extra A will pair with a T, resulting in an insertion.

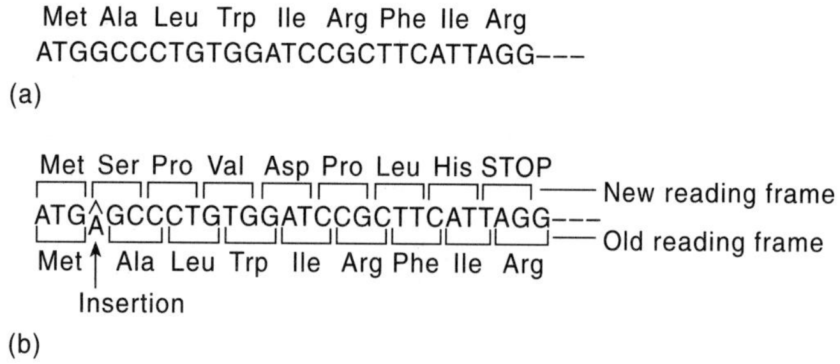

Met Ala Leu Trp Ile Arg Phe Ile Arg
ATGGCCCTGTGGATCCGCTTCATTAGG---
(a)

Met Ser Pro Val Asp Pro Leu His STOP — New reading frame
ATGAGCCCTGTGGATCCGCTTCATTAGG---
Met Ala Leu Trp Ile Arg Phe Ile Arg — Old reading frame
Insertion
(b)

FIGURE 9.2
A frame-shift mutation. (a) the normal nucleotide sequence codes for the amino acid sequence shown. (b) Between the first and second codons, an A is inserted. From that point on, the normal codon sequence is scrambled, resulting in a new amino acid sequence.

opposite type follows the first. For instance, an insertion might be followed by a deletion, or vice versa. Between the two changes, the sequence is garbled, but before the first change and after the second change the reading frame is normal. Of course, the more nucleotides between the two mutations the more damage the polypeptide will suffer because more information is garbled. And the greater the damage to the polypeptide the greater the chance its biological activity will be affected.

If deletions or insertions occur in groups of three (the number of nucleotides in a codon) no frameshift occurs, although an amino acid will be removed or added to the polypeptide and that change may affect the molecule's function. How seriously the function is affected would depend in large measure on the location in the polypeptide of the amino acid lost or gained. For example, an enzyme's catalytic site is a more important location than say a site at some distance from the catalytic site.

It should be understood that, although insertion mutations can revert to the original nucleotide sequence by removal of the extra base, deletions are not as easily restored. A simple explanation is that removal of an extra base is always 100% correct, but insertion of a base where a deletion has occurred may be correct only 25% of the time since any one of the four bases may be inserted; three out of four times the wrong base could be inserted. Deletions, therefore, are much more troublesome to the organism than are insertions.

Regardless of how a change comes about, the nucleotide sequence of the genome is altered; therefore, a mutation has occurred. If such a change is expressed in a critical segment of a protein or RNA molecule, an altered biochemical activity, an enzyme reaction, may occur.

## Base Substitutions: Tautomeric Shifts

The causes and mechanisms of naturally occurring, or spontaneous, mutations are not completely understood. But the study of induced mutations (laboratory-produced mutations) offers insights into naturally occurring base sequence changes. Spontaneous mutations are thought to arise as a result of such natural phenomena as changes in the chemical structure of nitrogenous bases, called **tautomeric shifts.** Examples include shifts from the normal **keto tautomer** to the unusual **enol tautomer** by thymine or the shift of adenine from the amino form to the imino form. In either case, a faulty base pairing results (Figure 9.3).

### *Transitions and Transversions*

Base substitutions can be one of two kinds: transitions or transversions. In **transitions,** the more common of the two, pyrimidines replace pyrimidines and purines replace purines so that, for example, A-T ↔ G-C. In this example, the left-strand nucleotides replace one another (A ↔ G, a purine replaces a purine) as do the right-strand nucleotides (T ↔ C, a pyrimidine replaces a pyrimidine). All deamination-caused mutations are transitions.

In the other form of base substitution, **transversion,** pyrimidines replace purines or purines replace pyrimidines—for example, G-C ↔ T-A. The left-strand nucleotides replace each other (G ↔ T, purine and pyrimidine replace one another), and the right-strand nucleotides also replace each other (C ↔ A, pyrimidine and purine replace one another).

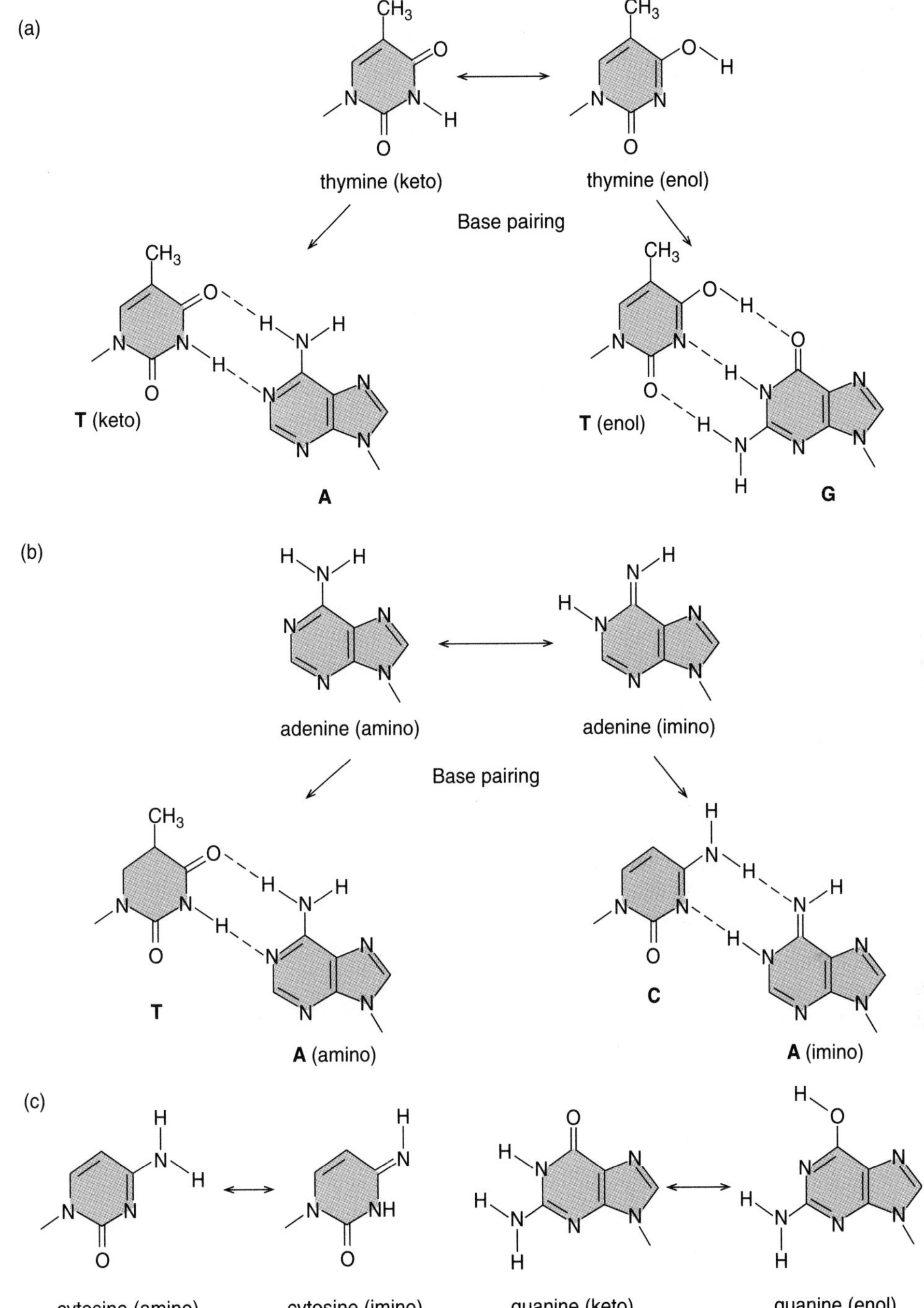


FIGURE 9.3
Tautomeric shifts. The normal tautomers are the keto and amino forms. Shifts to the enol and imino forms result in faulty base pairings: T=A T=G and T=A C=A. Cytosine and quanine can also convert to unusual tautomers; cytosine would then pair with adenine and guanine with thymine.

In either transitions or transversions it is important to look at the same-strand position of the bases in the parental pair and in the progeny pair, as was just done in the example. Transversions are much more difficult to explain than are transitions.

One explanation of transitions is the **alkylation** (the addition of an alkyl group such as methyl or ethyl to a hydrogen-reactive site in an organic compound) of guanine. The addition of such alkyl groups to the hydrogen-bonding oxygen of guanine blocks a base pairing bonding site. This blockage can cause guanine to base pair with thymine instead of cytosine, and at the next replication the thymine pairs with adenine; a transition has occurred.

Alkylation of guanine can also cause its removal from DNA in a reaction called **depurination.** The purine is detached from its deoxyribose, leaving a site that is said to be apurinic—without a purine. When this happens, DNA replication stops, and a special repair system called the SOS system is activated and replication continues. However, in the daughter strand, at the position opposite the apurinic site, an adenine is almost always put in where normally a cytosine ought to have been placed (to pair with the original guanine). Then, at the next round of replication, where there should be a G-C pairing we get, instead, a T-A pairing; a transversion has occurred. The treatment of some phage with buffers at pH 4 also produces depurinations, and subsequent replication of these treated phage produces many transversions of this type.

## Proofreading Errors

Spontaneous mutations may also arise as a result of a mutation that causes the proofreading activity of DNA polymerases (Chapter 2) to inefficiently detect and repair incorrect base pairs during DNA replication.

In *Escherichia coli*, at least, mutations in particular genes give rise to frequent and obvious phenotypic expressions. The term *mutator genes* has been applied to these genes. In fact, however, they are not special genes but rather genes that code for normal cellular products whose function is to assist in the replication and repair of DNA. Because of these functions, any defect in the products of these genes would result in an unusually high rate of what appear to be spontaneous mutations in other genes.

A number of these mutator gene effects have been resolved. For example:

1. mutation of the exonuclease activity of DNA polymerase I, in particular, a reduction in the $3' \rightarrow 5'$ exonuclease activity: Reduces proofreading efficiency.
2. mutation of the methylase enzyme, whose activity is required to mark DNA strands so that parental and progeny DNA strands can be distinguished by the mismatch repair system (more on this later in the chapter): If the strands cannot be distinguished, mismatch repair cannot take place.
3. mutation of the excision enzyme of the mismatch repair system: A defective enzyme means repairs cannot be completed.
4. mutations in the SOS repair system so that the system is not, or is incompletely, turned off: This is an error-prone repair mechanism in which the altered sequences to be repaired may provide little if any information to the repairing mechanism. So, if the mechanism is "on" when it ought not to be, indiscriminate "repair" occurs, which alters what previously were normal sequences. The alterations are then the mutations (more on this system later in the chapter).

# MUTATION RATES

Rates of spontaneous mutation differ for various loci within the same organism as well as among different organisms (Table 9.2). At the moment, there is no really accurate measurement of the mutation rate among eukaryotes, but it is assumed to be about the same as for prokaryotes.

If mutation rates are calculated as inactivation of genes, within a given species most genes show fairly similar rates of mutation relative to gene size. (One must take into account the size of genes: the larger the gene, the larger the target for a mutagen and, therefore, the greater the likelihood the gene will be mutated.)

However, if the question posed is Are all base pairs of a gene equally susceptible to mutation? the answer is No. It turns out that some sites within a nucleotide sequence may mutate 10 or even 100 times more often than would be expected if the mutation of base pairs were a random phenomenon and base pairs were equally susceptible to change. Sites that mutate at a rate significantly greater than statistically probable are referred to as **hotspots.** Hotspots may be sites of either spontaneous or induced mutations. Different mutagens can have different hotspots, and particular hotspots are not hotspots for all mutagens.

An example of a hotspot is found in *E. coli*. This particular hotspot is a major site of spontaneous mutation in this bacterium. The mutation involves the modified base 5-methylcytosine (5-methyl-C) produced by a methylase enzyme after the incorporation of cytosine into DNA during its replication. This modified base undergoes spontaneous deamination at appreciable frequency. Deamination of 5-methyl-C replaces the amino group, with a keto group, thereby converting 5-methylcytosine to thymine (Figure 9.4).

This conversion results in the mismatched base pair G-T. The second round of replication yields a normal G-5-methyl-C pairing and a mutant A-T pairing.

Spontaneous deamination of cytosine (as opposed to 5-methylcytosine) does not have the same effect. Cytosine deamination produces uracil, which normally base pairs with adenine. *E. coli*, however, contains an enzyme, uracil-DNA-glycosidase, that removes uracil from DNA, leaving an unpaired G, which is then repaired with cytosine by a repair system.

table 9.2

RATES OF SPONTANEOUS MUTATIONS AT VARIOUS LOCI IN DIFFERENT ORGANISMS

| ORGANISM | CHARACTER | GENE | RATE | UNITS |
|---|---|---|---|---|
| Bacteriophage T2 | Lysis inhibition | $r \rightarrow r^+$ | $1 \times 10^{-8}$ | Per gene replication |
| | Host range | $h^+ \rightarrow h$ | $3 \times 10^{-9}$ | |
| Bacterium *Escherichia coli* | Lactose fermentation | $lac^- \rightarrow lac^+$ | $2 \times 10^{-7}$ | Per cell per division |
| | Lactose fermentation | $lac^+ \rightarrow lac^-$ | $2 \times 10^{-6}$ | |
| | Phage T1 sensitivity | $T1\text{-}s \rightarrow T1\text{-}r$ | $2 \times 10^{-8}$ | |
| | Histidine independence | $his^- \rightarrow his^+$ | $4 \times 10^{-8}$ | |
| | Histidine requirement | $his^+ \rightarrow his^-$ | $2 \times 10^{-6}$ | |
| | Streptomycin dependence | $str\text{-}s \rightarrow str\text{-}d$ | $1 \times 10^{-9}$ | |
| | Streptomycin sensitivity | $str\text{-}d \rightarrow str\text{-}s$ | $1 \times 10^{-8}$ | |
| | Radiation resistance | rad-s →rad-r | $1 \times 10^{-5}$ | |
| | Leucine independence | $leu^- \rightarrow leu^+$ | $7 \times 10^{-10}$ | |
| | Arginine independence | $arg^- \rightarrow arg^+$ | $4 \times 10^{-9}$ | |
| | Tryptophan independence | $try^- \rightarrow try^+$ | $6 \times 10^{-8}$ | |
| Alga *Chlamydomonas reinhardi* | Streptomycin sensitivity | $str\text{-}s \rightarrow str\text{-}r$ | $1 \times 10^{-6}$ | |
| Fungus *Neurospora crassa* | Inositol requirement | $inos^- \rightarrow inos^+$ | $8 \times 10^{-8}$ | Mutant frequency among asexual spores |
| | Adenine requirement | $ade^- \rightarrow ade^+$ | $4 \times 10^{-8}$ | |
| Corn *Zea mays* | Shrunken seeds | $Sh \rightarrow sh$ | $1 \times 10^{-5}$ | Mutant frequency per gamete per generation |
| | Purple | $P \rightarrow p$ | $1 \times 10^{-6}$ | |
| Fruit fly *Drosophilia melanogaster* | Yellow body | $y^+ \rightarrow y$, in males | $1 \times 10^{-4}$ | |
| | | $y^+ \rightarrow y$, in females | $1 \times 10^{-5}$ | |
| | White eye | $w^+ \rightarrow w$ | $4 \times 10^{-5}$ | |
| | Brown eye | $bw^+ \rightarrow bw$ | $3 \times 10^{-5}$ | |
| | Ebony body | $e^+ \rightarrow e$ | $2 \times 10^{-5}$ | |
| | Eyeless | $ey^+ \rightarrow ey$ | $6 \times 10^{-5}$ | |
| Mouse *Mus musculus* | Piebald coat color | $S \rightarrow s$ | $3 \times 10^{-5}$ | |
| | Dilute coat color | $D \rightarrow d$ | $3 \times 10^{-5}$ | |
| Human *Homo sapiens* | Hemophilia | $h^+ \rightarrow h$ | $2 \times 10^{-5}$ | |
| | Huntington disease | $Hu^+ \rightarrow Hu$ | $5 \times 10^{-6}$ | |
| | Retinoblastoma | $R^+ \rightarrow R$ | $2 \times 10^{-5}$ | |
| | Epiloia | $Ep^+ \rightarrow Ep$ | $1 \times 10^{-5}$ | |
| | Aniridia | $AN^+ \rightarrow An$ | $5 \times 10^{-6}$ | |
| | Achondroplasia | $A^+ \rightarrow A$ | $5 \times 10^{-5}$ | |

Source: R. Sager and F. J. Ryan, Cell Heredity. Copyright © 1961 John Wiley & Sons, Inc., New York, NY.

Some geneticists have suggested that the presence of thymine in DNA compared with uracil in RNA helps to ensure a measure of nucleotide sequence stability. Because of the presence of thymine, any deamination of cytosines (generating uracils) would quickly be recognized and replaced because uracil is not normally present in DNA. Of course, this does not hold true for the deamination of 5-methylcytosine because the resulting base is a normal DNA constituent—thymine.

## ISOLATING MUTANTS

In 1927, Herman J. Muller reported the first results of induced mutations on **Drosophila** by x-rays. This report was followed in 1928 by Lewis J. Stadler's report of the same effects in barley. Since then, a great variety of agents have been identified as capable of inducing mutations. The agents have been given the names mutagens or mutagenic agents. As one might suspect, the rates at which the individual mutagens produce changes in the genotype vary depending on such factors as the dosage, the type of test organism used, and when the agent is administered during the life of the animal (prenatal or postnatal).

The use of artificial means to bring about mutations poses a problem. The investigator, in using physical or chemical mutagens, usually has little control over precisely which genes are mutated unless sophisticated techniques are used (Chapter 11). The genes are hit or affected randomly. As a result, mutant isolation techniques become

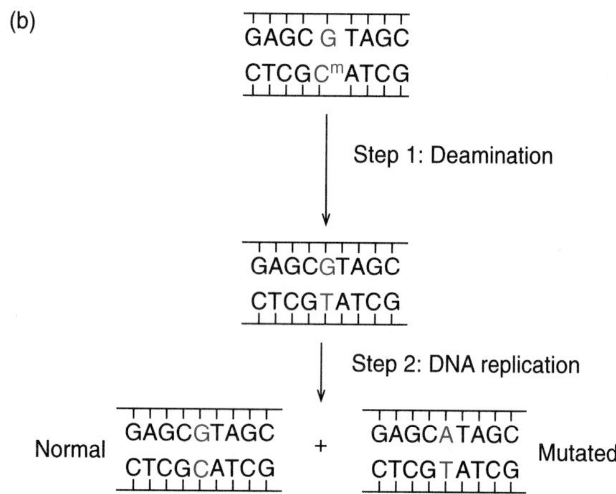


FIGURE 9.4
Mutation resulting from deamination of 5-methylcytosine. Deamination produces thymine, which pairs with adenine. A $C^m$=G pairing is converted to a T=TA pairing.

critical. Of the many mutations created, the investigator must be able to isolate the desired mutant.

Experiments are normally designed to isolate individual mutants, not all mutants. For example, if an auxotrophic mutant (i.e., one incapable of synthesizing a particular nutritional compound) is wanted, the organism must be grown after the agent is applied, under conditions such that the expression of the mutation can occur and the mutant can be detected and isolated. One technique to do this is called replica plating (Figure 9.5).

If a nutritional mutant incapable of synthesizing a particular amino acid is wanted, a mutagen is applied, and the organisms, in this case a bacterial culture, are spread over the surface of an agar plate containing a medium with the amino acid. The plate is incubated to allow colonies to grow. Next, felt cloth is gently pressed to the agar to pick up some cells from each colony. The felt is then gently pressed to the surface of another agar plate. The felt is applied to the second plate in such a way that the position of each colony appearing on this plate is exactly the same as on the original plate. This plate, however, contains a medium that does not have the amino acid.

Colonies appearing on the second plate are still capable of synthesizing the amino acid since they are growing in its absence. Colonies from the first plate that do not appear on the second plate, however, are those not able to synthesize the amino acid and so grow only in its presence (i.e., on the first plate). Then culturing the colonies from the first plate that do not appear on the second plate isolates the mutants, and they can be maintained as stock cultures for later use and study on a medium containing the amino acid.

With the advent of much more sophisticated techniques of genetic engineering, it is now possible to mutate a specific gene at a specific nucleotide. This more complex method is known as site-specific mutagenesis (see Chapter 11).

## SILENT MUTATIONS

An assumption we make in discussing mutations is that they can be detected. That is, that the phenotype is sufficiently altered to make detection feasible and relatively straightforward. Otherwise, we must rely on nucleotide sequence analysis, and these techniques are still not simple to use (Chapters 11 and 12). Analysis of phenotypes is still most frequently employed in the study of mutations. It is possible to screen for a relatively large number of mutations in a short period of time.

Some mutations, however, have no apparent phenotypic consequences. These are referred to as **silent mutations.** For these mutations, phenotypic expression either does not occur or is undetectable using standard methods.

Whether a particular physical or chemical agent is a mutagen is important when, for example, human and animal food additives or pesticides are concerned. Relying on phenotypic expression may miss the effects of some of these agents.

Silent mutations may go undetected using phenotypic analysis because:

1. The genetic code is redundant or degenerate, so the new triplet may code for the same amino acid as the original codon.
2. The change causes an amino acid substitution that does not appreciably alter the biological activity of the protein.
3. The change occurs in a gene that is not expressed or whose protein is not necessary under the experimental conditions being used for detection of the mutation.
4. The change is not expressed because of the simultaneous presence of a suppressor mutation that allows the protein to assume its normal conformation and restores biological activity (Figure 9.6).

Detection of silent mutations may require nucleotide sequence analysis, if not of an entire genome at least of sus-

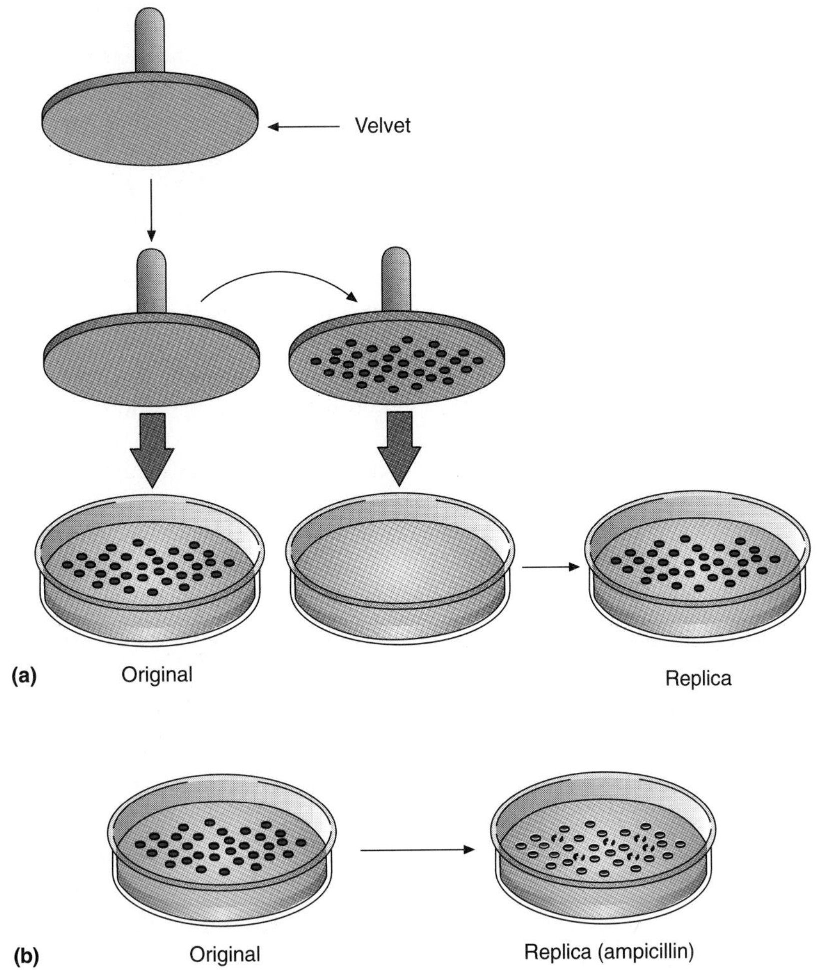


FIGURE 9.5
Replica plate. By transferring cells from one plate (original) to another (ampicillin) cells resistant to the antibiotic, but which have not been exposed to it, can be isolated from corresponding colonies on the original plate. This procedure is used to demonstrate spontaneous mutation. Cells on the original plate corresponding to colonies on the replica plate have spontaneously mutated to the resistant state.

pected chromosomes. And, except for some viruses, this is still a daunting task.

Even if we could sequence without error at the rate of one million base pairs per day, which is a goal not yet achieved, the human genome, with a haploid number of 3 billion base pairs, would still require 10 years to sequence. And it should be kept in mind that mutagenic agents affect more than one site at a time, further complicating detection. Which mutation causes which effect may be extremely hard to determine until we know the location of genes.

## SOME MUTAGENIC AGENTS

A variety of agents have been shown to be mutagenic. They differ with regard to how each alters the nucleotide sequence of the genome, the potency of their actions, and the type of cells most grievously affected.

These agents are most conveniently grouped as biological, chemical, and physical agents. Here we shall merely scratch the surface and look at some representatives of each group: those most intensely studied and for which we have some understanding of their mechanisms of action.

### Biological Agents

The biological agents known to be mutagens can be divided into two groups: **transposable elements** and **viruses.**

#### *Transposable Elements*

Transposable elements are a group of DNA sequences that have been found in both prokaryotes and eukaryotes

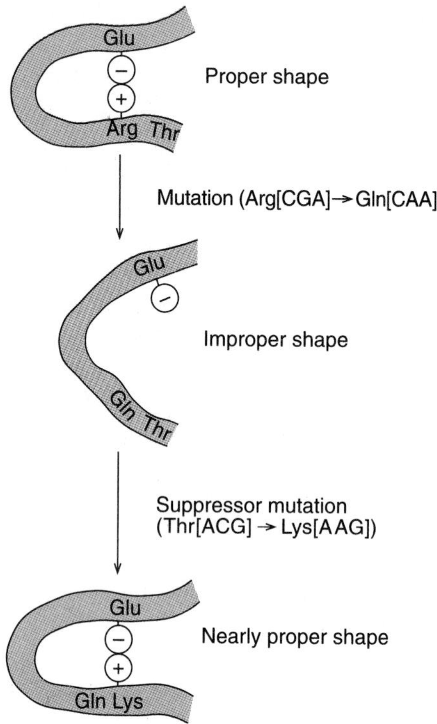


FIGURE 9.6
A suppressor mutation. The conformation of a protein is disrupted by the first mutation. A second mutation, the suppressor mutation, restores the protein's conformation and biological activity.

(*E. coli*, maize, yeast, *Drosophila*, *C. elegans*, and human beings, for example; also see Chapter 5). They were first reported in 1951 by Barbara McClintock (see also Chapter 5) in a study of pigmentation patterns in maize (Indian corn). Her reports were at first ignored because of the then strongly held belief that genes were fixed in chromosomes. She was later (1983) awarded the Nobel Prize for her work. The transposable elements, now known as transposons, she discovered were mobile sequences. And it would be another 20 years before these elements were discovered in another organism—*E. coli*.

Some transposable elements carry genes for enzymes that allow the elements to insert themselves into chromosomes. They also encode for genes that can influence the phenotype of the host, and over longer periods of time they may influence the evolutionary development of the host species. A number of observations suggest that they have, in fact, done so. First, *Alu* elements (a family of transposons) with significant similarities are found in two orders, primates and rodents, that diverged some 60 million years ago. Second, in human beings, for example, the *Alu* elements are found in the regions of the chromosomes that have the highest concentration of known genes. Third, transposable elements are known to foster gene mutations and chromosomal aberrations in every organism having transposable elements in an active form.

Transposable elements are known to promote inversions, deletions, and rearrangements of host cell DNA. The mechanism by which **transposition** (movement of a transposon) occurs is unknown, but apparently it does not require homology (sequence pairings) between donor and recipient DNAs. Transposition occurs fairly infrequently, at the rate of between $10^4$ and $10^7$ events per generation.

In one form of transposition, the intact element may remove itself entirely from the donor DNA and insert itself into the target DNA, leaving the donor DNA, without a transposon and the target DNA with a new sequence and nucleotide sequence disruption. This form is known as **conservative transposition.**

Another form of transposition, known as **replicative transposition,** requires the replication of the transposon followed by the insertion of the copy at the new site, where nucleotide disruption occurs.

Transposons can be classified on the basis of whether the elements transpose via an RNA intermediate (Class I) or directly as DNA (Class II). This classification scheme is based on the work of D. J. Finnegan and A. M. Weiner (Table 9.3).

### *Viruses*

An example of a virus capable of causing mutations is the bacteriophage Mu (mutator phage). This virus causes mutations when it inserts itself into the sequences of host genes. With the inserted nucleotides present, the normal sequence is disrupted and the gene affected is inactivated. Mu can cause a number of other types of mutations besides gene inactivation. For example, it can cause inversions, deletions, translocations, and fusions of independent DNAs such as the fusion of two plasmids. Mu replication requires repeated transpositions of Mu DNA to a number of sites on the host genome. So, in lytic replication, Mu must behave as a transposon during replicative transposition.

## Chemical Agents

A large number and variety of chemical agents are known to cause mutations. Chemical mutagens are substances that (1) upon being inserted into DNA during replication mimic a normal base but, via enhanced tautomeric shifting, cause inappropriate base pairing to occur or (2) can modify a base already part of DNA and as a result alter the base's hydrogen-bonding properties, resulting in a changed base pair.

Different chemical mutagens cause trouble in different ways. The following are some examples of chemical mutagens and the ways by which they produce mutations: 5-bromouracil, nitrous acid, ethyl methane sulfonate, and the acridine dyes.

1. 5-bromouracil: Resembles thymine, so the DNA-replicating machinery will usually base pair it with adenine. But it has an enhanced tendency to shift to

**table 9.3**

**CLASSIFICATION OF TRANSPOSABLE ELEMENTS**

Class I. Transpose by means of RNA intermediates
- A. Viral superfamily (retrovirus-like retrotransposons)
  - 1. These have the following characteristics:
    - a. Have long direct terminal repeats (LTRs)
    - b. Encode reverse transcriptase from ORFs (open reading frames) in DNA between LTRs
    - c. Able to generate 4- to 6-bp target site duplications
    - d. Have no 3′ terminal poly(A) track
    - e. Are dispersed in genome
  - 2. Examples
    - a. *Ty (Sacharomyces cerevisiae)*
    - b. *Copia*-like *(Drosophila melanogaster)*
    - c. *DIRS-1 (Dictyostelium)*
    - d. *BSI* (maize)
    - e. *IAP* (rodent)
    - f. *THE* (humans)
    - g. *VL30* (rat and mouse)
- B. Nonviral superfamily (nonviral retroposons)
  - 1. These may have the following characteristics:
    - a. Have no terminal repeats
    - b. Have ORFs
    - c. Do not encode enzymes responsible for their transposition (passive transposition)
    - d. Have 3′ terminal poly(A) tract.
    - e. Are dispersed in genome
  - 2. Examples
    - a. Transcripts of RNA PolII
      - *F* family *(Drosophila melanogaster)*
      - *Lines I* family (human, ape, monkey, mouse, and rat)
      - Processed pseudogenes (retropseudogenes)
    - b. Transcripts of RNA PolIII
      - *SINES*
      - 7SL RNA retropseudogenes
      - ***B1*** family (rodents)
      - ***Alu*** family (primates)
      - 7SK RNA retropseudogenes
      - tRNA retropseudogenes
      - Polymerase unknown
      - ING1/5RSI (Trypanosomes)

Class II. Apparently transpose directly: DNA → DNA by transposase

All have terminal inverted repeats (IRs)
- A. With short inverted repeats (SIRs)
  - *P; hobo (D. melanogaster)*
  - *Ac-DA; Spm/En* (maize)
  - *Tam (Antirrhinum majus*, snapdragon)
  - *Tc1 (Caenorhabditis elegans)*
- B. With long inverted repeats (LIRs)
  - *FB* (foldback) *(D. melanogaster)*
  - *TU (Strongylocentrotus purpuratus)*

Source: Data from D. J. Finnegan, "Eukaryotic Transposable Elements and Genome Evolution" in *Trends in Genetics*, 5, 103, 1989 and A. M. Weiner, et al., "Nonviral Retroposons" in *Annual Review of Biochemistry*, 55, 631–661, 1986.

its enol tautomer, in which case it then base pairs with guanine by mimicking cytosine (Figure 9.7).

2. nitrous acid: Converts amino groups to keto groups by oxidative deamination, resulting in altered base pairing (Figure 9.8). Nitrites used as a food preservative may be metabolized by the body to nitrous acid and therein lies the controversy of its use.
3. ethyl methane sulfonate: Alkylates the hydrogen-bonding oxygen of guanine in particular, but also of thymine. The guanine is thereby maintained in the unusual enol state, which readily base pairs with thymine rather than with cytosine (Figure 9.9).
4. acridine dyes: Acridine orange and proflavin are planar, three-ring structures whose dimensions approximate a purine-pyrimidine pair (Figure 9.10). Via a process called intercalation, these compounds can be set between base pairs, spreading apart the adjacent base pairs. Intercalation, a process not fully understood, causes deletions or additions of base pairs during DNA replication that result in frameshift mutations in the progeny strand. Most often, single bases are added or deleted, but occasionally two or more bases may be involved. Additions are more common than deletions.

## Physical Agents

Ultraviolet radiation at 260 nm, though relatively weak, can cause damage to DNA. Principally, this damage results from the cross-linking of adjacent thymines (Figure 9.11), which produces a so-called cyclobutane ring referred to as a **thymine dimer.**

Cytosine dimers and thymine-cytosine dimers can also form, but they cross-link less often than thymine. These pyrimidine dimers eventually produce a local distortion of the DNA molecule. The result is that, at these locations, DNA is neither a proper transcriptional sequence for RNA synthesis nor a proper template for DNA replication.

During DNA replication, thymine and other dimers stall synthesis at replication forks (see Chapter 2). Although the hydrogen bonding of the thymines to the adenines across the double helix can be brought about, a distorted section appears. When DNA polymerase III adds two adenines to base pair with the thymine dimer, the distortion signals that a mismatched base pairing has occurred even though there actually has been no mismatch. The polymerase thereupon removes the newly installed adenines and attempts to add two new adenines, with the same consequence. As adenines are added and then removed, the net result is replication stalls at the thymine dimers. The stalling of replication prevents cell division since DNA replication is a prerequisite of cell division. The effect on the organism, if the damage is not repaired, is obvious (see SOS Repair, later in this chapter).

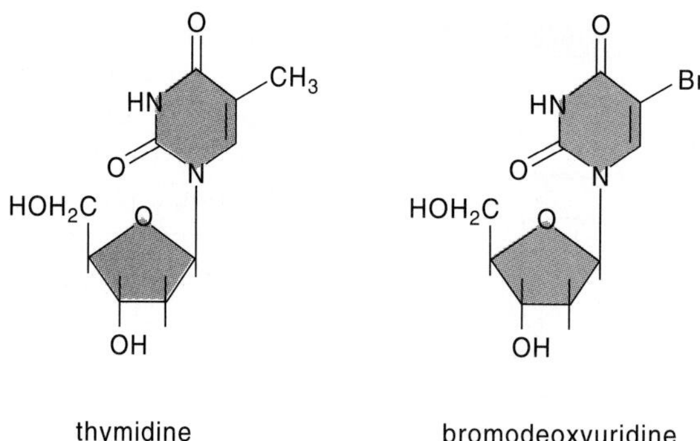


FIGURE 9.7
5-bromodeoxyuridine resembles thymidine but readily shifts to its enol tautomer. The enol form pairs with guanine.

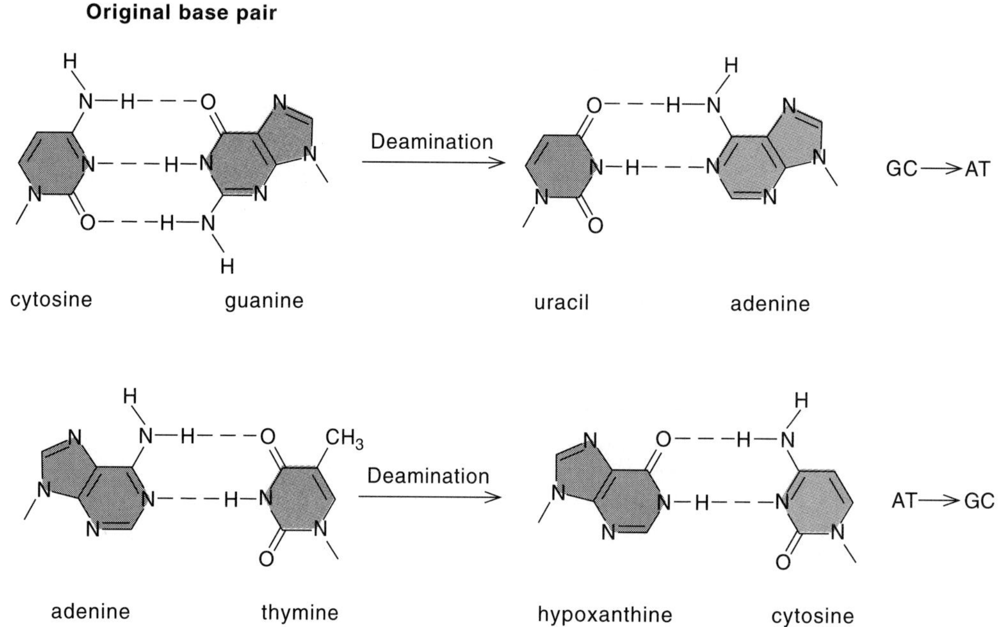


FIGURE 9.8
Nitrous acid deaminations. Both cytosine and adenine can be deaminated to uracil and hypoxanthine, respectively. During the next round of DNA replication the deaminated product pairs with the "wrong" base. The second round of DNA replication pairs the "wrong" base with its normal partner, setting the mutation.

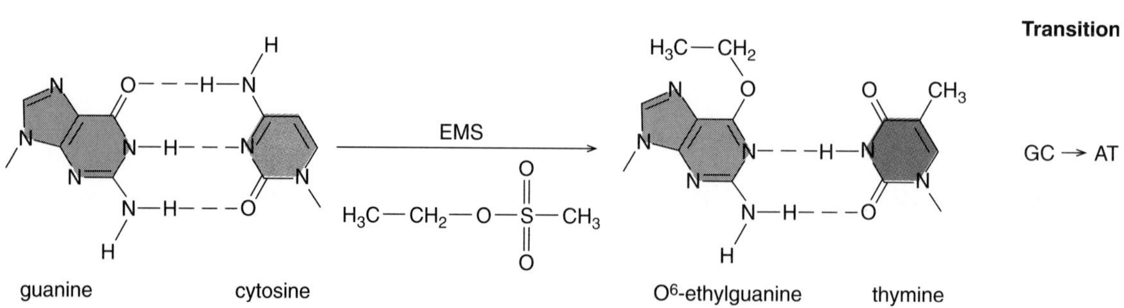


FIGURE 9.9
Alkylation of guanine by ethyl methane sulfate. The aklylated guanine will be paired with thymine and not with cytosine during the first round of replication. In the subsequent round of replication the thymine will be base paired with adenine. The result is that a GC base pair will be exchanged for an AT base pair.

FIGURE 9.10
Proflavin, one of the acridine dyes.

Gamma rays and x-rays are much more dangerous than UV radiation because their levels of energy are higher and, therefore, they have greater penetrating powers. These rays are referred to as ionizing radiation because they produce free radicals among the molecules, especially water molecules, that surround the DNA. The free radicals, having unpaired electrons, are extremely reactive. The free radicals that contain oxygen are particularly

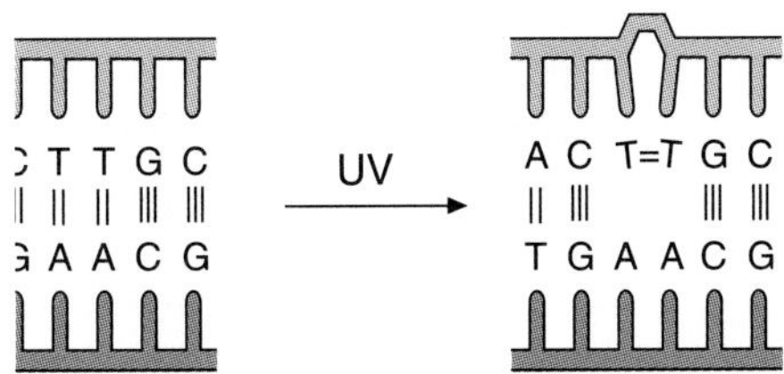


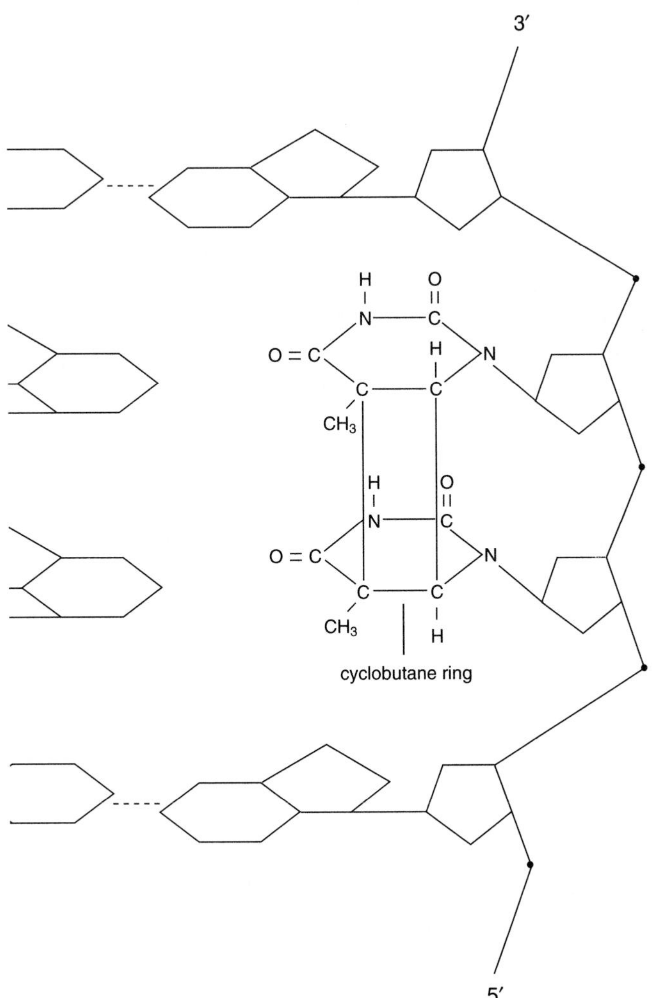


FIGURE 9.11
Thymine dimer (cyclobutane ring) produced by UV irradiation.

reactive. Free radicals such as hydroxyl and hydroperoxyl are potent oxidizing agents that can alter the structures of purines and pyrimidines, which then can form inappropriate base pairs. Formation of free radicals can cause arrested or abnormal mitosis, chromatid breaks (double-stranded breaks are much more difficult to repair than are single-stranded breaks), and increased mutation rates. These are some of the concerns plaguing the use of radiation as a means of sterilizing foods as an alternative to freezing.

## CONSEQUENCES OF MUTATIONS

### Cancer

Many mutagens are also powerful carcinogens, substances that induce uncontrolled cell growth. This fact suggests that there is a strong link between genes and cancer. **Carcinogens** are agents that damage DNA or that interfere with replication or repair of DNA. The agents may be chemicals, radiation, or certain viruses.

Cells and many viruses contain genes known as **proto-oncogenes** (in cells) and **oncogenes** (in cells and viruses) that can induce uncontrolled cell growth and differentiation. Oncogenes of viruses are thought to derive from host cell proto-oncogenes. Cellular proto-oncogenes seem to be genes whose products participate in cell division. So a relationship between oncogenes and cancers is quite plausible. Table 9.4 lists some oncogenes and proto-oncogenes associated with various animals and retroviruses (Chapter 3). Genetic changes associated with cancers included:

1. altered proteins: The oncogene thought to be responsible for human bladder cancer differs from its proto-oncogene counterpart by a single point mutation that results in a glycine codon to be changed to a valine codon (GGC ↔ GTC): a transversion (a purine is replaced by a pyrimidine) replaces a G with a T. This change affects a protein, p21, whose function is to hydrolyze GTP. This altered GTPase activity continues beyond the normal time limit and thereby contributes to the development of this epithelial tissue cancer. This particular cellular proto-oncogene, designated c-*ras* (c = cell), is the most common proto-oncogene implicated in human cancers.
2. altered regulatory sequences: A retroviral gene called v-*fos* (*v* = virus) and its cellular proto-oncogene, c-*fos*, differ primarily in their regulatory nucleotide sequences. Removal of the c-*fos* regulatory unit and its replacement by the v-*fos* regulatory unit results in the loss of control of degradation of the *fos* gene's mRNA. The c-*fos* gene's product is a nuclear phosphoprotein that associates with other proteins to form a transcription factor complex that binds DNA at a consensus sequence found in many promoters and enhancers. These protein complexes may be a part of a cellular regulatory cascade.
3. chromosomal translocation: In the human cancer Burkitt's lymphoma, an 8:14 translocation occurs. A portion of the number 8 chromosome carrying the proto-oncogene c-*myc*, which codes for a nuclear protein, is moved to the number 14 chromosome near where a cluster of antibody genes is located. These antibody genes are normally very active, in keeping with the roles of the genes' products in immunity. Translocation leads to the loss of the first of c-*myc*'s three introns (noncoding nucleotide sequences; more about these in Chapter 10). This intron, though it does not code for an amino acid sequence, may be important as a regulatory region. In this new site, c-*myc* becomes active and appears not to be subject to normal control; this activation is associated with the tumorigenic state. Exactly how tumorigenesis occurs is still open to question, however.
4. proto-oncogene amplification: In this situation, a gene such as c-*myc* is copied over and over, often as sequential repeats. This repeated copying produces more and more gene product. Gene amplification has been observed in a number of animal and human cancers.
5. insertion of viral DNA into chromosomes: Avian lymphoma induced by the avian leukosis virus, a retrovirus lacking an oncogene, is brought about by viral insertion near the c-*myc* gene. The insertion, it appears, brings the cellular proto-oncogene under the transcriptional control of the viral DNA.
6. loss or inactivation of anti-oncogenes: In at least one case, that of the human eye cancer retinoblastoma, there is clear evidence that the cancer results from a deletion of a portion of chromosome 13. Then a mutation of the allele of the missing gene, the *Rb* gene, causes a defective protein to be produced. The normal product of the *Rb* gene somehow prevents uninhibited cell proliferation. The mutant product no longer fulfills this role, and an eye tumor results. The *Rb* gene has been given the designation anti-oncogene. The product of this gene is a 105 kD DNA-binding protein found in the nucleus of normal retinal cells. The protein is not found in retinoblastome cells. Adenoviruses that transform cells to the cancerous state have an oncogene that codes for a protein that does not bind DNA but does inactivate Rb protein, strong evidence for the role of the *Rb* gene and its protein product in the prevention of cellular transformation (Voet and Voet 1995).

table 9.4

Some Oncogenes of Retroviruses

| Oncogene | Species of origin | Type of tumor | Proto-oncogene in vertebrate DNA | Oncogene product: Protein kinase | Phosphorylates tyrosine | Located on plasma membrane |
|---|---|---|---|---|---|---|
| *v-src* | Chicken | Sarcoma | Yes | Yes | Yes | Yes |
| *v-fps* | Chicken | Sarcoma | Yes | Yes | Yes | Yes |
| *v-yes* | Chicken | Sarcoma | Yes | Yes | Yes | ? |
| *v-ros* | Chicken | Sarcoma | Yes | Yes | Yes | ? |
| *v-myc* | Chicken | Carcinoma, sarcoma, leukemia | Yes | ? | ? | ? |
| *v-erb* | Chicken | Leukemia, sarcoma | Yes | ? | ? | ? |
| *v-myb* | Chicken | Leukemia | Yes | ? | ? | ? |
| *v-rel* | Turkey | Lymphoma | Yes | ? | ? | ? |
| *v-mos* | Mouse | Sarcoma | Yes | ? | ? | ? |
| *v-bas* | Mouse | Sarcoma | Yes | ? | ? | ? |
| *v-abl* | Mouse | Leukemia | Yes | Yes | Yes | Yes |
| *v-ras* | Rat | Sarcoma, leukemia | Yes | Yes | ? | Yes |
| *v-fes* | Cat | Sarcoma | Yes | Yes | Yes | ? |
| *v-fms* | Cat | Sarcoma | Yes | Yes | ? | ? |
| *v-sis* | Monkey | Sarcoma | Yes | ? | ? | ? |

From "Oncogenes" by J. M. Bishop. 

## Inborn Errors of Metabolism

We have been looking at the process of mutation in bacteria for the better part of this chapter. As mentioned at the beginning of the chapter, bacteria are used as study tools because of their relatively simple chromosomal structure, metabolism, and ease of manipulation. Nonetheless, we are aware of literally thousands of genetic defects, particularly in animals and human beings. The existence of these defects has been determined by a variety of techniques.

Congenital defects are called inborn errors of metabolism, a phrase first used by Archibald Garrod at the turn of the century. Garrod was an English physician who became interested in human disorders that appeared to be inherited. Among the disorders that caught his interest were albinism, cystinuria, and alkaptonuria. He published his findings in 1909 in a book entitled *Inborn Errors of Metabolism*. This treatise is the earliest attempt at an explanation of the biochemical genetics of human beings or any other species.

By studying the patterns of inheritance of **alkaptonuria,** Garrod concluded that this disease is inherited as a simple recessive trait. The disease, though not serious, is persistent and can lead to a benign arthritic condition caused by a buildup of an intermediate in the metabolism of the amino acids phenylalanine and tyrosine. The intermediate, homogentisic acid, accumulates because of the deficiency of the enzyme homogentisic oxidase, and the acid is excreted in the urine. Rapid oxidation of this compound when exposed to air causes the urine to blacken.

Not so fortunate are those individuals who also cannot metabolize phenylalanine but whose metabolic block occurs sooner in the metabolism of the amino acid. Here the block is at the first metabolic step. A deficiency of the enzyme phenylalanine hydroxylase results in the disease **phenylketonuria,** or **PKU,** which was first described in 1934. In this disease phenylalanine accumulates, and some may be converted to phenylpyruvate, phenylacetate, and hydroxyphenylacetate. These substances spill into the cerebrospinal fluid, and their levels in the brain are elevated. The occurrence early in the development of the fetus is thought to lead to the severe mental retardation seen within months after birth if preventive measures are not taken. Retardation can be prevented by early detection and a diet modified to either eliminate phenylalanine or reduce it to a low level. Screening of newborns is now routine. The frequency of PKU is about 1 in 11,000 births, and the disorder accounts for almost 1% of patients in mental institutions.

*table 9.5*

SOME GENETIC DISORDERS

| DISORDER | ENZYME INVOLVED | IN UTERO DIAGNOSIS POSSIBLE |
|---|---|---|
| Acid phosphatase deficiency | Acid phosphatase | |
| Albinism | Tyrosinase | |
| Alkaptonuria | Homogentistic acid oxidase | |
| Cholesterol ester deficiency | Lecithin: cholesterol acyltransferase | |
| Fabry's disease | α-galactosidase | Yes |
| Fructosuria | Fructokinase | |
| Galactosemia | Galactose-1-phosphate uridyl transferase | Yes |
| Gaucher's disease | Glucocerebrosidase | Yes |
| Glycogen storage disease I (Von Gierke's disease) | Glucose-6-phosphate (liver) | |
| Glycogen storage disease V (McArdle's disease) | Phosphorylase (muscle) | |
| Hemolytic anemia (at least 10 different hemolytic anemias are known) | Glucose-6-phosphate dehydrogenase | |
| Homocystinuria | Cystathionine synthetase | Yes |
| Hypoglycemia | Fructose-1, 6-*bis*phosphatase | |
| Hypophosphatasia | Alkaline phosphatase | |
| Ketoacidosis (infantile) | Succinyl~CoA: 3-keto acid CoA transferase | |
| Krabbe's disease | Galactocerebrosidase | Yes |
| Lesch-Nyhan syndrome | Hypoxanthine-guanosine phosphoribosyl transferase | Yes |
| Metachromatic leukodystrophy | Arylsulfatase A | Yes |
| Niemann-Pick disease | Sphingomyelinase | Yes |
| Ornithinemia | Ornithine ketoacid amino transferase | |
| Orotic aciduria | Orotate phosphoribosyl transferase + orotidylic decarboxylase | |
| Phenylketonuria | Phenylalanine hydroxylase | |
| Porphyria, erythropoietic | Uroporphyrinogen III cosynthase | |
| Refsum's disease | Phytanate α-hydroxylase | |
| Tyrosinemia | Parahydroxyphenylpyruvate oxidase | |
| Wolman's disease | Acid lipase (lysosomal) | |
| Xeroderma pigmentosum | Endonuclease (UV-specific) | Yes |

Sources: Data from K. O. Raivo and J. E. Seegmiller, "Genetic Diseases of Metabolism" in *Annual Review of Biochemistry*, 41, pp. 543–576, 1972; In utero data from V. Woodward, *Human Heredity and Society*, 1992, West Publishing Co., St. Paul, MN.

Table 9.5 lists a few diseases associated with mutations and notes a few for which in utero diagnosis is possible. The locations of some of the defective genes are listed in Table 10.4.

Mutations are not necessarily always harmful to the individual. A case in point is human sickle-cell mutations. Individuals suffering this condition may be of two genotypes: homozygous recessive (the person carries the defect in both alleles) or heterozygous (the person carries one normal allele [gene] and one defective allele [gene]). In the first case, the individual rarely reaches maturity unless modern medicine intervenes; the individual suffers from **sickle-cell anemia.** In the second case, individuals are, for the most part, asymptomatic; they are said to exhibit sickle-cell trait.

The mutation affects the number 6 amino acid of the beta chain of hemoglobin (Hb). The normal amino acid in this position is glutamic acid, but in sickle-cell hemoglobin (HbS), the amino acid in position 6 is, instead, valine. The genotype changes are point mutations: either transversion GAA → GUA or transversion GAG → GUG.

Under low oxygen tension, HbS is less soluble than normal hemoglobin. Presumably, the change from an acidic amino acid to a neutral one (i.e., glutamic acid →

valine) produces a conformational distortion of the protein that leads to the sickling of red blood cells. Cells containing HbS become defective oxygen carriers.

However, sickle-cell genes are distributed more widely in Africa and Asia in precisely those regions where the distribution of malaria is high. Sickle-cell heterozygotes are much more resistant to malaria than are normal individuals. (Homozygotes in these areas usually die in infancy.) Resistance appears to be the result of the sickling process. Upon infection by the mosquito-borne protozoan *Plasmodium falciparum*, the pH of the erythrocytes is lowered by about 0.4 pH units. This lower pH causes an increase in sickling from about 2% of the erythrocytes to about 40%. Sickling results in a change in the red blood cell's membrane permeability to potassium ions. Normal red blood cells maintain a potassium ion internal concentration that is higher than that of the surrounding blood. Upon sickling, the red blood cell loses potassium. At this stage in the parasite's development, a high potassium concentration is required. The loss of the ion upon sickling deprives the parasite of the ion and the parasite dies.

## Chromosomal Rearrangements: The Immune System

We will end this section by taking a very brief look at an absolute requirement for mutations in vertebrates, human beings in particular. Changes in the nucleotide sequences of vertebrates' somatic cells can be brought about in two ways. First are the random and accidental changes of the genome that may occur at any stage in the development of the fetus or the adult. Second are the programmed changes occurring according to a set of instructions found in the DNA. Two examples of this type of change are the inactivation of the X chromosome in the female (Chapter 10) and the genetic rearrangement necessary for the functioning of the defense system of the individual.

In mammals, the body is ultimately protected by the immune system, which can generate a huge variety of immunoglobulins, or antibodies, against a wide assortment of foreign materials called antigens such as microorganisms and their products, toxins. This ability is dependent on mutations.

Mammals have in B lymphocytes (the precursors to the antibody-producing plasma cells) a large number of genes that encode the information for the amino acid sequences in the so-called variable regions of the antibody molecule. It is this variable region that recognizes and reacts with the foreign material, antigen, introduced into the body, thereby destroying or neutralizing the material.

During B cell development, variable-region genes readily mutate. The mutations lead to different B cell populations, each of which can make a limited number of different types of variable regions. These, in turn, will be built into new antibodies as they are synthesized. Also during B cell development, chromosomal recombination of DNA segments close to the variable-region genes can occur to make up new combinations of variable regions. So, by mutation and recombination of the different genes responsible for antibody synthesis, the number of antibodies of different specificities (i.e., abilities to react with different antigens) that can be produced is astoundingly large.

The importance of the immune system to the survival of the individual cannot be overstated. The immune system is the court of last resort in protection against infectious and some noninfectious diseases. If this system fails, the individual is in mortal danger. In fact, babies born with a severely impaired immune system are doomed to life lived in isolation from other human beings. The common cold or contact with microorganisms that normally pose no problems to other individuals could kill them.

# DNA REPAIR

Nature has devised a number of ways to deal with damage to DNA nucleotide sequences. An ability to restore damaged genomes can almost be considered a sine qua non of life. Damage to the master molecule, if serious enough, must be repaired or the organism's very life is threatened.

At any one time, there are perhaps 25–100 molecules of repair enzymes present in each cell. In addition, there are operons coding for enzymes needed to protect the cell against heat (heat shock proteins) and oxidative damage. For example, the so-called *oxy*R regulon of *Salmonella typhimurium*, another intensively studied bacterium, includes the genes for catalase-peroxidase, superoxide dismutase, glutathione reductase, and alkyl hydroperoxide reductase. Any of these enzyme systems can be induced by exposure to the oxidizing agent hydrogen peroxide.

The ability to repair damaged DNA is based on the fact that, in all cellular forms of life, DNA is a double-stranded molecule. This means that an error in one strand can be repaired (bases removed and new ones added) by using the other strand as a template. In this way, the correct bases are put in place to maintain a proper nucleotide sequence. In the absence of a double-stranded molecule in which only certain base pairs are allowed (i.e., Chargaff's pairing), repairs, more often than not, would incorporate the wrong nucleotides. Indeed, in three out of four chances, the wrong base would be incorporated, since any one of the four nucleotides could be incorporated. It is of interest in this regard that single-stranded viruses, such as HIV, have a very high rate of mutation.

Among the repair systems found to be most active are: SOS, photoreactivation, mismatch, excision, and recombination. These repair mechanisms either undo the damage per se or remove the damaged nucleotide and replace it with an undamaged nucleotide. Some of these repair systems are more efficient and accurate in repairing damaged DNA than others. Most DNA repair systems are constitutive (e.g., proofreading), though some, such as the SOS response, are induced.

## SOS Repair

The SOS repair system is a complex cellular response mechanism activated by certain types of DNA damage. Because the repairing, is done in the absence of proper template instructions, many errors are made, so it is referred to as an **error-prone mechanism.**

The **SOS response** is induced by DNA damage that blocks replication (e.g., UV-produced thymine dimers). In *E. coli,* the response involves the activation of some 20 SOS-response genes. Under normal conditions (i.e., no DNA damage), expression of the SOS genes is at a low constitutive level (see Chapter 7). A simplified account of this system's functioning is as follows.

The genes of the SOS response are negatively controlled by the protein LexA. The protein binds the SOS box, which comprises about 16–20 nucleotides. This box is found in the regulatory sequences of the *lex*A gene, the *rec*A gene, and other genes of the SOS system.

Upon UV damage, a replication fork stalls, resulting in ssDNA. Production and activation of RecA protein, a protease or protein-cleaving enzyme, is induced by some as-yet unknown process. The binding of RecA to LexA causes a conformational change in the LexA protein, promoting the self-cleavage of LexA. The destruction of LexA frees the genes of the SOS system that code for proteins with various functions in repairing DNA.

As we saw earlier, thymine dimers cause distortions of DNA, and the molecule is then not a proper template. The SOS response base pairs the thymine dimer with any bases to get around the distortion, which stalls DNA synthesis. The trade-off is that DNA replication continues, and the cell survives. But mutations are present (the error-prone mechanism). In fact, the rate of mutations at these points is higher than usual. It may be argued, of course, that this repair process may also produce new cells better adapted to living under the conditions that produced the mutations in the first place by eliminating thymines at the mutation site during the next round of replication. Both transitions and transversions can result from this type of repair.

Once DNA repair is completed, RecA ceases to catalyze LexA self-cleavage. Newly synthesized LexA is then available to once again repress the transcription of the SOS-response genes and the cell returns to normal.

## Photoreactivation Repair

Another mechanism by which thymine dimers can be repaired is photoreactivation. This mechanism requires **photoreactivating enzyme,** also called **DNA photolyase.** Figure 9.12 shows such repairing of UV-damaged DNA.

This method of repairing the thymine dimers restores the thymine monomers and is found in cells ranging from bacteria to animal cells. The repairing mechanism depends on the absorption of visible light by the repair enzyme. All photolyases contain two chromophores: one is the reduced form of the coenzyme flavin adenine dinucleotide (FAD) (the reduced form is $FADH_2$; an electron carrier also involved in ATP generation via the cytochrome system); the second absorbs a wavelength of light characteristic for the particular organism. The energy of the absorbed light, 300–500 nm for *E. coli,* is used by the enzyme to split the dimer, thereby restoring the thymine monomers.

Once again, distortion of the DNA produced by the thymine dimer may be the signal that repair is necessary. This repair system, however, splits only thymine dimers, other types of pyrimidine dimers are not repaired by photolyase.

## Mismatch Repair

As noted at the end of Chapter 2, replication of DNA normally proceeds with a high degree of fidelity. On occasion, however, errors of base pairing do occur. The errors of incorporation are usually dealt with by the proofreading functions of the DNA polymerases. These enzymes are not altogether foolproof (they too are subject to mutations), and now and again a mismatched base pair is incorporated into DNA.

The recognition and repair of mismatches have been observed in a variety of organisms including *E. coli,* yeast, and mammals. Though exactly how repair is accomplished in higher organisms is as yet unknown.

To repair the mismatched base pair, the system must somehow be able to distinguish between the template strand and the newly synthesized strand containing the wrong base. If it could not, the repair system might "repair" the template base, setting the error permanently.

In *E. coli,* parental DNA is methylated at an adenine residue found in the sequence 5′ GATC 3′ , which, by the way, is a palindrome, so the other strand also reads the same in the 5′ → 3′ direction. A newly synthesized DNA strand containing this sequence will also be methylated at the adenine. The key to this recognition of template DNA versus newly made DNA is that for several minutes after synthesis the new strand is unmethylated. During this time, mismatched bases are recognized and repaired.

Repair is accomplished by a group of seven proteins, including endonuclease, helicase, single-stranded binding proteins (SSB), and DNA polymerase III. One of the other proteins locates a mismatch and is then joined at the site by two other proteins. The three proteins then move along the DNA in either direction, for several thousand base pairs if need be, until an unmethylated GATC sequence is

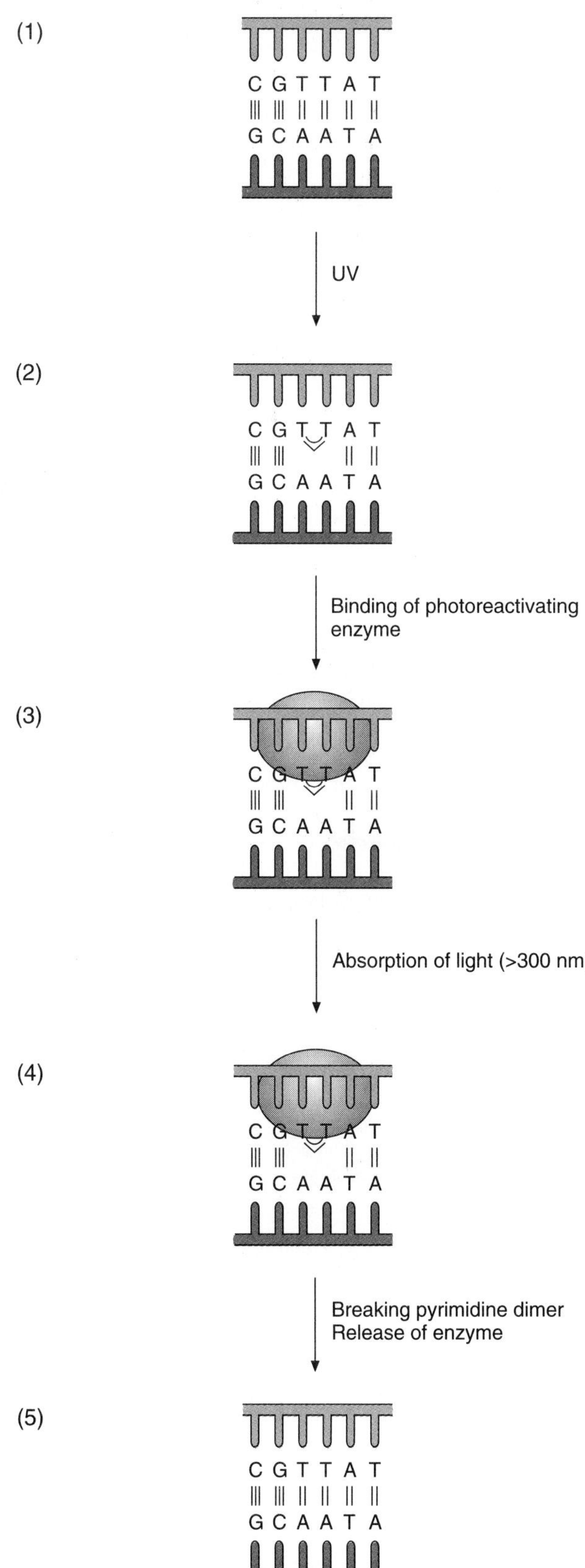


FIGURE 9.12
Photoreactivation by the photoreactivating enzyme (DNA photolyase).

encountered in either strand. At that sequence, one of the proteins cuts the unmethylated strand. Then, with the help of the exonuclease, helicase, and SSB, the unmethylated strand containing the mismatched bases is removed. The gap is filled in by DNA polymerase III.

Again, a distortion of DNA is the signal that a problem exists. In this case, the helical distortion resulting from mismatched bases is exceedingly small but recognized. Repairs are made with great efficiency.

Interestingly, it has recently been suggested that the mismatch repair system is what prevents the successful cross-breeding of species. Offspring of cross-breeding usually die or are sterile because maternal and paternal genetic material cannot be combined. The problem appears to center on recombination. Under normal conditions, recombination via crossing-over is the usual way by which maternal and paternal chromosomes are scrambled. This mechanism produces nonparental combinations and, perhaps, genetically better progeny. The process of recombination is complex but requires the pairing of homologous chromosomes.

Experimental evidence suggests that it is the mismatch repair system that somehow prevents the pairing of nonhomologous chromosomes. The two bacteria, *E. coli* and *Salmonella typhimurium* apparently diverged some 150 million years ago, and their DNAs differ by about 20 percent. These two species cannot conjugate (mate sexually) very successfully under normal conditions, but they can be made to conjugate at frequencies up to 1000 times more than normal. Success depends on a defective mismatch repair system. Other workers have shown that mammalian genes can be made to recombine with *E. coli* DNA if the bacteria have a defective mismatch system.

## Excision Repair

Perhaps the most common repair system is the **excision repair** system. In all organisms, this repair mechanism can handle a wide variety of structural DNA defects. In *E. coli*, the best understood of the systems, the process requires a number of enzymes most of which are nucleases of one type or another.

Figures 9.13 and 9.14 show the prokaryotic excision of a methylated base (i.e., alkylation) and of a thymine dimer (photolyase is not used here). Recognition of the defects is via DNA distortion.

In mammalian cells, the excision repair systems are much more complex and require many more enzymes. This complexity may be directly attributable to the fact that chromosomes of eukaryotes are more complex than are the chromosomes of prokaryotes. The incision step itself in eukaryotes may require the products of at least 10 genes. A number of human diseases have been associated with defects in excision repair systems (Table 9.6).

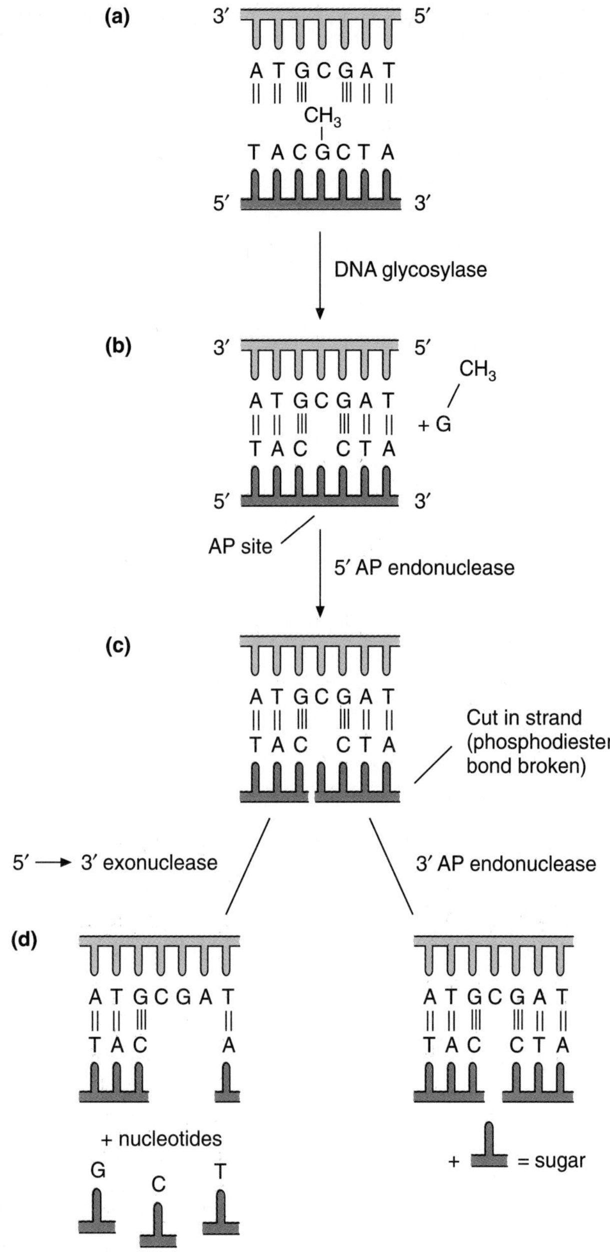


FIGURE 9.13
Removal of a methylated guanine by apurination. A 5′ endonuclease cuts the strand on the 5′ side of the AP site. This is followed by the removal of either several nucleotides or only the apurinic sugar. In either case, a gap results that must be repaired by DNA polymerase I and DNA ligase.

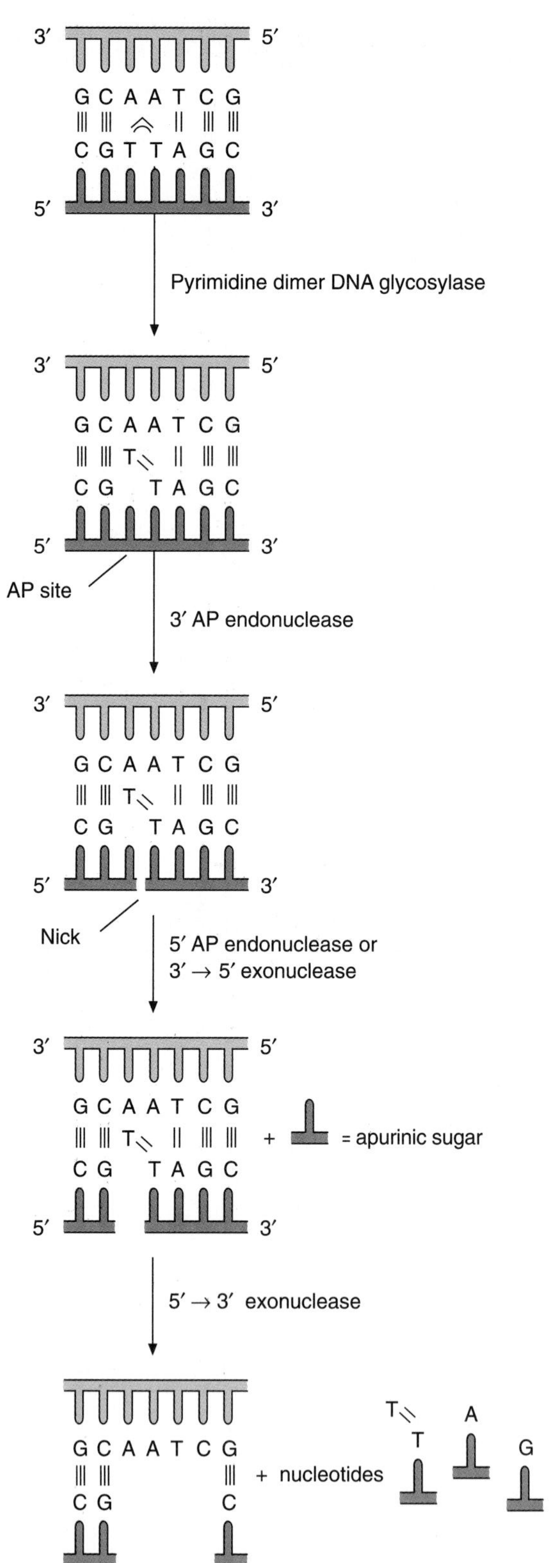


FIGURE 9.14
Thymine dimer removal. The DNA glycosylase releases one of the thymines from its sugar. A 3′ AP endonuclease cuts the strand on the 3′ side of the AP site. The sugar is then cut out by an endonuclease or an exonuclease and, last, another exonuclease removes the dimer. The gap is filled by DNA polymerase I and DNA ligase.

## Recombination Repair

One way by which damaged DNA that inhibits replication can be dealt with so that replication can continue is a form of **postreplication repair.** This mechanism allows replication to proceed, reminiscent of the SOS system, even though all the damage has not been repaired. It may take many hours to remove all the thymine dimers, or there may be damage to the DNA that does not signal a problem by distortion.

*table 9.6*

SOME HUMAN DISEASES WITH DNA-REPAIR DEFECTS

| DISEASE | SENSITIVITY | CANCER SUSCEPTIBILITY | SYMPTOMS |
|---|---|---|---|
| Ataxia telangiectasia | Gamma irradiation | Lymphomas | Uncoordinated muscle control; dilation of blood vessels in skin and eyes; chromosome aberrations; immune dysfunction |
| Bloom's syndrome | Mild alkylating agents | Carcinomas, leukemias, lymphomas | Photosensitivity; facial telangiectases; chromosome alterations |
| Cockayne's syndrome | Ultraviolet irradiation | | Dwarfism; retinal atrophy; photosensitivity; progeria; deafness; trisomy 10 |
| Fanconi's anemia | Cross-linking agents | Leukemias | Hypoplastic pancytopenia; congenital anomalies |
| Xeroderma pigmentosa | Ultraviolet, chemical mutagens | Skin carcinomas and melanomas | Skin and eye photosensitivity; keratoses |



This system, however, is not truly a repair mechanism, because when the process is completed the damage is still present. The advantage is that the cell has divided and repair can take place later on by another method. In other words, at the moment, division is more important to the cell than is correcting DNA damage.

Figure 9.15 shows how the mechanism works. Note that DNA replication is required, as is the protein RecA (Rec = recombination) and an exchange of sections of chromatid. The exchange is needed since, as we saw earlier, adenines can not be base paired with thymine dimers by DNA polymerase III. The adenines are continually removed, and replication is thereby stalled. RecA, however, can place a pair of adenines, cut from the dimer strand's complement, opposite the dimer. The gap in the donor strand can then be filled and closed by DNA polymerase and DNA ligase.

## THE AMES TEST

At present, human beings add some 1000 new synthetic compounds yearly to a stock currently estimated to be over 60,000. Testing all of these compounds by using the usual mouse toxicity tests becomes prohibitively expensive and time-consuming; it takes 2–3 years to test each compound.

We have, so far, discussed mutation in the context of the deleterious effects produced by changes in the genome's nucleotide sequence. In 1974, Bruce Ames developed a rapid screening technique for mutagens and carcinogens and an organism to be used in place of mice. The **Ames test** is based on mutation, and it is rapid (completed in 2 days) and less expensive than the mouse toxicity method.

This test is a technique designed to show the probability that a substance is a mutagen or carcinogen and the relative mutagenicity or carcinogenicity of the substance. The procedure depends on the fact that particular mutants of the bacterium *Salmonella typhimurium* spontaneously revert, or back-mutate, to the wild type at a given, but low, frequency.

Ames isolated cells that are auxotrophic histidine (*his*) mutants with either a base substitution or a frameshift mutation in the *his* operon. The cells also lack the lipopolysaccharide layer at their surface that would normally make the cells impermeable to many of the compounds of interest to industry. The isolated cells were also deficient in excision repair.

Approximately $10^8$–$10^9$ *S. typhimurium* cells are grown on an agar medium containing barely enough histidine to initiate cell division of individual cells but not enough histidine to allow colonies to develop. And, since many compounds are not themselves mutagenic or carcinogenic but are enzymatically converted to these forms during normal metabolism, a small amount of rat liver extract is added to the medium. (A normal function of the liver is to metabolize a great variety of substances, and there is no liver counterpart in bacteria; hence, the necessary enzymes are not found in bacteria.) Spontaneous reversion of the *his* operon will result in the production of a colony. Half the plates inoculated will contain the substance of interest, and the

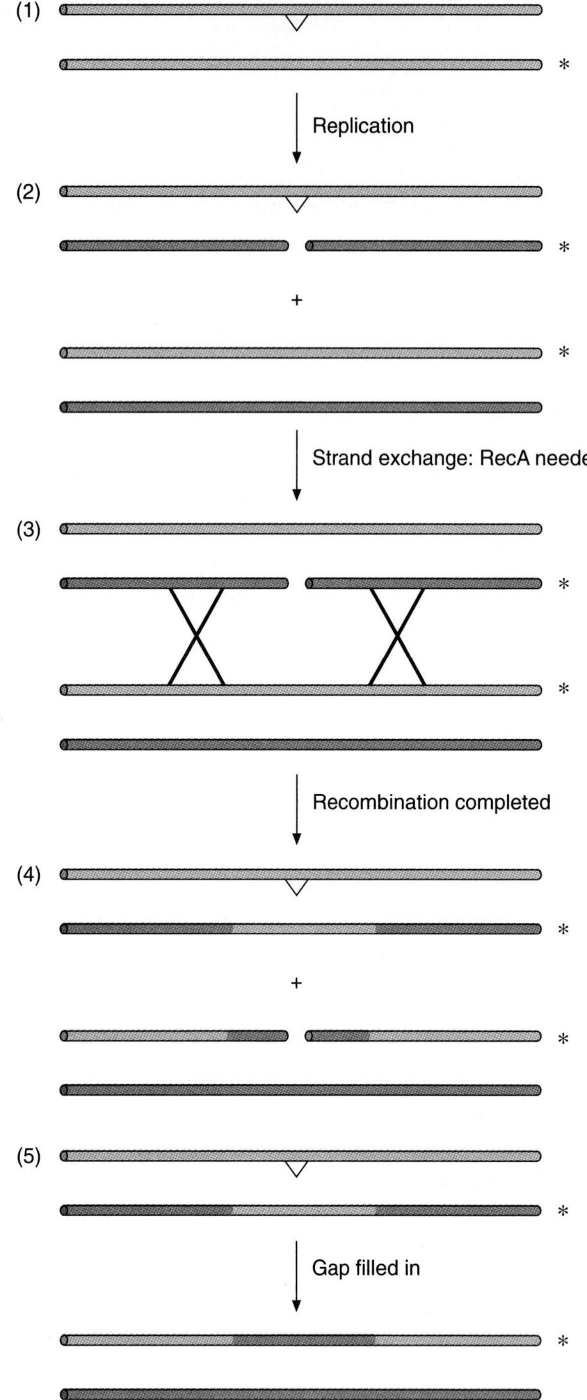


FIGURE 9.15
Recombination repair. In this mechanism, the damaged region (V) is skipped during replication. An exchange then occurs between homologous strands (*) in which RecA participates. The gap is filled by using its complement strand as the template.

other half will not contain the substance (these are the controls to check the number of spontaneous reversions).

The number of colonies on the plates containing the test substance should be statistically greater than the number of colonies appearing on the plate without the test substance. There is a correlation between the concentration of the test substance added to the medium and the number of colonies that appear. For known carcinogens, there is a significant correlation between the number of colonies produced on the plates containing the test material and the effectiveness of the carcinogen in producing cancer.

It should be noted that a positive result on an Ames test is not absolute proof that a substance is a mutagen or a carcinogen. Rather, a positive test indicates a probability that the substance is a mutagen or carcinogen. In the case of carcinogens, direct evidence of cancer-causing properties requires testing for tumor formation in laboratory animals. Nor should it be assumed that a compound is absolutely safe because the results of the Ames test were negative.

## Summary

Mutations are defined as changes in the nucleotide sequence of an organism's genetic material. Two rounds of DNA synthesis are needed to set the mutation.

These changes affect the organism's genotype but are most easily detected by analysis of the phenotype, which is the expression or manifestation of the genotype and includes such things as biochemical and morphological characteristics.

At the chromosomal level, mutations include changes that are deletions, duplications, inversions, and translocations of large segments of chromosomes.

At the gene level, mutations can involve single nucleotides; these are the mutations most intensely studied. Single nucleotide mutations are referred to as point mutations. These mutations can be brought about by deletions or insertions of bases, by base substitutions in nucleotide sequences, by proofreading errors, and by chemical modifications of bases.

Mutations can be naturally occurring, the result of faulty DNA or RNA synthesis. These types of mutation are called spontaneous mutations. Those mutations produced in the laboratory are referred to as induced mutations. A study of this type of mutation is the way in which spontaneous mutations and their consequences are explored.

Because not all mutations bring about detectable phenotypic expressions and because of the difficulty in doing nucleotide sequence analysis of entire genomes, some mutations remain undetected and are called silent mutations.

Substances known to cause mutations can be assigned to one of three groups: biological, chemical, or physical mutagens. These substances range from viruses to a great variety of chemicals such as 5-bromouracil and nitrous acid to UV radiation. These mutagens cause disruption of reading frames, mismatched base pairs, additions or deletions, and pyrimidine dimers.

Cancer is considered to be a problem of DNA replication or of DNA repair or of both. Many mutagens are also powerful carcinogens. A number of genetic changes have been associated with cancers: for example, altered proteins or regulatory sequences, chromosomal translocations, and proto-oncogene amplification.

Human beings are subject to thousands of genetic errors. Many are present at birth and are, therefore, called inborn errors of metabolism. Since the early 1900s, many of these errors have been identified at the biochemical level and are associated with a faulty enzyme or a lack of an enzyme that results in the accumulation of toxic biochemical intermediates.

Human beings are protected, as are all mammals, from some forms of disease by an exquisite immune system that is itself dependent on mutations for its effectiveness.

All cells possess the capacity to repair DNA damage, but not all cells use the same mechanisms. The repair systems utilize an array of enzymes and other proteins to either restore the original nucleotide sequence or repair the damage in such a way that cell replication can continue even though the original nucleotide sequence has not been restored.

Many of these repair systems are induced by damage that distorts the DNA molecule. Others are constitutive systems such as polymerase proofreading activity.

## Study Questions

1. Define and give an example of a mutation at the genotypic level, and give an example of a phenotypic expression of a mutation.
2. List the types of mutations that affect chromosomal structure.
3. What is a point mutation? List three types of point mutations.
4. Give three examples of different categories of mutations, such as auxotrophic mutations.
5. What is the difference between spontaneous and induced mutations? How are these mutations similar?
6. How might base sequence changes come about?
7. Briefly describe four reasons why silent mutations might go undetected.
8. Explain what transitions and transversions are.
9. Give the groupings of mutagenic agents and at least one example for each group.
10. Explain the connection between mutation and cancer.
11. Briefly explain at least three genetic changes associated with cancers.
12. Five DNA repair systems were discussed in this chapter. Briefly explain how each of them works.
13. Describe how the Ames test is done and the significance of positive and negative results.
14. What are inborn errors of metabolism? Give a few examples.

## Readings and References

Benson, P. F., and A. H. Fenson. 1986. *Genetic biochemical disorders*. New York: Oxford University Press.

Bishop, J. M. 1982. Oncogenes. *Sci. Am.* 246:80–92.

Cavenee, W. K., and R. L. White. 1995. The genetic basis of cancer. *Sci. Am.* 272:72–79.

Cohen, S. N., and J. A. Shapiro. 1980. Transposable genetic elements. *Sci. Am.* 242:40–9.

Datta, A., and S. Jinks-Robertson. 1995. Association of increased spontaneous mutation rates with high levels of transcription in yeast. *Science* 268:1616–19.

Echols, H., and M. F. Goodman. 1991. Fidelity mechanisms in DNA replication. *Annu. Rev. Biochem.* 60:477–511.

Epstein, C. J., and M. S. Golbus. 1977. Prenatal diagnosis of genetic diseases. *Am. Sci.* 65:703–11.

Finnegan, D. J. 1989. Eukaryotic transposable elements and genome evolution. *Trends Genet.* 5:103.

Freifelder, D., and G. M. Malacinski. 1993. *Essentials of molecular biology*. Boston: Jones and Bartlett.

Grindley, N. D. F., and R. R. Reed. 1986. Transpositional recombination in prokaryotes. *Annu. Rev. Biochem.* 54:863–96.

Hanawalt, P. C. 1994. Transcription-coupled repair and human disease. *Science* 266:1957–58.

Haseltine, W., and F. Wong-Staal. 1988. The molecular biology of the AIDS virus. *Sci. Am.* 259:52–62.

Kornberg, A., and T. A. Baker. 1992. *DNA replication*. 2d ed. New York: W. H. Freeman.

Lewin, B. 1997. *Genes VI*. New York: Oxford University Press.

Marx, J. 1994. DNA repair comes into its own. *Science* 266:728–30.

McClintock, B. 1951. Chromosome organization and gene expression. *Cold Spring Harbor Symp. Quant. Biol.* 16:13–47.

Modrich, P. 1994. Mismatch repair, genetic stability, and cancer. *Science* 266:1959–60.

Muller, H. J. 1927. Artificial transmutation of the gene. *Science* 66:84–87.

Raivo, K. O., and J. E. Seegmiller. 1972. Genetic diseases of metabolism. *Annu. Rev. Biochem.* 41:543–76.

Rayssiguier, D., S. Thaler, and M. Radman. 1989. The barrier to recombination between *Escherichia coli* and *Salmonella typhimurium* is disrupted in mismatch-repair mutants.

Roitt, I., J. Brostoff, and D. Male. 1993. *Immunology*. 3d ed. London: Gower Medical.

Sancar, A. 1994. Mechanism of DNA excision repair. *Science* 266:1954–56.

*Science*. 1991. *Science* 254 (November), Cancer Issue.

Shapiro, J . A. 1995. Adaptive mutation: Who's really in the garden. *Science* 268:373–74.

Stadler, L. J. 1928. Mutations in barley induced by x-rays and radium. *Science* 68:186–87.

Voet, D., and J. G. Voet. 1995. *Biochemistry*. 2d ed. New York: John Wiley & Sons.

Wagner, R. P., M. P. Maguire, and R. L. Stallings. 1993. *Chromosomes: A synthesis*. New York: Wiley-Liss.

Wallace, D. C. 1992. Diseases of mitochondrial DNA. *Annu. Rev. Biochem.* 61:1175–212.

Weaver, R. F., and P. W. Hedrick. 1997. *Genetics*. 3d ed. Dubuque, Iowa: Wm. C. Brown.

Weiner, A. M., P. L. Deininger, and A. Efstratiadis. 1986. Nonviral retroposons: Genes, pseudogenes, and transposable elements generated by the reverse flow of genetic information. *Annu. Rev. Biochem.* 55:631–61.

Woodward, V. 1992. *Human heredity and society*. St. Paul, Minn.: West.

CHAPTER
TEN

# EUKARYOTIC GENES: BASIC STRUCTURE AND FUNCTION

## OUTLINE



## CHAPTER OBJECTIVES

*This chapter will discuss:*

- The structure of eukaryotic DNA, its packaging into chromosomes, and how chromosomes are distinguished one from another
- How genes are dispersed among the chromosomes and how individual genes are structured
- How genes control development and how methylation is a way of controlling gene expression
- How the various RNA polymerases, the RNA processings and modifications, and the regulatory proteins needed for transcription are distinguished
- How one vertebrate system uses the same genetic information in different combinations to synthesize different proteins (the immune system)

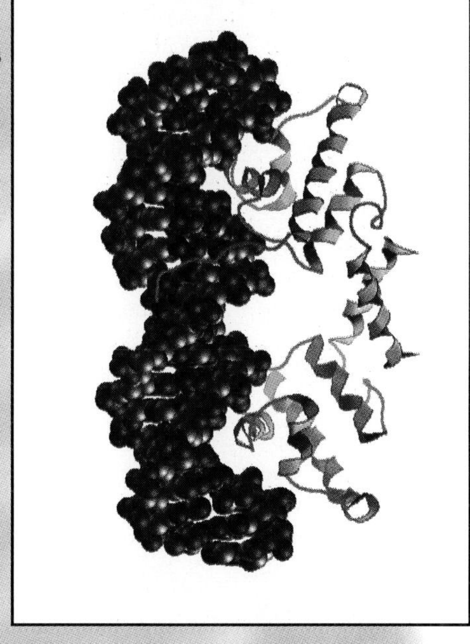

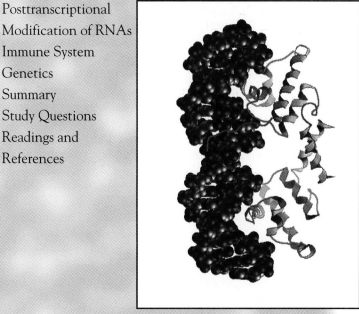

## INTRODUCTION

Compared with prokaryotes, the eukaryotes have many more genes, and the structure and organization of these genes are much more complex. The genes are located on structures called **chromosomes.** The complexity of structure and organization imposes a complexity on the mechanisms by which the information of the genome is transmitted to the next generation during DNA replication and to the cell during protein synthesis.

Because of the enormous length of eukaryotic chromosomes relative to the nucleus, packaging of the DNA is necessary. Packaging reduces the volume taken up by the chromosomes. Special proteins called histones are used for this purpose.

Unlike prokaryotic genes, which have uninterrupted nucleotide sequences, eukaryotic genes are split. The nucleotide sequences that bear information are separated by nucleotides that do not carry information for protein or RNA synthesis. The processing of mRNA after transcription removes noninformational ribonucleotides specified by noninformational deoxyribonucleotides of DNA. The other species of RNA (i.e., rRNA and tRNA) are also modified after their synthesis. The eukaryotes also differ from the prokaryotes in that they use different RNA polymerases to produce mRNA, tRNA, and rRNA.

Although, in general, protein synthesis in eukaryotes is very similar to the process in prokaryotes, there are differences. These differences involve, among other things, transcription factors, RNA polymerases, and RNA processing.

## EUKARYOTIC DNA AND CHROMOSOMES

### Eukaryotic DNA

Like that of the prokaryotes, the genetic information of eukaryotic cells is stored in the nucleotide sequences of DNA. But there are major differences between the prokaryotic and the eukaryotic genomes.

Prokaryotes have a single chromosome. Higher organisms, on the other hand, display a wide range of chromosome numbers. In fact, the variation is from 1 to 630 for the haploid number. Males of the Australian jumper ant *Myrmecia pilosula* have a single chromosome, and the worker ants have two chromosomes. And for the fern *Ophioglossum reliculatum* the haploid number is 630. The mammal that has the distinction of having the largest haploid number, 46, is the rodent *Anotomys leander*, and the Indian deer called muntjac (*Muntiacus* sp.) has the lowest haploid number, 3. Table 10.1 shows the haploid chromosome numbers of some common organisms.

*table 10.1*

**Organisms and Their Haploid Chromosome Numbers**

| Common name (scientific name) | Haploid chromosome number |
|---|---|
| **Mammals** | |
| Dog (*Canis familiaris*) | 39 |
| Horse (*Equus caballus*) | 32 |
| Donkey (*Equus asinus*) | 31 |
| Chimpanzee (*Pan troglodytes*) | 24 |
| Human (*Homo sapiens*) | 23 |
| House mouse (*Mus domesticus*) | 20 |
| Cat (*Felis catus*) | 19 |
| **Other animals** | |
| Carp (*Cyprinus carpio*) | 52 |
| Chicken (*Gallus domesticus*) | 39 |
| Frog (*Rana pipiens*) | 13 |
| Fruit fly (*Drosophila melanogaster*) | 4 |
| **Plants** | |
| Tobacco (*Nicotiana tabacum*) | 24 |
| Pine (*Pinus* spp.) | 12 |
| Corn (*Zea mays*) | 10 |
| Garden pea (*Pisum sativum*) | 7 |
| **Fungi** | |
| Baker's yeast (*Saccharomyes cerevisiae*) | 17 |
| Black bread mold (*Aspergillus nidulans*) | 8 |
| Pink bread mold (*Neurospora crassa*) | 7 |

Figure 10.1 shows the range of haploid DNA contents for various organisms. These numbers nicely illustrate the so-called C-value paradox (Chapter 5). The C-value is the amount (content) of DNA found in the haploid genome. One would expect the cells of more-complex organisms to have more DNA than the cells of simple organisms. That is, one would expect the DNA content of organisms to increase as we move from prokaryotes to eukaryotes and from single-celled organisms to multicellular organisms. It would not be surprising to find more DNA in mammals than in amphibians, and more in amphibians than in insects. But that is not always the case. Salamanders, for example, have more DNA than do mammals. And the eukaryotic alga *Pyrenomas salina* has less DNA than the bacterium *Escherichia coli*, a prokaryote ($6.6 \times 10^5$ bp and $4.2 \times 10^6$ bp, respectively).

The biological significance of this phenomenon remains to be explained. Like chromosome numbers, chromosome sizes show a large variation among organisms. Higher organisms usually have larger chromosomes than

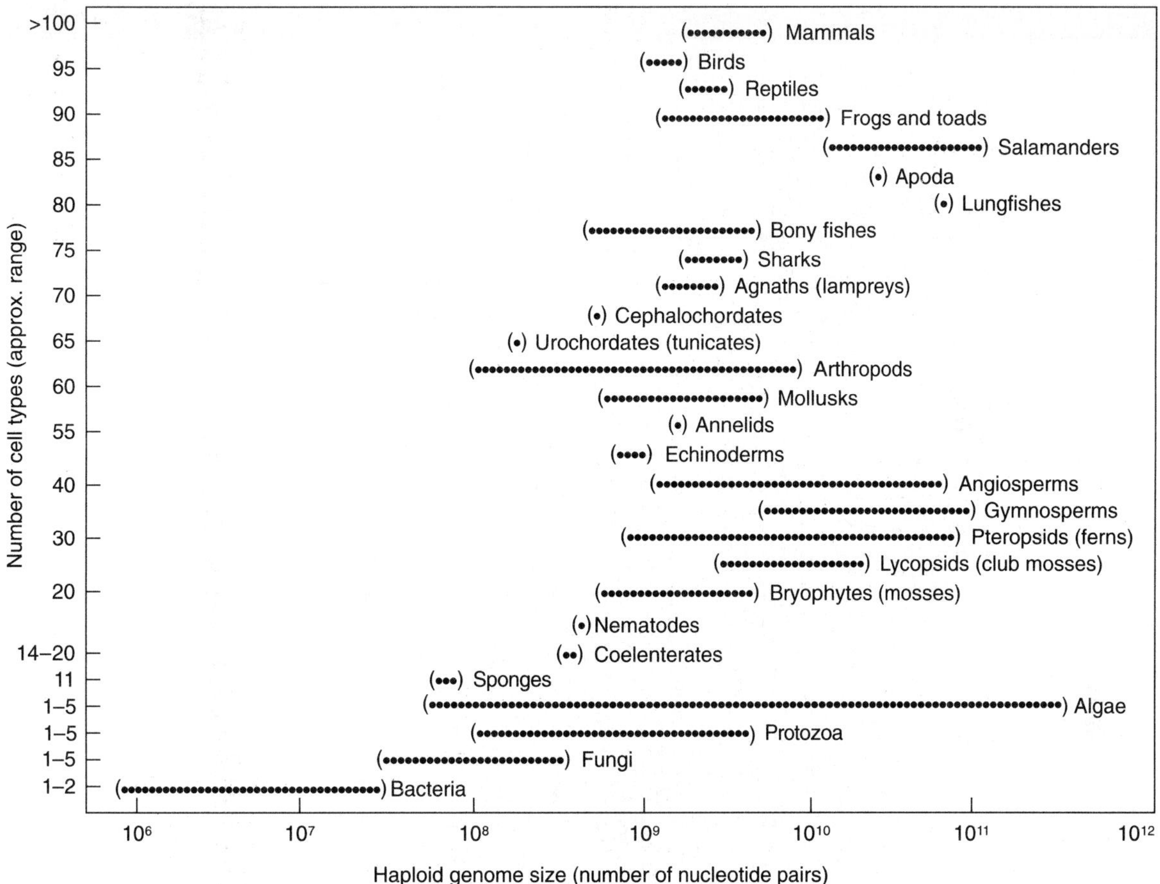


FIGURE 10.1
C-value paradox: DNA content of haploid cells is not directly related to complexity of the organism.

lower organisms, but not always. Some of the largest chromosomes have been found among the amphibians, the liliaceous plants, and grasshoppers, and some fungi and algae have some of the smallest chromosomes.

This variation in size is found even among closely related species with the same diploid number. And the chromosomal complement of some species of animals, birds, and lizards may contain both large and small chromosomes. Table 10.2 shows the variation in the size of human chromosomes.

Eukaryotes also show a large range of DNA per diploid nucleus. For example, *Drosophila* contains 0.2 pg (picogram) DNA per nucleus, and the salamander *Amphiuma* has 168 pg DNA per nucleus. The liliaceous plant *Fritillaria* has about 197 pg DNA per nucleus, and human beings have approximately 6.4 pg DNA per nucleus.

Interestingly, only a very small percentage of this DNA appears to actually code for proteins or RNA. Some estimates place the amount as low as 2% of the total. Though the human genome contains 1000 times the DNA of *E. coli,* it is estimated to have only 50 times as many genes. What function most of the other DNA serves is still a mystery. Table 10.3 shows the classes of eukaryotic DNA.

# Eukaryotic Chromosomes

To function as a chromosome, the DNA molecule must not only be able to act as the template for RNA synthesis, it must also be able to direct its own synthesis. For this latter function, each eukaryotic DNA molecule must have three specialized sequences. These sequences are the sites of attachment for special proteins required for replication and segregation of chromosomes. Each chromosomal DNA molecule must contain a centromere, two telomeres, and a replication origin.

All eukaryotic chromosomes have the same general structure. They have a short arm; designated **p (petite) arm,** and a long arm, the **q arm.** The arms are separated by a **centromere,** a constricted region. Each arm ends in a region called the telomere.

Centromeres vary in size within a single species as well as between species. Yeast centromeres are on the order of a few hundred base pairs, and human centromeres range from about 300 to 5000 kbp (kilobase pairs).

Centromeric DNA from mammals is simple-sequence DNA. In human beings, a 170–base pair sequence is repeated over and over again anywhere from 1700 to 29,000

table 10.2

## Average Lengths of Human Chromosomes as a Percentage of the Total Nuclear Genome of Somatic Cells

| Chromosome Group | Chromosome No. | Average length (in % of autosomal genome) of: Long arm | Short arm | Chromosome Group | Chromosome No. | Average length (in % of autosomal genome) of: Long arm | Short arm |
|---|---|---|---|---|---|---|---|
| A | 1 | 4.68 | 4.57 | D | 13 | 3.29 | — |
| | 2 | 5.28 | 3.35 | | 14 | 3.12 | — |
| | 3 | 3.80 | 3.32 | | 15 | 2.89 | — |
| B | 4 | 4.85 | 1.84 | E | 16 | 1.93 | 1.34 |
| | 5 | 4.66 | 1.75 | | 1·7 | 2.07 | 0.96 |
| | | | | | 18 | 2.04 | 0.76 |
| C | 6 | 3.87 | 2.36 | F | 19 | 1.32 | 1.11 |
| | 7 | 3.54 | 2.04 | | 20 | 1.30 | 1.05 |
| | 8 | 3.45 | 1.63 | | | | |
| | 9 | 3.23 | 1.72 | G | 21 | 1.26 | — |
| | 10 | 3.22 | 1.54 | | 22 | 1.38 | — |
| | 11 | 2.90 | 1.88 | | Total autosomes | 100.00 | |
| | 12 | 3.38 | 1.32 | | | | |
| | | | | | X | 3.26 | 2.02 |
| | | | | | Y | 1.64 | — |

*— short arms not measurable.

Revised by K. Patau (unpublished).

From E. Therman and M. Susman, *Human Chromosomes: Structure, Behavior, and Effects*, 3rd edition. Copyright © 1993 Springer-Verlag, New York, NY. Reprinted by permission.

table 10.3

## Classes of Eukaryotic DNA

Protein-coding genes
1. Single copies
2. Duplicated (most duplicates are not exact)

RNA-coding genes (most are tandomly duplicated)

Repetitious DNA
1. Simple sequences
2. Dispersed intermediate repeats (short, 150–300 bp; long, 5–7 kbp)
   - Mobile genetic elements
     - Transposons
     - Reverse transcription copies

Unclassified spacer or connecting DNA

times depending on the centromere. This DNA is often referred to as satellite DNA because in density-gradient centrifugation it settles above the main DNA band.

Associated with the centromeres are protein-rich regions called kinetochores. These are the sections of chromosomes to which the spindle microtubules attach when cells undergo mitosis.

Because the eukaryotic chromosome is a linear molecule, there are special problems in replication. The newly synthesized lagging strands are shortened by the length of the primer RNA that is removed (see Chapter 2, Semidiscontinuous Replication of DNA). This problem is overcome by telomeres and the enzyme **telomerase.**

**Telomeres** are made up of short nucleotide sequences that are also repeated from several hundred times in yeast to several thousand times in human beings. The human telomere repeated sequence is 5′ TTAGGG 3′. These sequences are similar in such diverse organisms as protozoa, fungi, plants, and mammals. The complementary strand (3′ to 5′—the lagging strand during DNA replication) of telomeric DNA ends a bit short of the 5′ to 3′ strand (the parental or template strand) so that each chromosome has a single-stranded tail at each end. Interestingly, because of this phenomenon, normal chromosomal duplication should result in successive shortening of telomeres with replicative age, and indeed it does because the shortening and restoration of telomere sequences are only in approximate balance. The special enzyme telomerase (a reverse transcriptase), containing template RNA for DNA synthesis, prevents the shortening. This enzyme recognizes the telomere sequence of the 5′ → 3′ strand and attaches to this strand. It then elongates the 3′ end of this strand. The enzyme contains its own template strand, which is used to produce the telomere sequences. The elongated sequences are then used by DNA poly-

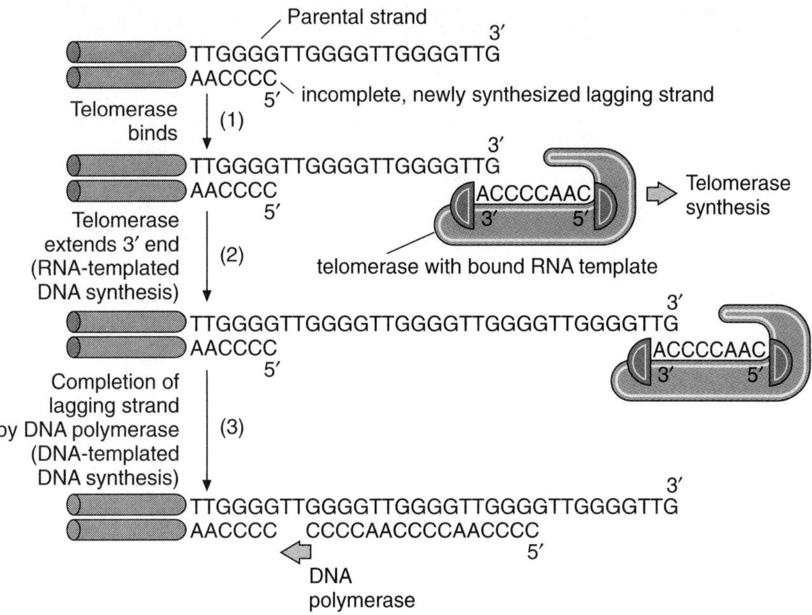


FIGURE 10.2
Telomere production. Binding of telomerase is followed by extension of the 3′ end using the enzyme's RNA template. The lagging strand is completed by DNA polymerase in the 5′ → 3′ direction.

merase as the template to extend the 5′ end of the lagging strand (Figure 10.2).

The telomeric DNA long strand (5′ → 3′) folds back on itself in such a way that the guanines are aligned, and an unusual hydrogen bonding between paired guanines takes place. In this way, both ends of the chromosome are capped.

One role of telomeres appears to be to prevent chromosomes from joining together end to end as broken chromosomes can and do. Another role may be to anchor chromosomes in the nuclear membrane. Interestingly, recent findings (Counter et al. 1994) have suggested that, in human beings, telomerase is active in metastatic cells but not detectable in normal somatic cells and that, as these cells age, telomeres become progressively shorter. In vitro immortalization of cells appears to activate telomerase and to stabilize the shortening of telomeres. The findings also show that in human ovarian cancer cells, telomeres are significantly shorter than in corresponding nonmalignant cells. These cancer cells show telomerase activity not present in the nonmalignant cells. Counter and co-workers suggest that the continuation and progression of malignancy may depend on the activity of telomerase and that an effective anticancer treatment may involve telomerase inhibitors.

## *Chromosomal Banding*

Chromosomes are at their most condensed state during the M phase of the cell cycle, at which time mitosis occurs. During the rest of the cell cycle, called interphase, chromosomes are much less condensed and are dispersed throughout the nucleus. Because the less-condensed, dispersed interphase chromosomes are much more difficult to study, metaphase chromosomes are the most often characterized. They appear as duplicates because replication has been chemically arrested before the centromeres have replicated and the sister chromatids pulled apart. The chromosomes are then chemically treated, with trypsin, for example, to elongate them a bit. Then they are fixed to preserve their structures. Finally, they are stained with quinicrine (a fluorescent dye) or Giemsa to give Q- or G-banding patterns, respectively.

If Giemsa staining is used, euchromatin stains lightly and heterochromatin stains darkly, giving the G-banding patterns. With Q-banding produced by quinicrine and revealed by UV light, the bright bands correspond to the dark G-bands and the nonfluorescing or dark Q-bands correspond to the light G-bands.

**Euchromatin** is that portion of the chromosome not densely packed and so accepts less stain, resulting in a lightly stained (G) or nonfluorescing (Q) segment. **Heterochromatin,** on the other hand, is densely packed, so it stains intensely, giving dark G-bands and fluorescing (bright) Q-bands.

The G light bands (Q-dark) contain more genes than do the G dark bands (Q-bright). We now know that within this euchromatin are found all the so-called **housekeeping genes.** These are the genes whose products are required by all cells of an organism to maintain normal cellular functioning such as energy production and protein synthesis. Also found in the euchromatin are the tissue-specific genes, those genes whose products give particular cells, such as nerve cells, special physiological functions.

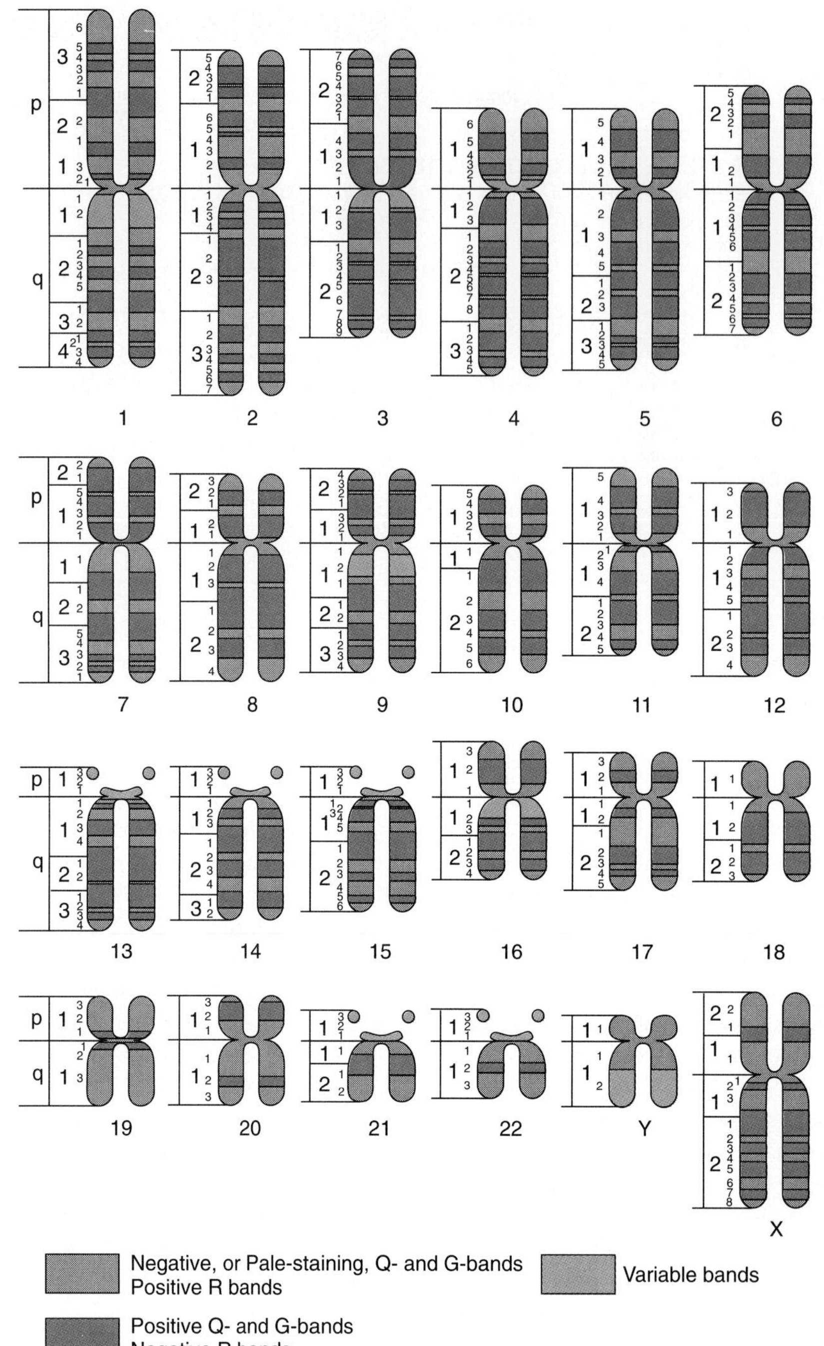


FIGURE 10.3
Banded human chromosomes.

Heterochromatin is of two types: constitutive and facultative. Constitutive heterochromatin consists of simple sequence-repeated DNA, contains no genes, is never transcribed, and is replicated late during DNA synthesis. Facultative heterochromatin is essentially euchromatin that is inactive (perhaps by condensation) during particular stages of development of higher organisms. Figure 10.3 shows the G- and Q-banding patterns of human chromosomes.

Another technique, known as prophase banding, increases band resolution from about 300 bands on a metaphase chromosome using quinicrine and Giemsa staining to anywhere from 500 to 2000 bands. This procedure arrests dividing cells during S phase (the DNA synthesis phase) with amethopterin (i.e., methotrexate), a substance that prevents the conversion of uracil into thymine.

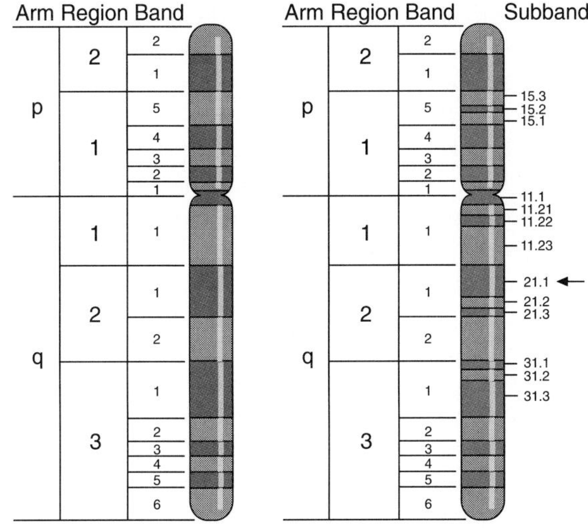


FIGURE 10.4
Example of G-banding designations: human chromosome 7.

The block is terminated with a thymine-rich medium, and a large proportion of the cells are in synchrony. Treating and fixing the chromosomes results in large chromosomes and more bands. A drawback, however, is that there is more band overlap and, consequently, analysis is more difficult.

Regardless of how the banding is achieved, the bands are used to localize genes on chromosomes. A system for identifying locations is based on the position of the centromere and the relation of the bands to the centromere.

The nonsex chromosomes, called **autosomes,** in humans are numbered 1–22 according to size, from the largest to the smallest. The **sex chromosomes** X and Y are grouped separately from the autosomes. Each chromosome is then divided at the centromere into the short arm p (petite) and the long arm q. Each arm is then divided into regions and each region into bands and subbands, if necessary. For example, the designation 7q21.1 means chromosome 7, the long arm (q), region 2, band 1, subband 1 (subband numbers follow the decimal point) (see Figure 10.4).

Table 9.5 lists some genetic disorders. The locations of some of the defective genes responsible for these conditions have been determined. Table 10.4 lists the chromosomal location of a sampling of these and other defective genes.

It should be pointed out that often sensational announcements are made in the news media concerning the discovery of the location of an important gene. A closer reading of the article, or better still, a reading of the scientific paper on which the news article was based, would reveal that what has been found is the location of the gene in a band on a specific chromosome where dozens if not hundreds of other genes may also be located. A more precise location would require nucleotide sequence analysis of the band. For example, the discovery of the location of the neurofibromatosis, or NF, gene (on chromosome 17) and of colon cancer genes (on chromosomes 2 and 3) were justly hailed as great discoveries. The initial announcements, however, did not specify where among the thousands of other genes on these chromosomes the genes were located. The NF gene was finally assigned to chromosomal location 17q11.2, while one of the cancer genes was located at chromosome 3p21.

*table 10.4*

CHROMOSOME LOCATION OF SOME DEFECTIVE GENES

| DISORDER | GENE LOCATION |
|---|---|
| Albinism | 11q14–q21 |
| Colon cancer | 2q15–q16 |
| Fabry's disease | xq22 |
| Fructosuria | 9q22 |
| Galactosemia | 9p13 |
| Gaucher's disease | 1q21 |
| Hemolytic anemia | 9q34.1 |
| Homocystinuria | 21q22–q23 |
| Huntington's disease | 4p16.3 |
| Krabbe's disease | 14q21–q31 |
| Lesch-Nyhan syndrome | xq26–q27.2 |
| McArdle's disease | 11q13 |
| Niemann-Pick disease | 11q15.1–q15.4 |
| Porphyria, erythropoietic | 10q25.2–q26.3 |
| Xeroderma pigmentosum | 2q21 |

Eukaryotic chromosomes are a complex of DNA, RNA, and proteins. Each chromosome usually contains a single molecule of double-stranded DNA. Here, too, however, there are some notable exceptions. The giant chromosomes of dipteran salivary glands are the best-known examples. These chromosomes may replicate dozens of times, with the new chromosomes remaining attached in parallel to the older ones. The giant chromosomes may have hundreds to thousands of neatly aligned DNA molecules. These types of chromosomes are referred to as polytene chromosomes.

Because of the size of even normal chromosomes, packaging becomes critical. The complex of DNA, RNA, and protein, called **chromatin,** must be reduced to a volume that will allow it to fit within the confines of the nucleus.

The total length of the entire human genome (diploid number of 46 chromosomes) is approximately 2 m (about 6 ft). The average length of a human chromosome is about 5 μm. Packed into this 5 μm is a length of DNA that, when unpacked, is almost 5 cm long. The DNA is about 10,000 times as long as the chromosome into which it is packed. Human chromosomes have a nucleotide content ranging from 48 million base pairs in the smallest autosomal chromosomes, 21 and 22, to about 240 million base

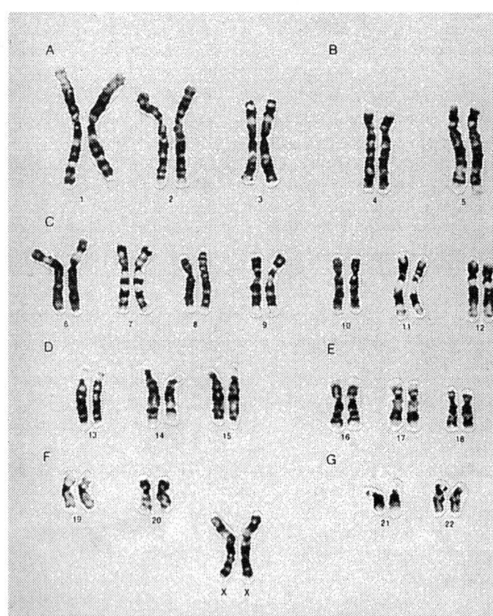

FIGURE 10.5
Karyotype of a normal female.
© Leonard Lessin/Peter Arnold, Inc.

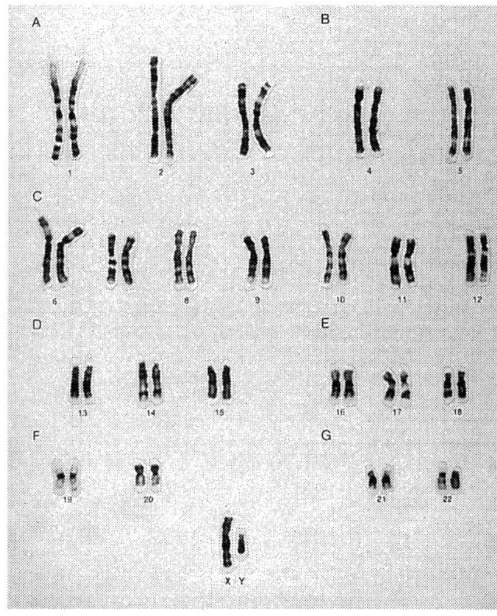

FIGURE 10.6
Karyotype of a normal male.
© Leonard Lessin/Peter Arnold, Inc.

pairs for the largest chromosomes, 1, 2, and 3. Figures 10.5 and 10.6 show the normal female and male chromosomal complement, respectively.

# HISTONES AND NONHISTONE PROTEINS

## Histones

A chromosome has about twice the mass of protein as the mass of DNA, and less than 10% of the mass is RNA. Chromosomal proteins are of two types: histones and nonhistones. Chromosomal RNA is, for the most part, mRNA, tRNA, and rRNA in various stages of completion.

The **histones** contain large amounts of the basic amino acids, particularly lysine and arginine. As a group, then, these proteins have a strong basic, or positive electrical, charge. This is not surprising considering their function in packaging the highly acidic DNA molecule (remember, the phosphate groups along the DNA backbone have a negative charge). Histones are divided into five classes designated H1, H2A, H2B, H3, and H4 (Table 10.5).

In virtually all eukaryotes studied thus far, the same five classes of histones have been found. A few exceptions have been noted, especially among the unicellular algae and yeast. For example, some algae have proteins identifiable as H1, H3, and H4, but they also have a single protein whose properties resemble those of both H2A and H2B. Some yeast, such as *Saccharomyces cerevisiae* (baker's yeast), have either no H1 or a functionally similar but structurally distinct protein equivalent.

H3 and H4 are among the most highly conserved proteins found in nature. Identical amino acid sequences have been found for these two proteins in organisms as different as the cow and the pea. H2A and H2B, although found in all eukaryotes, show significant species-specific variations in amino acid sequences. H1 proteins, however, have a good deal of amino acid variation between species and even between tissues of the same species.

The histones are also among the most highly modified proteins found in the cell. These modifications may include the covalent addition of acetyl and methyl groups, phosphate, ADP-ribosyl groups, and entirely different proteins such as ubiquitin. Some modifications have been correlated with important cellular mechanisms; for example, DNA replication is associated with the methylation of lysine and arginine. The role of other modifications, such as addition of ubiquitin, is still something of a mystery.

The relative constancy of histone structure, in general, and the highly conserved nature of H3 and H4, in particular, suggest that these proteins have identical functions in all eukaryotes. We shall discuss these functions in the next section.

## Nonhistone Proteins

The nonhistone proteins are all the other proteins associated with the DNA of chromosomes. These proteins represent a smaller proportion of the chromosomes' mass than do the histones. Nonhistone proteins are a highly varied group in both structure and function. Most of these proteins are either negatively charged or neutral at physiological pH. And, because their chemical properties closely resemble the properties of other cellular proteins, it has been very difficult to determine the numbers and kinds of these proteins.

Depending on the isolation and purification techniques used, there may be as few as 20 or as many as several

*table 10.5*

HISTONES

| CLASS | BASIC AMINO ACIDS LYSINE (%) | ARGININE (%) | ACIDIC AMINO ACIDS (%) | BASIC/ACIDIC RATIO | NUMBER OF AMINO ACIDS | MOLECULAR WEIGHT (Da) |
|---|---|---|---|---|---|---|
| H1 | 29 | 1 | 5 | 6 | 200–265 | 23,000 |
| H2A | 11 | 9 | 15 | 1.3 | 129–155 | 13,960 |
| H2B | 16 | 6 | 13 | 1.7 | 121–148 | 13,774 |
| H3 | 10 | 13 | 13 | 1.8 | 135 | 15,342 |
| H4 | 11 | 14 | 10 | 2.5 | 102 | 11,282 |

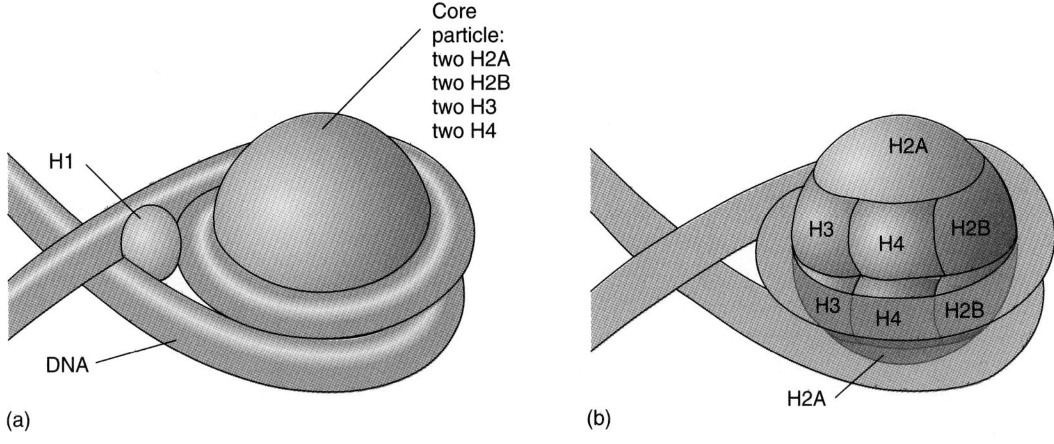


FIGURE 10.7
Nucleosome structure. (a) Approximately two turns of DNA are wrapped around the histone octamer with H1 presumably "locking" the DNA in. (b) x-ray diffraction positioning of individual histone molecules.

hundred of these proteins, with molecular weights ranging from 7000 to over 200,000 kD.

These nonhistone proteins do, however, appear to fall into about four major categories (Wolfe 1993):

1. Proteins required to regulate rRNA and tRNA synthesis.
2. Enzymes and other proteins functioning in transcription, replication, recombination, DNA repair, and modification of chromosomal proteins.
3. Proteins temporarily associated with the RNA products of transcription and needed for RNA modifications.
4. Proteins needed for stabilizing chromosomal structure or required for the conversion of chromatin from extended to condensed states during various phases of the cell cycle (e.g., mitosis).

## PACKAGING DNA

As noted earlier, because of its length, DNA must be packaged to fit the confines of the nucleus. To achieve this goal, DNA is condensed so that the packing ratio of the original DNA length to the shortened length is between 8,000 and 10,000. Three levels of organization are used to condense DNA: nucleosomes, solenoids, and radial loops.

### Nucleosomes

The DNA of each eukaryotic chromosome is first packaged into a series of discrete structures called **nucleosomes,** which are themselves separated by linker DNA. Nucleosomes are composed of a histone octamer containing two copies each of H2A, H2B, H3, and H4, around which is wrapped the double-stranded DNA (Figure 10.7). This is referred to as the core particle.

This structure is produced by the digestion of chromosomes with a micrococcal nuclease. A short digestion time releases a nucleosome: a histone octamer and a length of DNA of about 200 bp. Longer digestion produces the core particle, which is relatively resistant to further digestion. Associated with the core and removed by the nuclease is linker DNA made up of 8–114 bp between nucleosomes. The DNA of the core particle is made up of 146 base pairs wrapped around the histone octamer approximately twice. Therefore, nucleosomes along the length of a DNA molecule are like beads on a string.

Nucleosomes can be assembled in vitro with only DNA and the five histones. In fact, not only eukaryotic DNA but viral and bacterial DNA can also form nucleosomes in the presence of the histones as can laboratory-produced DNA. This ability suggests that the DNA itself plays little if any active role in nucleosome assembly.

On the other hand, H3 and H4 alone can force DNA to coil into a preliminary nucleosome particle that will assume the nucleosome structure upon addition of H2A and H2B. H1 then functions to "lock" the DNA in position around the octamer.

The histones at the center of the core particle, H3 and H4, are arranged in such a way that a tetramer is formed, while two molecules of H2A are positioned at roughly the top-center of the core particle and two molecules of H2B are positioned at about the bottom-center of the particle.

Apparently, as genes are transcribed and repressed, changes in nucleosome structure occur by some as-yet unknown mechanism. When genes are active for transcription or replication, nucleosome DNA becomes more sensitive to nuclease digestion, suggesting at least a loosening of the nucleosome structure during transcription and a complete loss of histones in regions where replication is occurring. One suggestion is that the loss of H1 may allow histone octamers to move along the DNA as transcription proceeds. Loosening and movement of nucleosomes may be dependent on histone modification. The formation of nucleosomes results in a sixfold packaging or condensation of DNA.

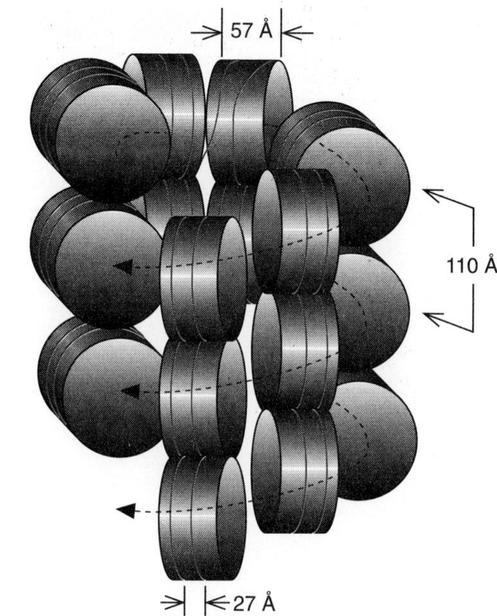


FIGURE 10.8
A solenoid with six nucleosomes in each turn. Solenoids are sometimes referred to as the 30 nm fibers.

## Solenoids

The next level of packaging involves the turning of the nucleosomes around themselves to form a hollow coil structure called a **solenoid.** Each turn of the chromatin contains approximately six nucleosomes. Solenoids increase the packing ratio to about 40 (each micrometer along the solenoid axis contains about 40 μm DNA). Figure 10.8 represents a portion of chromatin in solenoid formation.

Solenoid formation requires the presence of H1. The location of the protein and the exact nature of its function in solenoid formation are still unresolved.

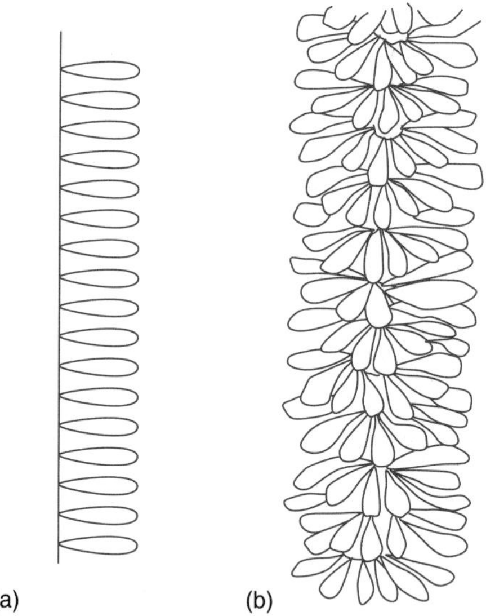


FIGURE 10.9
Radial looping of chromatin. (a) individual loops. (b) three-dimensional representation of loop arrangement. All the loops are a part of the same DNA molecule.

## Radial Loops

A final packing ratio of about 1000 for interphase euchromatin and of about 10,000 for highly condensed mitotic chromosomes is achieved by supercoiling of solenoids into loops that are held closed by special proteins (Figure 10.9).

This packaging from nucleosomes to radial loops allows a human diploid cell to take its chromosomal complement of some 6 billion base pairs with a total length of about 400 cm (the longest chromosome is about 10 cm) and condense the material so that it fits the volume of the cell nucleus with room to spare. This condensation of the genetic material may also help protect it from chemical and physical damage by making sensitive sequences relatively inaccessible.

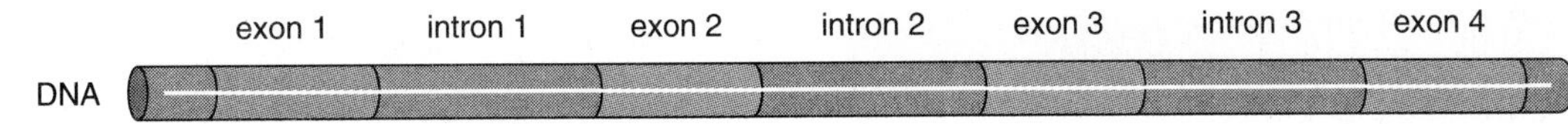


FIGURE 10.10
A representation of exons (coding) and introns (non coding) in a gene of a eukaryote.

## EUKARYOTIC GENES

Eukaryotic genes range in length from several hundred nucleotides (e.g., the human insulin gene) to half a million or more nucleotides (e.g., the human genes for Duchenne muscular dystrophy and neurofibromatosis). In the vast majority of cases, the information contained in a single gene codes for a single protein end product. But a very unusual situation has been discovered. The genes for Factor VIII hemophilia and for neurofibromatosis contain nested genes: that is, genes within genes.

The Factor VIII gene, located on the human X chromosome, has another gene embedded in its sequence, and the NF gene may have three or more other genes nested within its nucleotide sequence (recall the overlapping genes of ØX174, Chapter 5).

Along the eukaryotic DNA, genes may occur on either strand of the double-stranded DNA molecule. That is, each strand has both coding and noncoding segments. All the genes in a eukaryote can be placed in one of three categories, or classes: Class I, II, or III.

Class I genes: Those genes whose information codes for rRNAs (18S and 28S RNA) needed for ribosome production.

Class II genes: Those whose end product is a protein. The direct transcripts of these genes are precursor mRNA molecules, which are subsequently treated to form true mRNA molecules (we shall see how a little later in this chapter).

Class III genes: Responsible for the cell's tRNAs and 5S rRNAs.

As we will see further on in this chapter, each of these classes of genes requires its own DNA-directed RNA polymerase for transcription.

## SPLIT GENES

With very few exceptions, the genes of eukaryotes are composed of coding and noncoding nucleotide sequences called exons and introns, respectively (see Chapter 5). Among the vertebrates, the only known exceptions to this condition are the histone and interferon genes. These structural genes lack introns. Among the plants, the zein genes of corn (which code for storage proteins in maize kernels) also lack introns. Table 5.1 shows some examples of exons and introns in various genes. Note the wide range of both numbers and sizes of introns. Figure 10.10 is a schematic representation of the organization of exons and introns in a eukaryotic gene.

It should be noted here in discussing introns that they are not to be confused with spacer DNA. Introns are found within genes; **spacer DNA** is found between genes. And, although virtually all introns are noncoding segments of genes, there are exceptions. Certain mitochondrial genes carry introns that are coding sequences. For example, in S. *cerevisiae*, the gene for cytochrome b contains two introns (I2 and I4) whose nucleotide sequences code for two enzymes, maturases, needed for RNA processing. It is also thought that some introns may contain genetic information such as enhancer sequences and replication and chromosomal packing signals.

The current thinking on exons is that they carry information for particular sections (amino acid sequences) of polypeptides. Each of the sections, oftentimes referred to as **domains,** has a distinct function in the completed protein. For example, the genes for antibodies, hemoglobin, and a number of enzymes contain exons corresponding to particular domains of these proteins. One theory is that the spreading out of genetic information into discrete units (exons) makes possible the recombination of these units to form proteins with new functions. In other words, there would be an evolutionary advantage in this structural arrangement of genes. There is, however, the possibility that exons can be lost through gene rearrangement or recombination.

## MULTIGENE FAMILIES

In eukaryotes, most protein genes occur as single copies. There are, however, a number of exceptions to this generalization. (Genes for tRNA and rRNA may number in the thousands, from lower eukaryotes to the higher organisms.)

Some protein genes appear to have descended from a common ancestral gene. These groups of genes are referred to as gene, or multigene, families. The genes show sequence variations, particularly in exons, that result in protein end products that are different but related. The genes may be clustered together on one chromosome in a particular region; they may be scattered at various locations on a single chromosome; or they may be dispersed on different chromosomes.

Some families, such as those encoding histones, tubulins (the principal protein component of microtubules), and actins (the major protein component of microfilaments that

*table 10.6*

**SOME RELATED GENES AND PROTEINS**

Closely homologous in structure and function
- Actins
- α-globins
- β-globins
- Growth hormone/placental lactogen
- Opsins
- Steroid receptors
- Tubulins
- Zeins

Related, but divergent in structure and function
- α-globins/β-globins
- α-interferons/β-interferons
- Prolactin/growth hormone
- RNA polymerases
- Serum albumin/α-fetoprotein

Distant cousins
- Antithrombin III/ovalbumin/angiotensinogen
- Ceruloplasmin/factor V/factor VIII
- Haptoglobulin/serum proteases (chymotrypsin)

function to maintain cell structure and movement) are repeated regularly among the chromosomes of the genome. Histone genes, depending on the species, may be clustered, scattered singly and in small groups, and both clustered and scattered among the genome chromosomes. The tubulin genes, some 10–20 each of alpha and beta tubulin, in mammals are scattered singly throughout the chromosomal complement of the organism.

Genes of a given family are structurally and functionally related. That is, they share some common exons and introns, and the protein end products have related functions: for example, the genes for the digestive enzymes trypsin, chymotrypsin, and elastase (remember the domains mentioned earlier). Each of these enzymes hydrolyzes peptide bonds. Their specificities differ and are dependent on the amino acid side chains flanking the bond to be cleaved. These enzymes are synthesized by pancreatic acinar cells and are secreted, by the pancreatic duct, into the duodenum. Table 10.6 lists some related genes and proteins, and Table 10.7 gives some protein multigene families.

## GENE SUPERFAMILIES

Then there are the gene superfamilies containing many dozens of genes whose end products (such as hemoglobin and the immune system proteins) are related. For example, the immunoglobulin superfamily genes, the most extensive yet discovered, include:

1. Major histocompatibility Classes I and II (recognition sites) and Class III (complement components).
2. The five classes of antibodies: IgG, IgM, IgA, IgD, and IgE. There are also subclasses to each of these classes.
3. T-cell receptors.
4. Poly-Ig receptors, which bind IgA antibodies and transport them across epithelial cells.
5. CD4 and CD8 molecules found on T-helper and T-suppressor cells, respectively. These molecules serve a recognition function during the initial stages of the immune response. There are another 40 or so CD markers.

It should be pointed out that many families contain numerous faulty genes that are either not transcribed or, if they are transcribed and then translated, result in nonfunctional proteins. So, for example, of the 10 beta tubulin genes discovered in human beings, 5 code for nonfunctional proteins, 2 are not transcribed, and 3 code for fully functional end products.

The genes for rRNA occur anywhere from hundreds to thousands of times in eukaryotic genomes in near-perfect sequence copies. These genes, as many as 300 copies in the human being, are clustered in particular sections of five chromosomes called **nucleolar organizers.** This is the site around which the nucleolus forms when rRNA transcription and ribosomal subunit assembly are taking place. In maize, on the other hand, the rRNA genes are clustered on a single chromosome: chromosome 6.

Like the rRNA genes, tRNA genes are also found in clusters on one or several chromosomes. Clusters may contain multiple copies of a single gene or a number of different tRNA genes. Among this group of genes, pseudogenes are very common, especially in the higher eukaryotes. In human beings, there are approximately 2600 copies of tRNA genes in each diploid cell; they are found in clusters and as single copies dispersed among the chromosomes.

## HOMEOBOX GENES

For multicellular organisms to develop from a single cell to the adult requires the turning on and off of a preprogrammed set of genes. In other words, selective or differential transcription must occur, and it must occur at precise moments during development. Eventually this development will lead to the production of the many specialized tissues and organs of the adult—for example, of skin, muscles, nerves, brain, heart, blood cells, and so on.

But first, the organism, via selective transcription and cell differentiation, begins to divide into major regions—head, trunk, arms, legs. So, as development proceeds, individual cells begin to take on different sizes, shapes, functions, and potentials. The cells also develop specific spatial relationships with one another. In vertebrates, more than 200 cell types can be distinguished.

*table 10.7*

SOME POLYPEPTIDE MULTIGENE FAMILIES

| GENE FAMILY | ORGANISM | APPROXIMATE FAMILY SIZE (NUMBER OF GENES) | LINKED (L) OR DISPERSED (D) | CHARACTERISTICS |
|---|---|---|---|---|
| Acid phosphatase | Yeast | ≥4 | l, d | Individual genes are regulated in diverse ways (e.g., one is repressed by Pi, one is induced by Pi, another is constitutive) |
| α-amylase | Mouse | ≥3 | l | Multiple genes are expressed in pancreas, one in liver and salivary gland. Different mouse strains have variable numbers of pancreatic α-amylase genes |
| | Rat | ≥9 | ? | 4 encode pancreatic amylase |
| | Barley | ≥7 | ? | At least 3 functional genes |
| Collagens (interstitial) | Human | 4 | d | Interstitial collagens, types I (a heterotrimer), II, and III (homotrimers), are encoded by 4 genes dispersed to chromosomes 2, 7, 12, and 17 |
| Myosin heavy chain | Rat | ≥10 | | At least 1 gene for each type of muscle; at least 1 nonmuscle gene |
| Ovalbumin | Chicken | 3 | l | All 3 regulated by estrogen; the function of 2 is unknown |
| Ribulose-1,5-bisphosphate carboxylase, small subunit | Tomato | 5 | l, d | Expressed in different tissues at different times in development, dependent on light |
| Serum albumin | Mouse | 2 | l | Serum albumin and α-fetoprotein |
| Vitellogenin | Toad | 4 | ? | Pairs of genes diverge 20% from one another; within pairs, the genes are 5% divergent; all 4 are expressed |
| Zein | Maize | 100 | d, l | Encode a family of kernel storage proteins |

*Note:* Pi stands for inorganic phosphate. Linked (l) means that two or more genes are clustered in one locus, and dispersed (d) means that family members are in different loci. In some families, there are both linked and dispersed copies.

From Berg and Singer, *Genes and Genomes*, University Science Books, Sausalito, CA. Reprinted by permission.

What determines the pattern of development is the timing of activation and inactivation of spatially distinctive patterns of genes. This timing ensures the orderly development of the organism.

There is a hierarchical expression of genes. Some genes are sensitive to the nature of neighboring cells, and others are sensitive to the patterns of previous gene expression. Genes of individual cells are controlled by the products of other genes. In other words, the product of one gene or set of genes controls the expression of one or more other genes. The aim of the work in developmental biology and genetics is to bring together the timing of gene expression and the changes that occur in early embryonic development with the expression patterns of particular genes.

Much of this work has been done with the fruit fly *Drosophila melanogaster* and more recently with a roundworm (*Caenorhabditis elegans*), the mouse, and the South African clawed frog (*Xenopus laevis*).

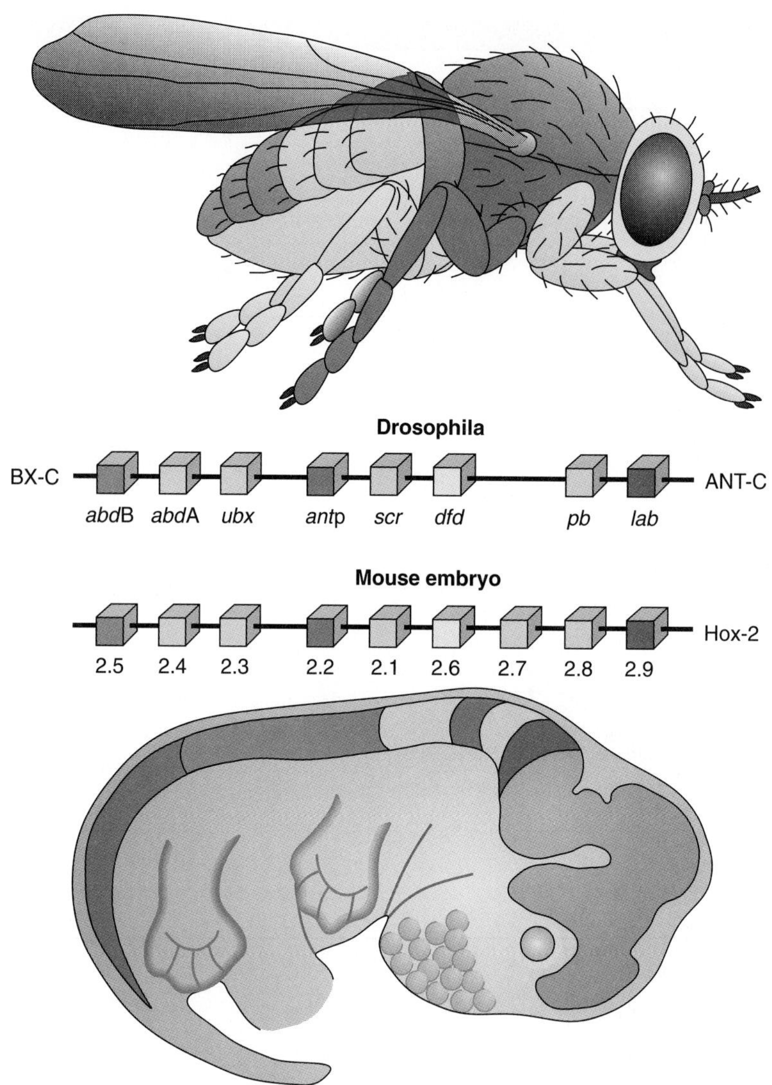


FIGURE 10.11
Homeobox genes and the head-to-tail development of *Drosophila* and mouse. The two organisms show same similarities in the genes that control some body segments. Colored boxes are related genes.

Early classical geneticists, using *Drosophila,* were able to produce mutants in which particular body parts developed in the wrong segments. For example, legs could be made to grow on the head where antennae ought to grow. The geneticists named the genes responsible for these mix-ups **homeotic genes.** The word *homeotic* derives from the Greek *homoio,* meaning "similar." The structures are produced just as the normal structures would be but in the wrong location.

In the *Drosophila* genome, two clusters of these homeotic genes are made up of 250,000 bp on chromosome 3. One cluster makes up what is called the **antennapedia complex (ANT-C),** which controls development of the head and three segments of the thorax. The other cluster, known as the **bithorax complex (BX-C),** regulates development of the eight abdominal segments and two of the thorax. A great deal of the anatomy of the fruit fly is controlled by these clusters. A mutation in the genes of these clusters can be severe enough to be lethal.

These genes in the fruit fly and the corresponding ones in the mouse divide the embryo into bands along a head-to-tail axis. Interestingly, these genes are found along the chromosome in relation to one another corresponding to where the particular gene is to be expressed along the head-to-tail axis (Figure 10.11).

The discovery of these genes in the nonsegmented, more primitive *C. elegans* has suggested to some investigators the possibility that the genes' function was at first to diversify cells along the head-to-tail axis and only later evolved into the genes' functioning to diversify body segments per se.

In the early 1980s, a number of homeotic genes were isolated and their nucleotide sequences compared. A startling discovery was made. Each gene contained highly conserved nucleotide sequences of about 180 bp that were virtually identical.

This conserved sequence is the **"homeobox."** The 180-bp sequence codes for 60 amino acids in the protein product. The 60 amino acids make up what is called the **homeodomain** of the proteins. The homeodomains of these regulatory proteins bind in the major groove of DNA. In this way, they influence the transcription of genes.

Amazingly, these same, almost identical, sequences have been found in fungi, amphibians, plants, and even human beings. In vertebrates, the homeotic genes occur in four clusters and have been given the name **Hox** (homeobox) **genes.** In the mouse, the genes are located on chromosomes 2, 6, 11, and 15. The human Hox genes are found on chromosomes 2, 7, 12, and 17. There may be as many as eight Hox genes in a vertebrate cluster.

The function of the protein products of homeobox genes is to regulate the activity of other genes. Specifically, they act as transcription factors that increase the expression of the target gene. In other words, the rate of selective transcription is increased.

This function, it turns out, is dependent on the homeodomain's recognizing a specific sequence of nucleotides on the gene being controlled. And, since these homeodomains have pretty much the same amino acid sequences, this means that the specific sequences of the target genes must be much the same. It has been found, in fact, that even though homeobox gene proteins differ greatly in their overall amino acid sequences, the homeobox amino acid sequences do not.

During development, different homeobox proteins are produced at different times as these genes respond to their own signals. The homeobox proteins bind to their target genes, which are thereby activated, facilitating cell differentiation and the development of the organism in a sequential pattern, from head to tail.

## METHYLATION OF GENES

As was noted in earlier chapters (see Chapter 2, Replication Fidelity), methylation of DNA at adenine and cytosine is very common in both prokaryotes and eukaryotes and in their viruses. Methylation is a process that occurs after DNA synthesis and is the result of the activity of specific enzymes, DNA methylases or methyltransferases. These enzymes transfer methyl groups from S-adenosylmethionine to adenines and cytosines at specific locations along the genomic DNA.

In *E. coli*, methylation, particularly adenine methylation, plays several roles (cytosine methylation is still not well understood):

1. Proofreading and repair of mismatched base pairs.
2. Methylation of oriC sequences is necessary to maintain synchronous initiation of replication at all sites and to bind the *ori*C site to the membrane. These steps may be necessary to apportion newly synthesized chromosomes to daughter cells.
3. Regulation of gene expression.
4. Methylation patterns distinguish host DNA from foreign DNA, which is degraded by restriction endonucleases. Phage infecting *E. coli* would have to either disable the host's protein-synthesizing machinery or destroy the host DNA. In either case, the virus methylation pattern becomes irrelevant or the phage may become a prophage, in which case its DNA is given the same methylation pattern as the host DNA.

In eukaryotes, however, methylation appears to have a much more restricted role. Its function is, primarily, to regulate gene expression and cell differentiation. For example, foreign DNAs (e.g., virus DNA) are not recognized and degraded, because eukaryotes lack restriction enzymes. DNA viruses infecting eukaryotic cells do not have methylated DNA, and, since eukaryotes do not have restriction systems, it would appear that methylation patterns play no role in protection against viral infection.

The sites of methylation in eukaryotes are also different from the sites in prokaryotes. Methylation of eukaryotic DNA, except in some unicellular organisms, is principally the modified base 5-methylcytosine. In fact, this is the only modified base in mammalian cells. Methylation patterns differ depending on the species and the tissue.

The evidence that methylation affects transcription is very strong. This evidence correlates methylation with gene inactivation and demethylation with transcription (however, the complete demethylation of a segment of DNA is not required for transcription). Of interest at this point is the fact that *Drosophila* DNA is not methylated, though, as we have seen, differentiation does take place. Methylation is, obviously, then, not a universal phenomenon.

A number of features of eukaryotic DNA methylation provide evidence of the correlation between methylation and transcription activity (Kornberg and Baker 1992):

1. Methylation of cytosines near the 5′ end of the gene changes the structure of DNA. Increased methylation is correlated with reduced transcription. The modification takes place close to, if not directly at, the promoter/operator sites. In fact, experiments have demonstrated that excess methylation can alter DNA at this site to such an extent that binding of repressor

protein is inhibited. Repressor binding normally inactivates gene expression, as does methylation, in general. But excess methylation changes the structure of DNA enough to prevent repressor binding, so DNA structure is critical. Just as excessive methylation affects the affinity of repressor protein for an operator site, so too, might excessive methylation affect the affinity of transcription factors for their binding sites on DNA.
2. Methylation of eukaryotic DNA is symmetrical on both strands, 90% of which are GC sequence base pairings.
3. Methylation patterns are somatically inherited (more on this shortly). That is, methylated sites on the parent or template DNA strand indicate which sites of the new strand are to be modified.
4. Methylation patterns differ among tissues of a given individual.
5. The cytidine analog 5-azacytidine is incorporated into DNA and can cause fibroblasts to differentiate into muscle cells and inactive X chromosomes to become active during mammalian cell differentiation. The explanation of these phenomena is that the analog is not recognized by the methylase and, as a result, it is not methylated; those genes that have incorporated the analog are, therefore, activated and transcribed.

## Fragile X Syndrome

An interesting effect of misdirected methylation is the fragile X syndrome, also known as Martin-Bell syndrome. The disorder gets its name, fragile X, from the fact that a region on the X chromosome fails to stain, giving the impression of a gap along the chromosome's length. The disease is characterized by mental retardation and is primarily sex-linked. The incidence of the disease is about 1/1500 males and 1/2500 females. These data have been difficult to explain. The disorder is second only to Down syndrome as a cause of mental retardation and occurs in all ethnic groups with equal frequency.

The fragile X site is located at Xq27.3. In a variety of banding procedures, including Q and G staining, the site appears as an unstained area or gap of varying size.

The disorder is one that is known as a genetic anticipation disease because it increases in severity as it is transmitted from one generation to the next. This increase in severity may be the result of an increase in gene size at location Xq27.3 due to the progressive amplification or duplication of a CCG sequence. At more than 52 copies of the CCG sequence, this section of the chromosome becomes unstable and can break. Mild retardation may occur when CCG is repeated between 50 and 200 times, with severe retardation occurring when the triplet has increased in number to between 230 and 1000.

In addition to size changes, the fragile X site has also been found to be relatively heavily methylated, which may not be surprising in view of the increased concentration of the CCG triplets there. This increased methylation may also result in decreased staining of the location and the appearance of a gap at the site.

## Inherited Methylation

Several investigators have suggested still another role for DNA methylation. It may be a mechanism by which gene activity patterns are inherited. This role would help to explain the order in which genes are to be transcribed once an egg has been fertilized by sperm. The ordering of gene transcription would be one control mechanism regulating cell differentiation. That is to say, during development, methylation may be responsible for initiating the transcription sequence of genes in individual cells so that the cells differentiate into particular types (i.e., nerve, muscle, skin, etc.). In the absence of this or another kind of regulation, all cells would follow the same developmental path and end up as identical cells, since all somatic cells have identical chromosomes (minus mutations, of course). Furthermore, information concerning this pattern of development must be passed from one generation to the next to ensure that the progeny will develop into the type of organisms their parents were.

Another way to look at this is to think of methylation as an inheritance of gene activities. Inheritance, then, would involve the transmission from one generation to the next of both genes and gene activity patterns. This idea is in contrast to arguments that organisms inherit not just DNA but a DNA-protein (nonhistone) complex, and it is the proteins that determine which gene pattern of transcription will be followed.

Methylation, per se, however, does not explain the mechanisms by which genes are switched on or off during development. That is, How is the protein-binding pattern necessary for transcription altered so that stem cells give rise to specialized cells?

Another intriguing idea concerning methylation has to do with cell aging and death. Here, it is proposed, the demethylation of certain genes, which ought to remain silent or inactive, results in their activation. That, in turn, upsets the cells' normal physiological processes and eventually leads to cell death.

Some experimental evidence can be interpreted to support such a proposal. For example, human cells are now known to replicate for a given number of generations independent of chronological time. As the number of generations increases, so does the number of unmethylated sites. Further, young cells treated with the cytidine analog azacytidine, which cannot be methylated, die out earlier than untreated cells. The fact that azacytidine is also a very potent carcinogen indirectly suggests that demethylation may by a contributing factor in cancer.

## IMPRINTING AND X CHROMOSOME INACTIVATION

### Imprinting

Imprinting is the differential modification of maternal and paternal genes. The X chromosome inactivation refers to the selective inactivation of one of the two X chromosomes in females. Both of these processes are critical to mammalian development and show a strong link to methylation. We shall take a brief look at both phenomena.

An article of faith since the very beginning of the study of genetics has been that it does not matter whether a gene comes from an individual's mother or father. The gene's effect was thought to be independent of its source. This, however, turns out not to be the case. There is very good evidence that selective marking of genes does occur, and a gene's effect is dependent on its source. The exact nature of the marking and the mechanism by which it is brought about are still to be discovered. In the mouse, characteristic modifications of maternal and paternal genetic material are required for normal development.

A major difference has been found in the 5-methylcytosine content of egg and sperm DNA. For example, the results of transgene experiments in mice, in which specific genes are introduced into newly fertilized eggs, demonstrate that genes from the mother and father are expressed and methylated differently. A gene derived from the father was expressed, but only in heart tissue; the same gene from the mother was not expressed at all, regardless of the tissue. Susceptibility to the endonucleases *Hpa*II and *Msp*I used as a measure of methylation at 5′ CCGG 3′ sites indicated that the transgenes taken from the father were hypomethylated and expressed, as noted above, while the maternal transgenes were hypermethylated and not expressed.

Imprinting must occur either before or during gamete formation (gametogenesis); exactly when is not known. And it must be stably inherited through many somatic generations. The germ line cells are not methylated until the time of egg and sperm development, when the methylation pattern for that particular organism is reestablished in the gametes.

When a zygote is formed, but before the embryo is implanted, both sets of chromosomes (maternal and paternal) are partly demethylated. Then, as fetal development takes place, chromosomes are remethylated, but maternal and paternal chromosomes are methylated in different patterns.

### X Chromosome Inactivation

Inactivation of the X chromosome in females appears to occur via methylation. This idea is supported by the fact that the inactivation can be reversed by treatment of the cells with 5-azacytidine, which is incorporated into newly synthesized chromosomal DNA but is not methylated. Inactivation seems to happen in three stages:

1. The X chromosome to be inactivated is selected by some unknown process.
2. Inactivation proceeds along the entire length of the chromosome.
3. The inactivation is somehow stabilized so that the same X chromosome remains inactive during subsequent cell divisions.

Inactivation ensures an equal contribution of X chromosome information in normal females (XX) and normal males (XY). In females, inactivation does not occur at the same time in all cells, nor is the same X chromosome always inactivated in different cells. For example, at about the fourth day after implantation, the paternal X chromosome in cells destined to be extraembryonic tissue is inactivated; after the sixth day, the female fetus develops as a mosaic because the X chromosomes inactivated are selected at random.

Individuals having more than two X chromosomes (i.e., aneuploidism) inactivate all but one. In diploid females and aneuploids the inactive X chromosome appears in cells as the Barr body, a highly condensed heterochromatic structure. Of some interest is the fact that active X chromosomes are replicated early in the S phase, whereas inactive X chromosomes are replicated late.

Inactivation is a very stable state. Reactivation of particular regions of the chromosome does occur rarely, but no examples of complete chromosomal reactivation are known, except in germ cells prior to meiosis.

A word of caution, however, is in order concerning methylation in imprinting and X chromosome inactivation. Whether methylation is a cause or an effect of these two processes still remains to be settled. Overwhelming experimental proof that methylation is a cause is lacking.

## DNA-DIRECTED RNA POLYMERASES I, II, AND III

In Chapter 8, we saw that prokaryotes make do with a single type of DNA-directed RNA polymerase for the manufacturing of the three species of RNA needed for protein synthesis: mRNA, tRNA, and rRNA. Eukaryotes, on the other hand, use three different RNA polymerases (except in their mitochondria, where one type is used): a different one for each type of RNA (Table 10.8). These enzymes are

*table 10.8*

**THE THREE EUKARYOTIC RNA POLYMERASES**

| POLYMERASE | RNA PRODUCT |
|---|---|
| I | rRNA precursor (18S + 28S + 5.8S) |
| II | mRNA precursor |
| III | tRNA precursor; 5S rRNA precursor |

also structurally more complex than the prokaryotic enzyme. And the eukaryotic genes transcribed by the different polymerases are sometimes referred to as Class I, II, or III genes, corresponding to the polymerases responsible for the transcription.

The transcription processes in which these polymerases are involved are more complex than in prokaryotes. For example, promoter regions must first be recognized by transcription factors; there may be dozens of different factors for different promoters in a human cell. These factors may in turn be bound by other proteins that facilitate the binding of the polymerases. The promoter sites themselves are not necessarily located immediately upstream (i.e., in the 3′ direction) of the transcriptional start site. In fact, they may be located as far as several thousand base pairs downstream (i.e., in the 5′ direction) of the transcriptional start site. The nuclear envelope of eukaryotic cells separates, or uncouples, eukaryotic transcription and translation. The nucleus is the site of transcription, and translation occurs in cytoplasm. And, last, the **primary RNA transcripts** are processed in the nucleus and then passed through the nuclear membrane to the cytoplasm as mRNAs, tRNAs, or ribosomal subunits. Let us now take a brief look at each of the three eukaryotic RNA polymerases.

## RNA Polymerase I

The enzyme RNA polymerase I is responsible for the synthesis of rRNA, which accounts for nearly half the total RNA found in the cell. The enzyme is tightly regulated so that ribosome synthesis keeps pace with the cell's protein requirements for growth, development, and division. RNA polymerase I can be distinguished from the other polymerases by its insensitivity to the potent mushroom poison alpha-amanitin. The complete enzyme includes two large polypeptide subunits and, depending on the source, from 4 to 10 smaller subunits. Some of these smaller subunits are common to the other two polymerases. The polymerase requires at least two **transcription factors** for activity. These are needed for binding of the polymerase at the promoter site and to initiate transcription.

## RNA Polymerase II

The enzyme RNA polymerase II produces all the pre-mRNA of the cell and is thus responsible for the transcription of the largest part of the genome. Polymerase II is sensitive to alpha-amanitin. It has been the most intensively studied of the three polymerases. The enzyme is composed of two large polypeptides and from 6 to 8 smaller polypeptides.

This polymerase recognizes three different elements of a gene:

1. A selector sequence containing a TATA box and a start sequence: the TATA box determines the start site, which is often found about 25 bp downstream, though start sites as far as 110 bp downstream of the box are known.
2. An upstream promoter sequence.
3. An enhancer sequence, which may be located at different sites in different genes: enhancers increase the rate of transcription presumably by allowing more-efficient binding of the polymerase to the promoter.

A large number of transcription factors used by polymerase II have been isolated from a variety of sources. These transcription factors are necessary for the specific binding of the enzyme to the DNA promoter and require other factors (protein-protein interactions), and the transcription factors are also necessary to initiate transcription. For example, in human cells, the following series of events has been deduced. First, TFIID (TF = transcription factor; the Roman numeral indicates the polymerase for which the factor is required) binds DNA at the promoter region; this association is then stabilized by the addition of TFIIA to the complex. The complex, TFIID-DNA-TFIIA is then recognized by polymerase II in the presence of other initiation protein factors. At least one of these other protein factors, TFIIH, has ATPase activity, which is required to phosphorylate polymerase II. Phosphorylation presumably causes a conformational change that releases the enzyme from the other transcription factors so that RNA synthesis can begin.

## RNA Polymerase III

The enzyme RNA polymerase III transcribes not only tRNA genes but also those genes for 5S RNA found on the large (60S) ribosomal subunit of eukaryotes and genes whose RNA end products (e.g., UsnRNAs) assist in the processing of pre-RNAs by spliceosomes (to be discussed later).

Polymerase III is the most structurally complex of the RNA polymerases. In yeast, the complete molecule is made

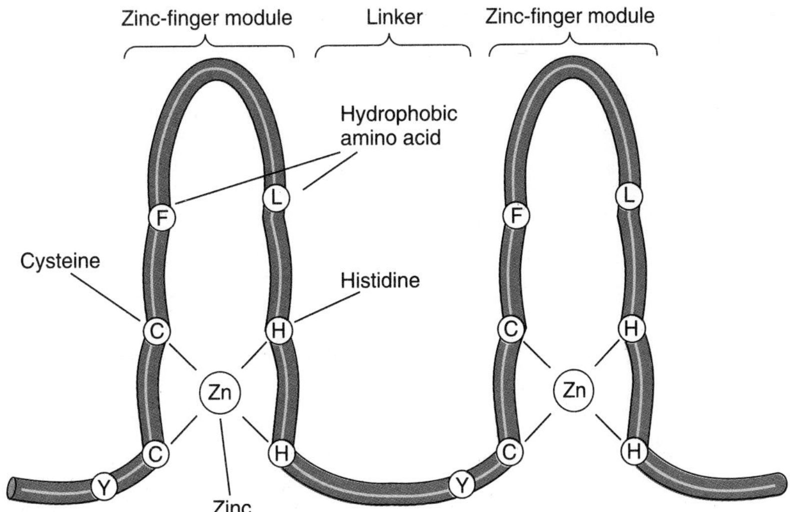


FIGURE 10.12
A representation of one of the two types of zinc fingers. The cysteine and histidine residues within the fingerlike projections complex a zinc ion.

up of 14 distinct polypeptides. As with the other enzymes, it has two large polypeptides associated with smaller subunits. A number of the smaller units are common to the other polymerases.

## REGULATORY PROTEINS

As we noted above, transcription (and DNA replication) requires DNA-binding proteins. These proteins function to activate or inactivate gene expression. There are four well-characterized types of structural patterns found among the hundreds of DNA-binding proteins. These types are: the helix-turn-helix, two kinds of zinc fingers, and the leucine zippers. Each type will be briefly described. Keep in mind, however, these are not the only motifs of protein structure found to affect transcription.

### The Helix-Turn-Helix Proteins

The **helix-turn-helix** types of positively charged regulatory proteins are represented by prokaryotic repressors and activators such as those of the *lac* and *trp* operons (e.g., repressors and CAP) and the lambda phage proteins (e.g., Cro and repressor).

The eukaryotic proteins are represented by those involved in regulation of the homeogenes of *Drosophila* and the genes encoding prolactin and growth hormone in mammals. Their designation, helix-turn-helix, derives from their three-dimensional structure. The molecules have two successive alpha helices (see Chapter 8) held at right angles to one another by a short sequence of amino acids. One of these helices fits the major groove of DNA. A given protein may have several helices so that successive turns of the double-stranded DNA can be occupied.

### Zinc Fingers

**Zinc fingers** are metalloproteins that have been found in nearly all eukaryotes. More than 100 such kinds of DNA-binding molecules have been identified. The first of these proteins to be recognized and studied was the RNA polymerase III transcription factor TF-IIIA, an activator of the 5S RNA gene (the Roman numeral indicates the polymerase for which the factor is required). This particular protein has nine zinc fingers. For different proteins, the number of fingers varies from 2 to 37.

These molecules get their name from the fingerlike projections of amino acid sequences that extend from the surface of the protein (Figure 10.12). Each finger contains at least one zinc ion. The fingers are thought to form alpha helices and beta sheets that allow the protein molecules to bind the major groove of DNA and thereby to affect transcription.

The two types of zinc fingers can be distinguished by the amino acids bound to the zinc ion: either there are a pair of cysteines on one side and a pair of histidines on the other or all four binding amino acids are cysteine (Figure 10.12). This last type of zinc finger is a common feature of yeast transcription factors and mammalian steroid receptor proteins that react with DNA after complexing with a steroid hormone.

Since the structure of all zinc fingers is essentially the same, what confers specific DNA binding must be the

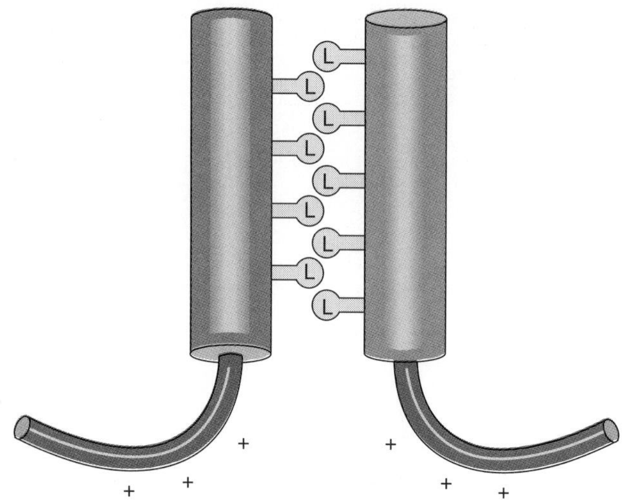


FIGURE 10.13
A representation of a leucine zipper. The leucines on α-helices of two proteins form the halves of the zipper. When the zipper is closed the positively charged regions of the proteins can then bind DNA.

sequence or arrangement of the fingers in each different protein so that different proteins bind different segments or sequences of DNA.

## Leucine Zippers

The **leucine zipper** proteins are so named because they have an alpha-helix domain containing four or five leucines spaced exactly seven amino acids apart (Figure 10.13). Pairs of these proteins can interact to form a dimer, which then positions positively charged domains of the protein to react with the negatively charged DNA.

Leucine zippers have been found that also contain zinc fingers. Presumably, the zipper binds the DNA in such a way that the fingers are then in position to fit the DNA's major groove. In other words, zippers and fingers have different functions.

The leucine zipper design appears to be of ancient origin. Two tobacco plant zippers have been found to be almost identical to those of animal zippers. This similarity suggests that leucine zippers may have had a common ancestor.

# PROCESSING RNAs

Each of the RNAs manufactured by eukaryotic cells is produced as a precursor molecule that must then be processed into the final product. Because of the existence of introns or the manner in which the genes are transcribed the RNAs must be modified. Table 10.9 gives some idea of the variety of RNAs the cell processes.

We shall take a brief look at the processing of each of the major types of RNA in the order corresponding to their respective polymerase number (e.g., RNA polymerase I—rRNA).

## rRNA

In the early 1980s, Thomas Cech and his colleagues made a startling discovery while studying an RNA molecule from the protozoan *Tetrahymena thermophila*. The RNA molecule was capable of cutting and splicing itself. Cech coined the word *ribozyme* for this enzymelike RNA. This was the first discovery of a nonprotein molecule with enzyme activity. One important implication of this finding is that in the earliest life forms, RNA may have served as both a genome and a catalyst (i.e., enzyme). This suggestion avoids the chicken-or-the-egg problem; that is, Did genes come first, or did proteins (enzymes)? Another interesting question, and one much more important than the first, is whether RNAs found in ribosomes have catalytic activity during protein synthesis (i.e., during translation).

The ribozyme of *Tetrahymena* is the large precursor RNA transcript of the rRNA genes. This molecule contains the RNA of the 40S and 60S ribosomal subunits, 17S and 26S RNAs, respectively, and a 5.8S RNA that becomes a part of the larger ribosomal subunit. (These RNAs are slightly smaller than the RNAs of other eukaryotes, 18S and 28S, respectively). In the precursor RNA, the 17S and 26S RNAs are joined together by a sequence of spacer RNA: when this spacer RNA is removed during processing, the 17S and 26S molecules are separated. Within the 26S sequence, there is an intron that must also be removed, and then the two segments must be spliced together. This splicing process is said to be autocatalytic because it occurs in the absence of any enzyme; the rRNAs are self-splicing.

The self-splicing 26S RNAs are divided into two groups based on the nucleotide that initiates the splicing and whether a so-called **lariat** configuration is formed during the splicing procedure.

Group I intron splicing requires a monovalent cation, a divalent cation, and a guanosine in the form of GMP. These introns are attacked at their 5′ end by the addition of the GMP, which breaks the exon-intron junction. Then the 3′ end of the exon attacks the 5′ end of the next exon, releasing the intron and joining the two exons. No lariat is formed.

Group II introns are removed when an adenine nucleotide that is a part of the intron's 3′ end is used to form a lariat and to remove an intron. The exons are then spliced together. Figure 10.14 summarizes the two methods of removing introns, and Figure 10.16 shows lariat formation in more detail.

## mRNA

In the nucleus of eukaryotes at all times, there is a class of large RNA molecules of various sizes known as **heterogeneous nuclear RNA (hnRNA).** These molecules have been shown to be the precursors of mRNAs. Their size variation is, of course, the result of the transcription of genes of various sizes. (Remember that eukaryotic transcription is most often monocistronic: one gene for each mRNA).

table 10.9

SOME IMPORTANT CELLULAR RNAS

| | APPROXIMATE NUMBER OF DIFFERENT KINDS IN CELLS | APPROXIMATE LENGTH (IN NUCLEOTIDES) |
|---|---|---|
| Transfer RNA | 40–60 | 75–90 |
| 5S ribosomal RNA | 1–2 | 120 |
| 5.8S ribosomal RNA* | 1 | 155 |
| Small ribosomal RNA | 1 | 1600–1900 |
| Large ribosomal RNA | 1 | 3200–5000 |
| Messenger RNA | thousands | vary |
| Heterogeneous nuclear RNA* | thousands | vary |
| Small cytoplasmic RNA | tens | 90–330 |
| Small nuclear RNA* | tens | 58–220 |

*Occurs only in eukaryotic cells.
From Berg and Singer, *Dealing with Genes: The Language of Heredity*, University Science Books, Sausalito, CA. Reprinted by permission.

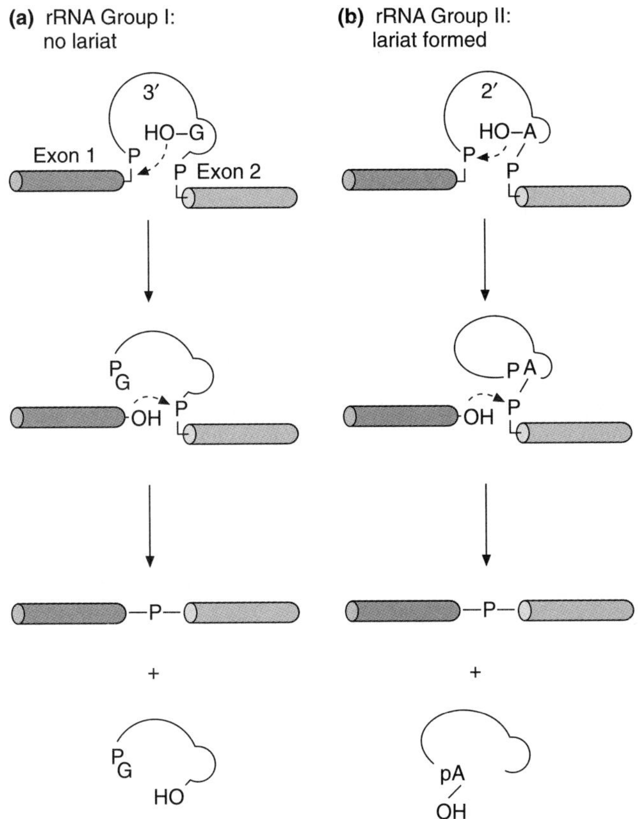


FIGURE 10.14
A summary of the self-splicing methods of the 26S RNA of *Tetrahymena*. Group I RNAs (a) do not form a lariat structure, whereas Group II RNAs (b) do.

When a gene is transcribed, its introns and exons are transcribed into a precursor mRNA molecule. This RNA is then processed to remove the introns, and then it is modified before passing through the nuclear membrane into the cytoplasm, where translation will take place (Figure 10.15).

Much as computers use a "cut-and-paste" program as an editing function, so too, the cell cuts and pastes precursor mRNAs to get mature mRNAs that are nearly ready for translation (some further modifications of the mRNA are needed as we shall see in the next section).

It is important to point out here the extremely critical nature of accurate cutting and pasting. As we saw in Chapter 9, single-base additions or deletions of gene nucleotides can result in garbled messages and faulty proteins (mutations). The same would happen if, in the process of removing introns, a base were to be inadvertently left in or removed; a reading frame shift would occur. So the importance of accuracy cannot be overemphasized! As it turns out, almost all introns, in precursor RNA have a GU at their 5′ end and an AG at their 3′ end. Figure 10.16 shows how introns are removed and exons are spliced in the lariat model.

Eukaryotes have other splicing mechanisms. For example, all or part of an exon can be treated as an intron. This mechanism allows a number of different mRNAs, and therefore different proteins, to be produced from a single gene. (There will be a unique amino acid sequence for each protein, because what has taken place is equivalent to a reading frame shift.) In this way, a cell can make more-efficient use of the limited amount of DNA it has. (Compare this arrangement with that of the overlapping genes of the ØX174; Chapter 3). In fact this technique, known as alternative splicing, can be regulated in such a way that at different times in different tissues a gene can be expressed (mRNA produced) in several different ways depending on the developmental stage. Female *Drosophila*, for example, are produced only if a given protein (SX1) is made (Figure 10.17).

SX1 binds one splicing site of a precursor RNA but not another, so the second splicing site is available. Splicing at this second site removes a larger nucleotide sequence, and the resulting mRNA codes for another protein that allows female adults to develop.

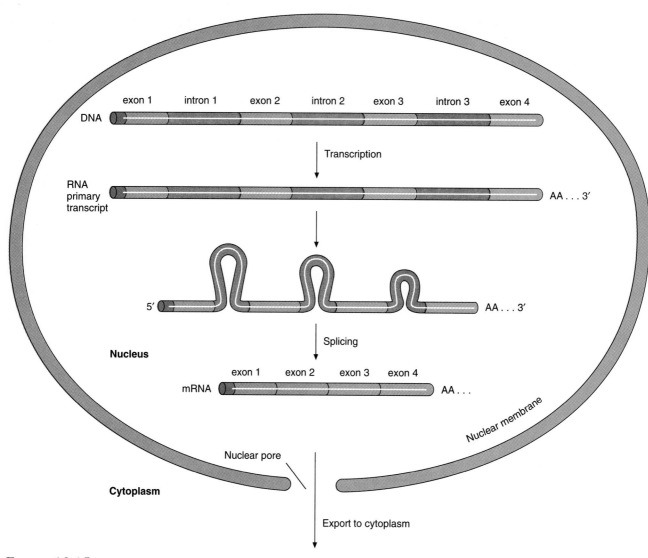

FIGURE 10.15
Splicing of hnRNA to produce mRNA by removal of introns. Capping the 5´ end and polyadenylation of the 3´ tail are also shown.

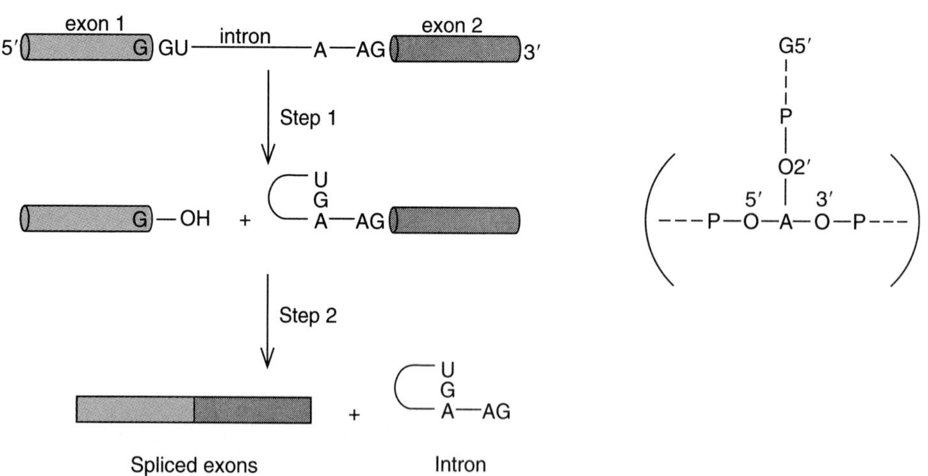


FIGURE 10.16
Lariat formation in the splicing of mRNA. The G at the 5´ end of the intron attaches an A within the intron to form the lariat at the intron's 3´ end. G attaches to the A at its 2´ OH position.

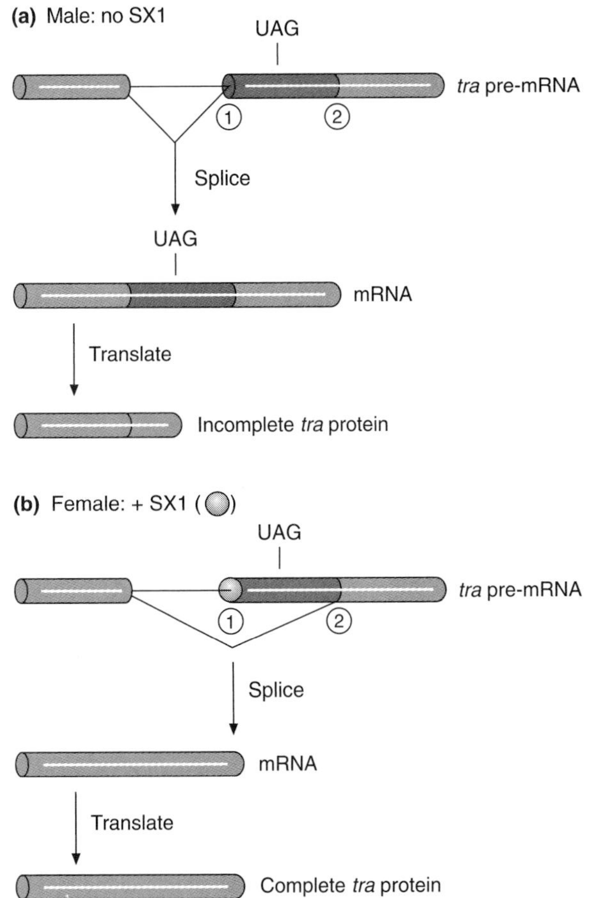


FIGURE 10.17
Regulation of sex determination in *Drosophila*. The product of the sex lethal gene (SX1) governs splicing of its own pre-mRNA and that of another gene's pre-mRNA-transformer (tra). The tra pre-mRNA can be spliced in two ways: (1) an incomplete tra protein results and (2) a complete tra protein can be made. In males, splicing occurs at site 1, no SX1 is available to block site 1. In females, SX1 is produced, which blocks site 1, so site 2 is used as the splice site. The mRNA formed is translated into complete tra protein. The fruit flies are female when complete tra protein is made. Maleness in the fruit fly is a result of the absence of complete tra protein. Note that the termination signal, UAG, is still present in the male mRNA. This is what causes premature stopping of translation. The codon is spliced from female pre-mRNA.

If the SX1 splicing site is free because no SX1 protein is made, splicing will occur there. A smaller nucleotide sequence is removed, leaving as part of the resulting mRNA a premature stop sequence (UAG); upon translation, an incomplete protein is made that is nonfunctional. As a consequence, a male fruit fly develops.

## *Splicing Precursor RNA*

The generation of nuclear mRNA from precursor RNA requires the activity of a structure called a **spliceosome.** These structures contain special proteins as factors and a ribonucleoprotein complex: **small nuclear ribonucleoproteins,** or **snRNPs** (pronounced "snurps").

The snRNPs of mammalian cells contain RNA molecules rich in uracil and its modified forms and have been designated URNAs. These molecules vary in size, uracil content, structure (e.g., single-stranded or base paired with one another), and activity (e.g., some are also needed for rRNA splicing; others have been implicated in the polyadenylation of mRNA—more on this later). The RNA molecules are identified by the number following the U, e.g., U1, U2, U3, and so on. The snRNPs needed for splicing hnRNA contain U1, U2, U5, and U4 and/or U6 and a common set of seven proteins plus several proteins characteristic of the type of snRNP. The complete spliceosome is about the size of the large ribosomal subunit, 50–60S.

Interestingly, the nucleotide sequences of the corresponding URNAs from vertebrates have some 95% homology. And even URNAs from diverse organisms such as yeast, sea urchin, cockroach, and *Tetrahymena* show a surprising degree of homology, though sequences distinct for each species are also present in the molecules.

Splicing requires the assembly of the spliceosome in an ordered sequence on the intron that is to be removed. Recognition of the intron occurs via the consensus sequences at the 5′ and 3′ intron ends. The intron is bound by at least five distinct types of snRNPs and several proteins. Spliceosome formation is dependent on several kinds of molecular interactions: base pairing between RNAs, binding of proteins to RNAs, and protein-protein associations. An example of these types of interactions is the base pairing between the 5′ end of U1 and the 5′ end of the intron—an interaction that involves at least 17 nucleotides. U1 is not directly a part of the spliceosome, however. Its role, though vague, may be to "prepare" the intron to receive the components of the spliceosome as it is assembled on the intron. Another example occurs at the intron's 3′ end, U5 and U4/U6 attach also via base pairing, and the binding is stabilized by protein interaction. Then hnRNA is converted to mRNA by intron removal within the assembled spliceosome—one spliceosome per intron.

At this point we have seen three different methods of intron removal: Group I, Group II, and hnRNA.

# tRNA

## *tRNA Introns*

The tRNAs found in cells are generally derived from a gene transcript precursor molecule containing more than one tRNA molecule. Introns of these transcripts are found to vary in size from 14 to about 60 ribonucleotides, but they always occur on the 3′ side of the anticodon (Figure 10.18).

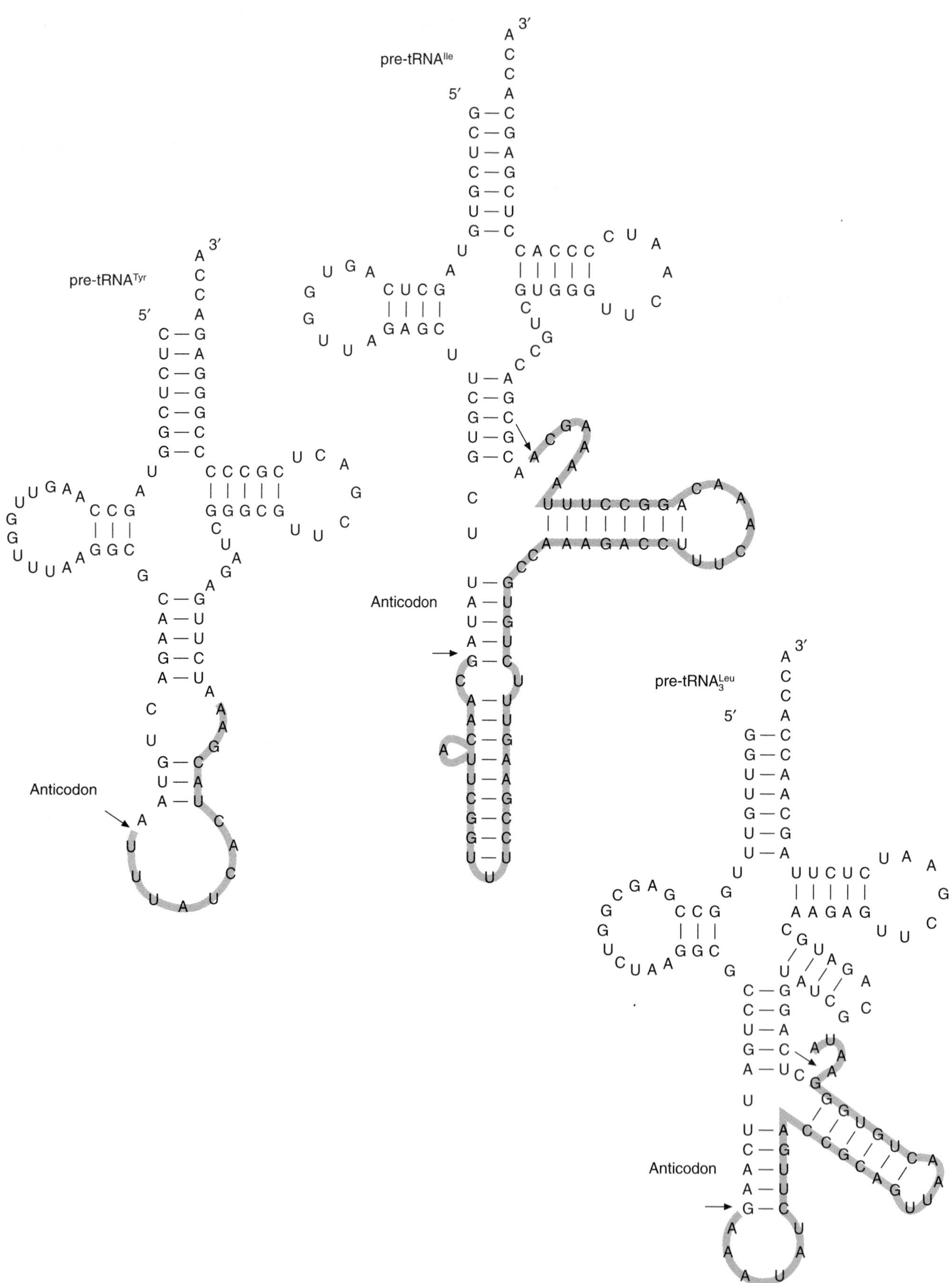


FIGURE 10.18
The introns of three different tRNAs. Introns occur after the anticodon triplets. The arrows indicate the splicing junctions.

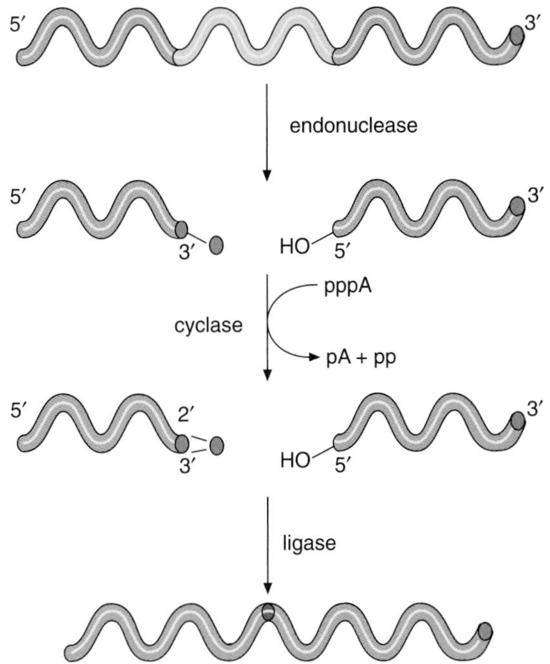


FIGURE 10.19
Splicing of tRNA in vertebrates.

All the enzymes from yeast cells needed to remove introns are known and have been highly purified. Three enzymes are required to excise introns:

1. a specific tRNA endonuclease
2. a cyclic phosphodiesterase
3. a tRNA-specific ligase

The reactions of the enzymes are shown in Figure 10.19.

### *Separation of tRNAs*

Although the separation of individual tRNA molecules from larger precursor molecules is best understood in *E. coli,* at least one of the four enzymes, RNAase P, needed for the separation of tRNAs in this bacterium has also been found in eukaryotes. So, in higher organisms, the processing of tRNAs may be very similar to that found in prokaryotes. The enzymes needed are:

1. RNAase P
2. an unidentified RNAase
3. RNAase D
4. tRNA nucleotidyl transferase

The procedure for producing mature tRNAs is shown in Figure 10.20.

RNAase P separates the individual tRNAs by digesting the 5′ leader and that stretch of ribonucleotides between the tRNAs. An unidentified RNAase removes a series of nucleotides from the 3′ ends of the separated tRNAs. Next the action of RNAase D releases bases at the 3′ end of the tRNA until a –CCA consensus sequence is reached. Type 2 tRNAs require the addition of the –CCA sequence, which is added by the tRNA nucleotidyl transferase.

A common feature of tRNAs is the high degree of base modification. Up to 20% of the bases are modified to produce such bases as psuedouridine, 4-thiouridine, 3-methylcytidine, 1-methyladenine, 7-methylguanosine, and others. Of course, not all organisms have all the modified bases, nor does any individual tRNA have all of the modifications.

## POSTTRANSCRIPTIONAL MODIFICATION OF RNAs

To acquire biological activity, the various RNAs of the cell must be modified in several different ways. Processing, then, does not end with intron removal. The RNAs can be modified by trimming, capping and methylation, and/or by polyadenylation.

### Trimming

Both rRNA and tRNA are modified by removing excess nucleotides, a process called trimming. In human beings, for example, the ribosomal RNA precursors contain 18S, 5.8S, and 28S segments embedded between spacer RNA sequences. Processing to remove these ribonucleotides takes place in the nucleolus. The steps to release the rRNA molecules are shown in Figure 10.21.

### Capping and Methylation

The 5′ end of mRNA is "capped" and methylated. The **cap** is the addition of a 7-methylguanosine group to this end and may also involve methylation of the second nucleotide at its 2′ hydroxyl group (Figure 10.22). In rare instances the 2′ hydroxyl group of the third nucleotide is methylated.

These modifications take place before the precursor mRNA is more than 20 nucleotides long. The cap is an important recognition site for eukaryotic translation initiation factor, eIF-4F, which has a cap-binding protein as one of its subunits. The binding of this protein to the cap then assists the binding of the mRNA to the 40S ribosomal subunit in preparation for complete ribosome assembly.

### Polyadenylation

Almost all eukaryotic mRNAs have a 3′ end to which 100–200 adenosines are enzymatically attached. Each added adenosine is derived from a molecule of ATP and is added to the tail by the enzyme **poly A polymerase.** This **polyadenylation** appears not to be required for mRNA translation but may serve to protect the mRNA from degradation.

As the mRNA ages, its **poly A tail** becomes progressively shorter. Unadenylated mRNAs have very short life

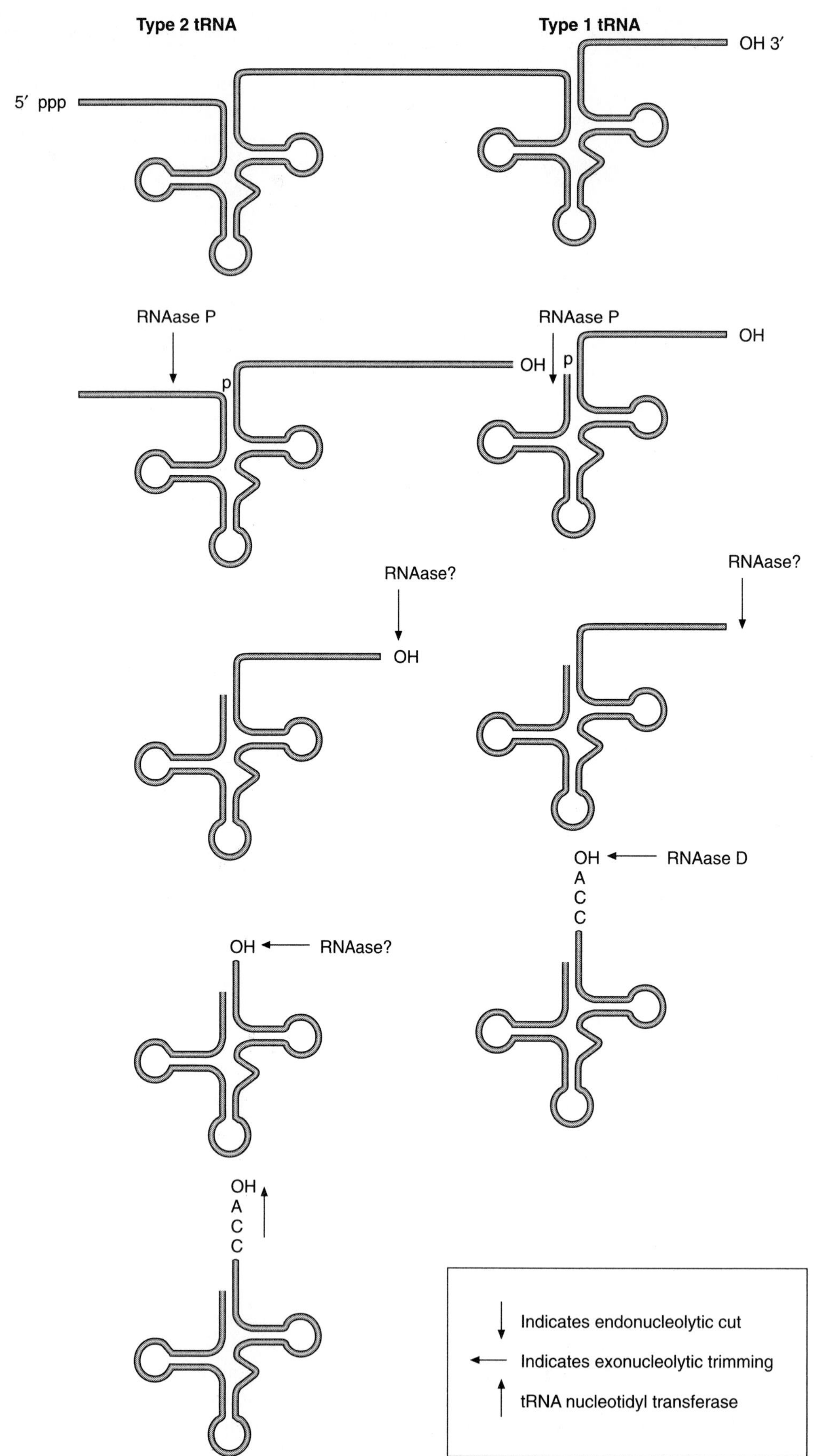


FIGURE 10.20
tRNA processing. The several enzymes needed are indicated. Type 1 tRNA already have the CCA triplet in place. Type 2 tRNA require the addition of the CCA triplet by the transferase enzyme.

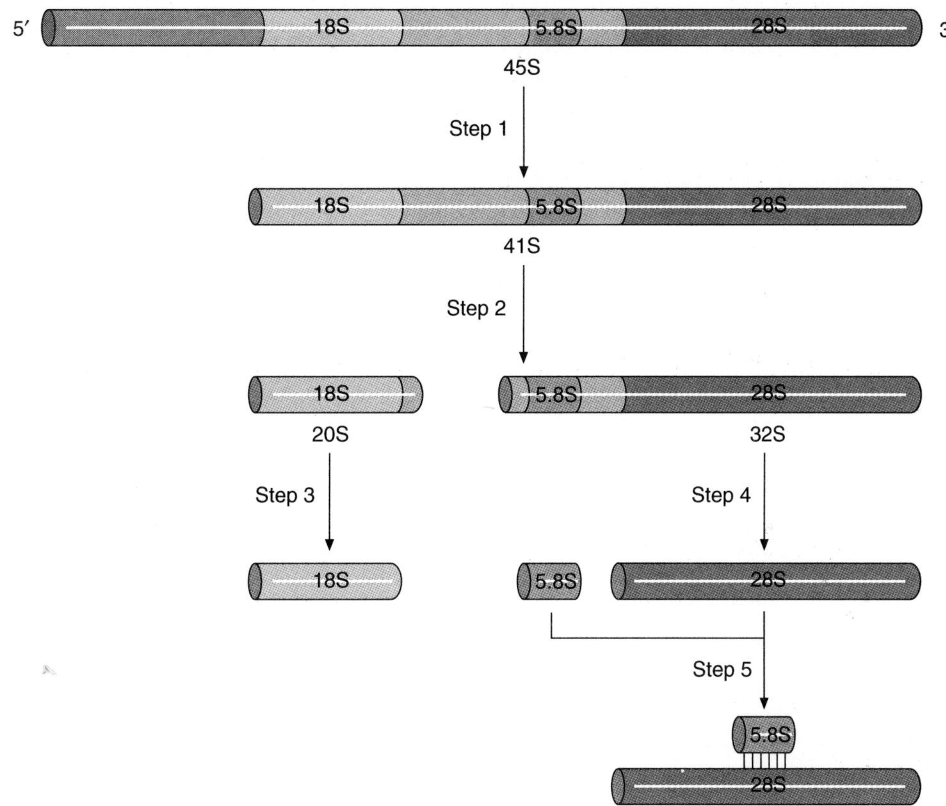


FIGURE 10.21
The processing of human rRNA to produce the RNA molecules for the ribosomal subunits.

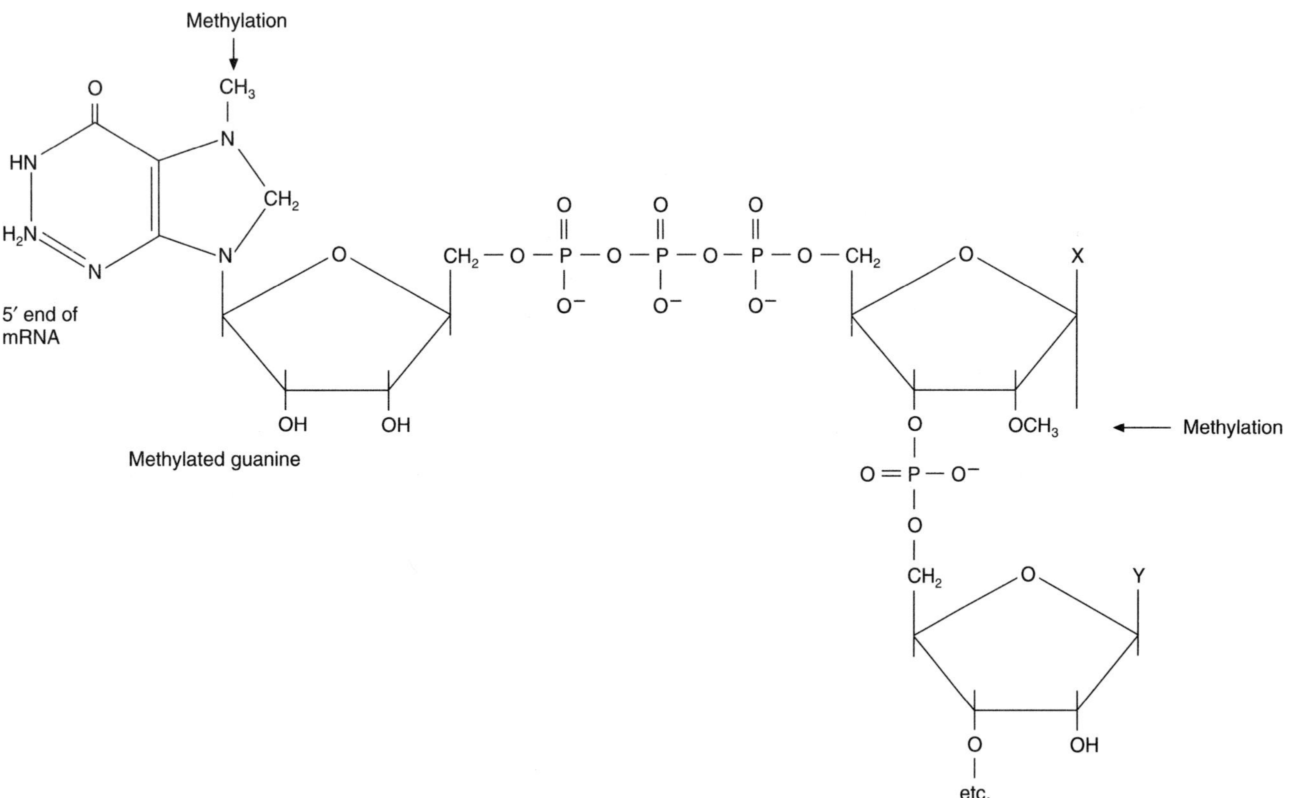


FIGURE 10.22
mRNA capping (guanine addition) and methylation at the 5′ end.

spans. Histone mRNAs, the only mature mRNAs that generally lack poly A tails, have life spans of less than 30 minutes. The vast majority of poly A–tailed mRNAs have life spans measured in hours and days. So polyadenylation may prevent degradation. Considering, once again, the energy expense in producing mature mRNA, the longer it lasts during protein synthesis the less energy the cell will have to expend making new mRNA: it is an efficient use of energy.

## IMMUNE SYSTEM GENETICS

The immune systems, of mammals in particular, are excellent examples of how eukaryotes are able to produce a rich diversity of protein molecules from a limited amount of genetic information. A detailed explanation of this ability is well beyond the scope of this book. In fact, the genetic complexity of the field is such that it has become a subspecialty of its own. We shall, however, take a brief look at the genetics of the immune system.

The **immune system** is designed to protect the individual against foreign materials, collectively called **antigens.** Antigens may be bacterial toxins (e.g., tetanus), bacteria and other cellular microorganisms, viruses, and even foreign tissue cells such as blood cells or lung and heart cells.

This protection can take one of two forms. In **humoral immunity,** special proteins called **antibodies,** or **immunoglobulins,** are produced. **Cellular immunity** involves specially altered lymphocytes. In both forms, the events leading to immunity occur only after activation by an antigen. Each antigen stimulates the production of a specific antibody (i.e., humoral immunity) or the alteration of lymphocytes (i.e., cellular immunity). The synthesis of antibodies is a variation on the theme of protein synthesis and requires all the stages of protein synthesis described in Chapter 8. All antibodies have a common four-chain structure: two light chains and two heavy chains.

The processes leading to antibody synthesis have been most intensively studied, so we will concentrate on those processes:

1. The antigen is first recognized as foreign by a group of special cells called antigen-presenting cells (APCs) located throughout the body.
2. The APCs ingest and process the antigen and present it to subsets of lymphocytes called T cells and B cells. The T cells are stimulated by antigen to produce chemical signal molecules called interleukins. The interleukins stimulate the B cells that were presented the same antigen by APCs.
3. The stimulated B cells first proliferate and then differentiate into plasma cells.
4. These plasma cells then synthesize, in huge quantities, antibodies that will react with the original antigen.

Each of these steps or stages requires the selective transcription of genes. In step 1, APCs must make and display surface proteins that react with foreign materials, and they make the enzymes to process the antigen. In step 2, antigen presented to T cells results in the induction of a set of genes whose protein end products are the interleukins, chemical signals. (Recall the induction of the *lac* operon brought about by lactose.) In step 3, the interleukins signal the B cells to proliferate and differentiate. Each of these complex processes requires the activation of new sets of genes. One set is needed for proliferation (e.g., for new DNA synthesis and cell division), and a different set is needed to get the B cells to differentiate into another type of cell, the plasma cell. Finally, in step 4, the genes responsible for proliferation and differentiation must be inactivated, and the genes for antibody synthesis must be activated, or turned on.

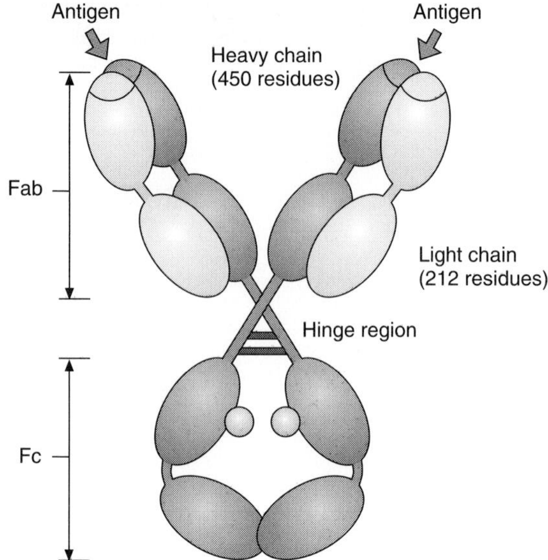


FIGURE 10.23
The structure of the most prevalent antibody: IgG. The molecule is composed of two identical light chains and two identical heavy chains. The Fab section contains the antigen binding sites, while the Fc section contains the cell binding site.

Depending on antigenic stimulation, a virtually endless variety of different antibodies can be produced. Estimates of the number of different antibodies a human being is capable of making range well into the hundreds of thousands if not into the millions.

At any given time, the body is exposed to and is reacting to numerous antigens, so it is continuously producing a variety of immunoglobulins. All this diversity of antibody molecules results from the various amino acid combinations that can be built into these proteins. This variation, of course, suggests that the genes of antibody-producing cells can be recombined in novel ways in response to antigen stimulation. Figure 10.23 is one example, among countless others, of how this rearrangement might occur to produce one of the polypeptide chains of the multichain antibody molecule.

## Summary

Though, in general, eukaryotes of a more complex nature, such as human beings, contain more DNA than less-complex organisms, such as *Drosophila*, there are exceptions. All organisms appear to have more DNA than seems necessary, causing one to wonder about the function of so much "extra" DNA. This extra DNA among lower eukaryotes, in particular, is referred to as the C-value paradox. The variations in genomes among eukaryotes include the total amount of DNA and the size and number of chromosomes.

The chromosomes are composed of DNA and proteins. Each chromosome is divided by a centromere—to which the mitotic spindles attach—into a long arm, designated the q arm, and a short arm, designated the p arm. The ends of the chromosomes are capped by special sequences called telomeres.

Various stains produce banding patterns in individual chromosomes. Lightly stained bands are composed of euchromatin-containing genes, and darker bands are composed of heterochromatin, which is normally nontranscribable DNA. Bands are used as a means of localizing individual genes.

The proteins found associated with DNA are histones and nonhistone proteins. There are five classes of histones, all of which are positively charged owing to their amino acid content. The nonhistone proteins are all the other proteins associated with chromosomal DNA. These proteins serve regulatory, enzymatic, and stabilizing functions.

The huge size of the chromosomal DNAs relative to the nucleus in which they are found necessitates a condensation and packaging of the molecules. Histones are formed into octamers around which DNA is wrapped. This configuration produces the first level of packaging, called nucleosomes. Nucleosomes are then gathered into so-called solenoids, and, finally, the solenoids are brought together to form radial loops. This packaging reduces the length of the DNA by as much as 8,000–10,000 times, and it produces the chromosome's structure.

Eukaryotic genes vary in size from a few hundred nucleotides to hundreds of thousands of nucleotides, and they may occur on either strand of the DNA. The genes themselves are divided into coding (exon) and noncoding (intron) sequences. Genes code for RNA or protein end products.

In many, if not all, cases genes evolved from other genes. The protein products have structural and/or functional similarities. These related genes are grouped into gene families or superfamilies. They may be found throughout the genome. A number of special genes have been discovered in multicellular organisms whose functions are to control the anatomical development of the organisms. These genes have been given the name homeotic genes.

A general characteristic of both prokaryotic and eukaryotic DNA is that both are heavily methylated. But methylation serves different functions in the different organisms. In eukaryotes, cytosine is preferentially methylated, and methylation functions primarily to regulate gene expression and cell differentiation. Methylation has been implicated in at least one disorder, fragile X syndrome. Methylation is also thought, by some investigators, to be an important inherited mechanism by which gene-activity patterns are controlled in early embryonic development. It may also play a role in cell aging, imprinting (the differential modification of maternal and paternal genes), and X chromosome inactivation in females.

Eukaryotes, unlike prokaryotes, use three different RNA polymerases to transcribe the genes of mRNA, tRNA, and rRNA. Transcription of eukaryotic genes has been found to require an assortment of protein transcription factors. Some of these factors have been extensively studied. Different structural patterns are evident among these factors, including a helix-turn-helix motif, two kinds of zinc finger proteins, and the leucine zipper arrangement.

Each of the three types of RNAs produced by eukaryotes undergoes modifications before being used by the cell. These modifications include excising introns, altering bases, methylation, capping, polyadenylation, and trimming.

In the case of rRNA, an amazing discovery has been made. In at least one example from a protozoan, rRNA has been found to be autocatalytic; it is capable of self-splicing. In other words, this species of RNA has enzyme properties. These RNAs are called ribozymes.

The immune system of mammals is a good example of how a limited amount of genetic information can be rearranged in such a way as to provide an almost limitless assortment of proteins.

## Study Questions

1. Explain the C-value paradox and give two examples.
2. Describe the basic structural features of a chromosome, and discuss the role each feature plays in chromosome behavior.
3. Describe the G- and Q-banding techniques, and discuss what banding reveals about chromosomes.
4. Describe the functions of the nonhistone proteins associated with DNA.
5. How is DNA condensation achieved, and what roles do the histones play in this process?
6. What are some major differences between prokaryotic and eukaryotic genes?
7. What are multigene families and gene superfamilies?
8. What role do homeobox genes play in the development of multicellular organisms?
9. Describe the different roles of gene methylation in prokaryotes and eukaryotes.

10. How might methylation contribute to fragile X syndrome, embryonic development, cell aging, imprinting, and X chromosome inactivation?
11. What are the major differences in structure and function of the eukaryotic RNA polymerases?
12. Describe the structural features of the regulatory proteins and how these features help to control gene expression and repression.
13. Explain the RNA processings that lead to lariat formation, and explain how these differ from the nonlariat-forming process.
14. How are individual tRNAs produced?
15. Describe the posttranscriptional modifications of the three species of RNA.
16. Explain how vertebrates are able to produce great numbers of different antibodies using a limited amount of genetic information.

## Readings and References

Alberts, B., D. Broby, J. Lewis, M. Raff, K. Roberts, and J. D. Watson. 1994. *Molecular biology of the cell*. 3d ed. New York: Garland.

Baskin, Y. 1995. Mapping the cell's nucleus. *Science* 268:1564–65.

Berg, P., and M. Singer. 1992. *Dealing with genes: The language of heredity*. Mill Valley, Calif.: University Science Books.

Biessmann, H., and J. M. Mason. 1992. Genetics and molecular biology of telomeres. In *Advances in Genetics*, ed. J. G. Scandalios and T. R. F. Wright, vol. 30, pp. 185–249. New York: Academic Press.

Cech, T. R. 1994. Chromosome end games. *Science* 266:387–88.

Counter, C. M., H. W. Hirte, S. Bacchetti, and C. B. Harley. 1994. Telomerase activity in human ovarian carcinoma. *Proc. Natl. Acad. Sci. USA* 91:2900–04.

DeRobertis, E. M., G. Oliver, and C. V. Wright. 1990. Homeobox genes and the vertebrate body plan. *Sci. Am.* 263 July:46–52.

Development: Frontiers in biology. 1994. *Science* 266:513–700.

Greenspan, R. J. 1995. Understanding the genetic construction of behavior. *Sci. Am.* 272 April:72–78.

Grunstein, M. 1992. Histones as regulators of genes. *Sci. Am.* 267 October:68–74B.

Holliday, R. 1989. A different kind of inheritance. *Sci. Am.* 260 June:60–73.

Kenyon, C., and B. Wang. 1991. A cluster of antennapedia-class homeobox genes in a non-segmented animal. *Science* 253:516–17.

Kornberg, A., and T. A. Baker. 1992. *DNA Replication*. 2d ed. New York: W.H. Freeman.

Lawrence, P. A. 1991. *Making of a fly: The genetics of animal design*. Cambridge, Mass.: Blackwell Scientific.

Lewin, B. 1997. *Genes VI*. New York: Oxford University Press.

Meyerowitz, E. M. 1994. The genetics of flower development. *Sci. Am.* 271 November:56–65.

Moyzis, R. K. 1991. The human telomere. *Sci. Am.* 265 August:48–55.

Nilisen, T. W. 1994. Unusual strategies of gene expression and control in parasites. *Science* 264:1868–69.

Papadopoulos, N. 1994. Mutation of a *mutL* homolog in hereditary colon cancer. *Science* 263:1625–29.

Parker, P. J., and M. Katan, eds. 1990. *Molecular biology of oncogenes and cell control mechanisms*. New York: Ellis Horwood.

Ptashne, M. 1989. How gene activators work. *Sci. Am.* 260 January:40–47.

Rhodes, D., and A. Klug. 1993. Zinc fingers. *Sci. Am.* 268 February:56–65.

Roitt, I., J. Brostoff, and D. Male. 1993. *Immunology*. 3d ed. London: Gower Medical.

Russo, V. E. A., S. Brody, D. Cove, and S. Ottolenghi. 1992. *Development: The molecular genetic approach*. New York: Springer-Verlag.

Service, R. F. 1994. Stalking the start of colon cancer. *Science* 263:1559–60.

Shi, Y., and J. M. Berg. 1995. Specific DNA-RNA binding by zinc finger proteins. *Science* 268:282–84.

Simpson, L., and D. A. Maslov. 1994. RNA editing and the evolution of parasites. *Science* 264:1870–71.

Singer, M. S., and D. E. Gottschling. 1994. TLC1: Template RNA component of *Saccharomyces cerevisiae* telomerase. *Science* 266:404–09.

Steitz, J. A. 1988. Snurps. *Sci. Am.* 258 June:56–63.

Therman, E., and M. Susman. 1993. *Human chromosomes: Structure, behavior, and effects*. 3d ed. New York: Springer-Verlag.

Tjian, R. 1995. Molecular machines that control genes. *Sci. Am.* 272 February:54–61.

Voet, D., and J. G. Voet. 1995. *Biochemistry*. 2d ed. New York: John Wiley & Sons.

Wagner, R. P., M. P. Maguire, and R. L. Stallings. 1993. *Chromosomes: A synthesis*. New York: Wiley-Liss.

Weaver, R. F., and P. W. Hedrick. 1997. *Genetics*. 3d ed. Dubuque, Iowa: Wm. C. Brown.

Wolfe, S. L. 1993. *Molecular and cellular biology*. Belmont, Calif.: Wadsworth.

Wolffe, A. P. 1994. Architectural transcription factors. *Science* 264:1100–01.

Wright, T. R. F., ed. 1990. Genetic regulatory hierarchies in development. In *Advances in Genetics*, vol. 27. New York: Academic Press.

Zaug, A. J., P. J. Grabowski, and T. R. Cech. 1983. Autocatalytic cyclization of an excised intervening sequence RNA is a cleavage-ligation reaction. *Nature* 301:578–83.

CHAPTER
ELEVEN

# GENE MANIPULATION: GENETIC ENGINEERING

## OUTLINE



## CHAPTER OBJECTIVES

*This chapter will discuss:*

- The conditions and elements necessary for the production of recombinant DNA
- How the amount of DNA can be increased in vitro using the polymerase chain reaction
- The various vectors and techniques used to transfect both prokaryotic and eukaryotic cells
- The ways in which genetic information can be stored in vectors as libraries and retrieved
- The ways by which genetic maps are constructed and gene nucleotide sequencing is done
- Some other techniques used in the study of gene structure and function

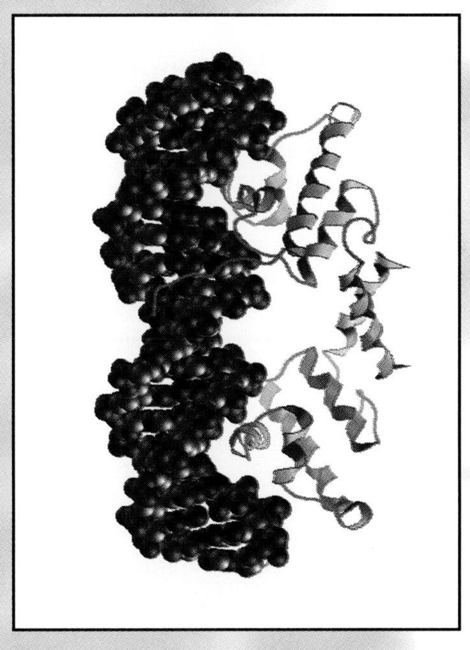

## INTRODUCTION

It is doubtful that anyone would dispute the claim that the dawn of molecular genetics came with the 1953 publication of the Watson and Crick papers on the structure of DNA. For the next several years our understanding of DNA structure was being refined, and, at the same time, scientists such as Jacob and Monod were working on the function of genes and their control (Chapter 7).

But as the work on the structure and function of genes was under way, other scientists were already thinking about and designing experiments in an effort to put this knowledge to use. They were beginning to toy with the idea of manipulating genes. The new field of gene manipulation, or genetic engineering, was developing and about to be born.

In 1972, Paul Berg and his colleagues D. A. Jackson and R. H. Symons reported that they had succeeded in splicing together, in vitro, the DNAs of two different viruses. Segments of DNA taken from the animal virus SV40 were spliced into the DNA of the *Escherichia coli* phage lambda.

The following year, Stanley Cohen, Annie Chang, and their colleagues H. W. Boyer and R. B. Helling published a report in which they described the construction of a biologically active hybrid DNA. They had joined segments of different bacterial plasmids in vitro. Both groups had produced what has come to be called **recombinant DNA.**

Using the knowledge gained since 1972, Michael Smith and C. A. Hutchinson developed a method to produce specific mutations via a process called site-directed mutagenesis. Further, a technique first developed in 1988 by Mario Capecchi and his colleagues S. L. Mansour and K. R. Thomas, referred to as gene targeting, now makes it possible to alter entire sections of genes and thereby alter a gene's function or inactivate the gene altogether. Then, by determining what anatomical or physiological feature of the organism is changed, researchers can assign a gene a given function in an organism.

We have come to a point today when the genes for human diseases are almost routinely being discovered and localized: Huntington's disease, colon cancer, Lou Gehrig's disease, adult-onset diabetes, glaucoma, cystic fibrosis, neurofibromatosis, severe combined immune deficiency (SCID), and others. Scientists, as a result of earlier successes in identifying and localizing some human genes, have taken a giant leap and have begun to sequence the entire human genome: all 6 billion base pairs of the human diploid cell (more about this in Chapter 12).

Remarkably, in a mere 20 years, we have reached the stage of attempting genetic cures for some of these diseases—e.g., cystic fibrosis and immune deficiency disorders. In the first case, a genetically engineered virus was used to carry the normal gene into the lung cells of a cystic fibrosis patient (see Chapter 12). In the second case, cells from a patient were removed, genetically altered, and returned to the patient. Time will tell how successful such gene manipulations will be.

If successful, these procedures would cure the immediate problem, but as new cells are produced from stem cells, they, too, would be defective and the "cure" would need to be repeated. One way around this problem is to isolate stem cells and genetically alter them. The progeny of these cells would then be "normal." An even more fundamental change could be accomplished if one could genetically alter germ cells— egg and sperm—without producing other undesirable changes in the rest of the chromosome. Individuals carrying a defective gene would have their germ cells genetically engineered, and an in vitro fertilization procedure would then be used to ensure the union of the desired egg and sperm. After implantation within the uterus and the normal gestation period, a baby would be born without the genetic defect. In this way, the transmission of inheritable diseases might be stopped. These procedures, however, pose enormous ethical questions because they could affect human evolution.

Laboratories have produced special proteins such as human insulin and human growth hormone. The idea is to implant the relevant genes into a microorganism such as a bacterium or yeast and then switch on the gene, forcing the cell to manufacture a protein it normally does not. These methods will help to ensure an adequate and relatively cheap supply of these proteins at least until a genetic cure for the diseases can be established.

In this chapter, we shall look at some of these techniques. Keep in mind that what is always the required outcome of any particular procedure or technique is a biologically active molecule, whether DNA or protein. The ability to produce a biologically active molecule is the ultimate test of the validity of any of the procedures and techniques.

## NECESSARY ELEMENTS

The gene-manipulation procedures used to produce recombinant DNA molecules that can be introduced into cells have four necessary elements:

1. A method of breaking and joining DNA fragments (genes) derived from different sources.
2. A suitable vector or carrier capable of replicating both itself and the foreign or passenger DNA linked to it. This product is the recombinant DNA.
3. A means of introducing the recombinant DNA into a bacterial, yeast, plant, or mammalian cell.
4. A procedure for screening or selecting from a large population of cells a clone of cells that has acquired the recombinant DNA molecule.

## Breaking and Joining Different DNAs

The obvious first step in moving pieces of DNA from one site to another is to be able to extract the designated segment from one molecule of DNA and then insert it into another molecule of DNA. For example, to clone a human gene requires the isolation of the gene. We must be able to locate the gene and separate it from the rest of the DNA. The separation must be precise. There must be no nucleotides missing from the gene's sequence, and there must be no additional nucleotides attached to it.

Fortunately, all DNAs are fundamentally alike. They are sequences of deoxyribonucleotides joined by phosphodiester bonds, and, for cellular life forms, all DNAs are double-stranded. The breaking and joining of DNA fragments is, then, the process of breaking and reforming phosphodiester bonds.

This process can be accomplished by chemical, physical, or enzymatic means. The first two methods are not useful for breaking the bonds precisely at selected sites; they would allow little if any control over where the DNA would be broken.

Enzyme reactions, on the other hand, are extremely specific and, one might add, very fast. Two classes of enzymes were found that could very precisely break phosphodiester bonds and reform them. These enzymes are the restriction endonucleases and the DNA ligases, respectively.

### *Restriction Endonucleases*

**Restriction endonucleases** are enzymes that, as the name implies, cleave phosphodiester bonds of DNA internally. They do not break the bond that joins a terminal nucleotide to a chain; exonucleases do that. The cleavage site lies within a specific nucleotide sequence of about 6–8 bp. Restriction endonucleases, or simply restriction enzymes, fall into three groupings. Two of the groups, Type I and Type III restriction enzymes also have methylase activity.

Type II restriction enzymes, however, do not have methylase activity. Apart from a particular recognition sequence, these enzymes require only magnesium ions for activity. We shall confine ourselves solely to Type II restriction enzymes, because these are the ones that have become indispensable to gene manipulation. They are, in effect, the molecular scissors of molecular genetics.

Restriction endonucleases were discovered by Hamilton Smith and his colleagues in 1968. Since then, more than 500 endonucleases have been discovered, and among them more than 100 specific cleavage sites are represented. The enzymes are derived from bacteria (remember, eukaryotes do not have these enzymes) and are named by using the first letter of the genus name followed by the first two letters of the species name. If a particular strain of organism is the source, the strain designation follows the species letters. And, finally, a Roman numeral is added to indicate a specific restriction enzyme of that strain. So, for example, Smith discovered endonucleases (Type II) in the bacterium *Haemophilus influenzae* strain Rd; the enzyme is thus named *Hin*dII (pronounced hin-dee-two). And, from *Escherichia coli* strain R, we get *Eco*RI (ee-co-r-one).

The sites recognized by most of these restriction enzymes are called **palindromes,** which in ordinary language are sentences that read the same forward and backward:

Able was I ere I saw Elba

In the "language" of DNA, palindromes are sequences that read the same on both strands in a $5' \rightarrow 3'$ or $3' \rightarrow 5'$ direction. For example, *Eco*RI's recognition palindrome is

5′ GAATTC 3′
3′ CTTAAG 5′

Each strand read in the $5' \rightarrow 3'$ direction has the same sequence GAATTC.

Besides their specificities, many restriction enzymes also have the advantage of making staggered cuts, leaving single-stranded ends, called sticky ends because they will easily base pair with their complementary single-stranded ends (see Figure 11.1). For *Eco*RI, digestion produces:

5′ G`AATTC 3′ ⟶ 5′ G + AATTC 3′
3′ CTTAA`G 5′ ⟶ 3′ CTTAA + G 5′

The break is at the accent mark on each strand. The phosphodiester bonds between the GA bases at the 5′ ends are cleaved, producing the single-stranded sticky ends.

Of course, the restriction enzyme chosen for a particular cut should not cut the gene itself internally. In other words, the gene and all its control elements must remain intact, otherwise the entire purpose of using the particular enzyme would be lost. Sometimes, however, only specific portions of a gene might be needed (e.g., a promoter sequence). In this case, cutting away the rest of the gene is of no consequence.

So, for *Eco*RI, the recognition sequence is 5′ GAATTC 3′ and the specific cleavage site within this sequence is 5′ GA 3′. Table 11.1 lists just a few restriction endonucleases and the recognition palindrome together with the specific cutting sites.

### *DNA Ligases*

The class of enzymes known as DNA ligases was first isolated in 1966 by B. Weiss and C. C. Richardson. Once the gene has been removed from flanking DNA and the vector DNA has been opened (also using the same restriction enzymes to break phosphodiester bonds), the two different DNAs (gene and vector) are brought together by **annealing,**

*table 11.1*

SELECTED RESTRICTION ENDONUCLEASES AND THEIR RECOGNITION PALINDROMES AND CUTTING SITES

| RESTRICTION ENDONUCLEASE | RECOGNITION PALINDROME* |
|---|---|
| *AluI* | A G↓C T |
| *BamHI* | G↓G A T C C |
| *BglII* | A↓G A T C T |
| *ClaI* | A T↓C G A T |
| *EcoRI* | G↓A A T T C |
| *HaeIII* | G G↓C C |
| *HindII* | G T Py↓Pu A C |
| *HindIII* | A↓A G C T T |
| *HpaII* | C↓C G G |
| *KpnI* | G G T A C↓C |
| *MboI* | ↓G A T C |
| *PstI* | C T G C A↓G |
| *PvuI* | C G A T↓C G |
| *SalI* | G↓T C G A C |
| *SmaI* | C C C↓G G G |
| *XmaI* | C↓C C G G G |
| *NotI* | G C↓G G C C G C |

*Source:* Weaver and Hedrick, 1997.
*Only one DNA strand, written 5′ → 3′ left to right, is presented, but restriction enzymes actually cut double-stranded DNA. The cutting site for each enzyme is represented by an arrow.

a process in which heating separates dsDNA (double-stranded) strands and subsequent cooling allows pairing of complementary strands. Then phosphodiester bonds are reformed, linking the two DNAs. A total of four such bonds must be reformed, two on each strand at the 5′ and 3′ locations. This process of phosphodiester bond formation is called ligation, and the enzymes that perform this function are called DNA ligases.

DNA ligases used in gene manipulations are derived from prokaryotes, eukaryotes, and viruses. These are the enzymes described in Chapter 2 that are needed to join the Okazaki fragments of newly synthesized DNA. The ligases require energy in a species-dependent manner supplied by the hydrolysis either of ATP or of $NAD^+$, as is the case in *E. coli*.

Ligation can be accomplished by three methods, depending on the nature of the cut ends of the DNA: sticky-end ligation, complementary-homopolymer ligation, and blunt-end ligation.

As we saw, the cut with the *Eco*RI enzyme yields single-stranded complementary ends, called **sticky ends.** So the foreign DNA and vector DNA can easily base pair at the cuts, and phosphodiester bonds can then be formed by DNA ligase (Figure 11.1).

The complementary-homopolymer ligation technique is used if the vector DNA and the passenger DNA have no common restriction sites at positions outside the vector acceptor site and outside the gene. Cut sites must not affect gene activity or the vector's ability to penetrate the cell and replicate within it. Restriction enzymes produce blunt ends either by cutting at different palindromes or by cutting away single-stranded regions.

Then an enzyme called terminal deoxyribonucleotidyl transferase, or simply terminal transferase, is used. It is a mammalian enzyme and the only known DNA polymerase not requiring a template. The two different DNAs may be cut by a restriction enzyme such as *Hind*II:

5′ GTC`GAC 3′ → 5′ GTC + GAC 3′
3′ CAG`CTG 5′ → 3′ CAG + CTG 5′
(Note the palindrome!)

Terminal transferase can now be used. The enzyme will use whatever triphosphate deoxyribonucleotides are available and add them randomly to the 3′ OH ends of DNA. If only a single type of nucleotide is present, a homopolymer tail is added.

So, for example, the vector DNA can be cut, and to its 3′ OH ends anywhere from 100–200 thymines are added. The passenger DNA is similarly cut, but to its 3′ OH ends 100–200 adenosines are added. The vector DNA would have poly T tails, and the passenger DNA poly A tails.

Now, during the annealing process, the two different DNAs are brought together. Any gaps resulting from uneven tail lengths are filled in by DNA polymerase I, and the phosphodiester bonds are reformed by DNA ligase (Figure 11.2).

A disadvantage of this technique is that the addition of the nucleotides to the 3′ OH ends of the passenger and vector DNAs changes the recognition site sequence. This may be an important consideration should it be necessary to recover the passenger DNA later. The problem, however, can be taken care of by the chemical synthesis of a restriction site, called a linker, which can be attached to the passenger DNA by a $T_4$ ligase then modified to match the vector restriction site (Figure 11.3).

The third method is blunt-end ligation. If the two DNAs to be joined have no regions where single-stranded complementarity exists and the other techniques cannot be used (e.g., too costly or too time-consuming), the molecules can still be joined by using $T_4$ DNA ligase. This enzyme will join the ends, provided the 5′ terminals have a phosphate attached and the 3′ ends have free hydroxyl groups. This ligase is the only enzyme known to be able to do this.

Regardless of the procedure used to join and seal two different DNAs, what is produced is recombinant DNA. A new DNA molecule composed of DNAs from different sources has been created.

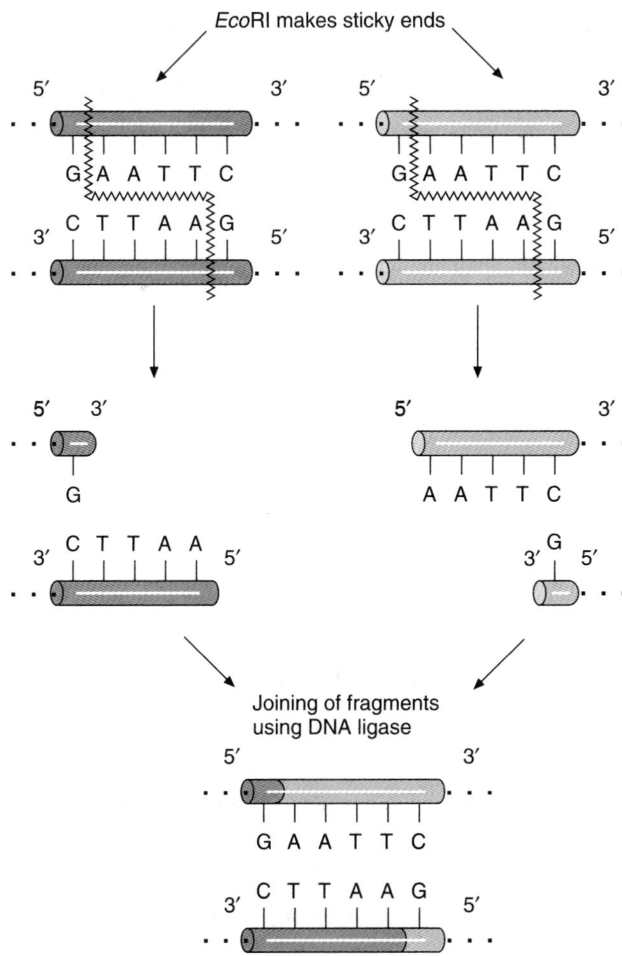


FIGURE 11.1
"Sticky-ends" produced by the restriction endonuclease *Eco*RI.

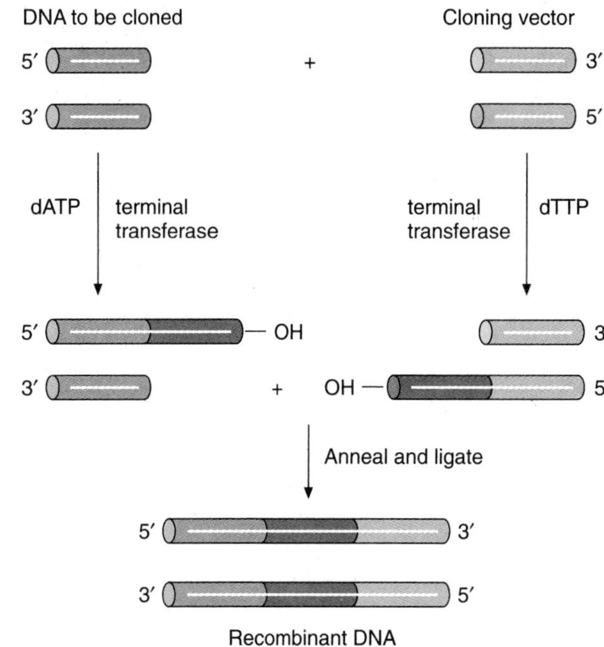


FIGURE 11.2
Complementary homopolymer ends used to join passenger DNA to vector DNA. A poly A tail binds a complementary poly T tail.

# A Suitable Vector or Carrier and a Means of Introducing the Recombinant DNA into a Cell

A number of different vectors, carriers, and techniques are used to place the recombinant DNA into a host cell. **Vectors** range from viruses and plasmids of bacteria to viruses of mammalian cells. Carriers include calcium phosphate coprecipitates of DNA and liposomes (synthetic lipid bilayers). Techniques include electroporation (electrical charges used to produce transient openings in cell membranes) and a method known as microinjection, which has had some success in transferring DNA into mammalian and frog eggs as well as into the eggs of *Caenorhabditis elegans* (the roundworm) and *Drosophila*. Vectors, however, are the most widely used systems for transfer of foreign DNA into a cell.

Where the more difficult plant cells are involved, DNA can be introduced into the cell preferentially by using the bacterium *Agrobacterium tumefaciens* or by using a "gene gun." In this latter technique DNA is coated onto the surface of gold or tungsten particles, which are then literally shot at high velocity through the cell wall using a blank pistol cartridge as the propellant. Other methods require the enzymatic digestion of the plant cell wall to produce protoplasts, which can then be directly injected with DNA or caused to take up the DNA by electroporation. The bacterial method utilizing *A. tumefaciens* is still the most widely used approach and the most successful.

Whatever method is used, the passenger DNA must either integrate into the host cell's DNA or be carried in the cell as part of a biologically active molecule that can replicate independently; otherwise, the inserted DNA (gene) will be lost in succeeding generations.

The biological molecule most commonly used for transferring genes into bacteria and yeast is the plasmid. This vehicle carries relatively small pieces of passenger DNA.

**Plasmids** are small, double-stranded, closed, circular, extrachromosomal DNA molecules. They are found in many species of bacteria and yeast and may carry one or more genes of their own. Most important for molecular geneticists is the fact that plasmids have an origin of replication site that allows them to multiply independently of the host cell's replication. The number of copies of a particular plasmid in a given cell is often controlled by the plasmid. A single bacterium may contain more than one type of plasmid. In general, plasmids do not integrate into host cell DNA.

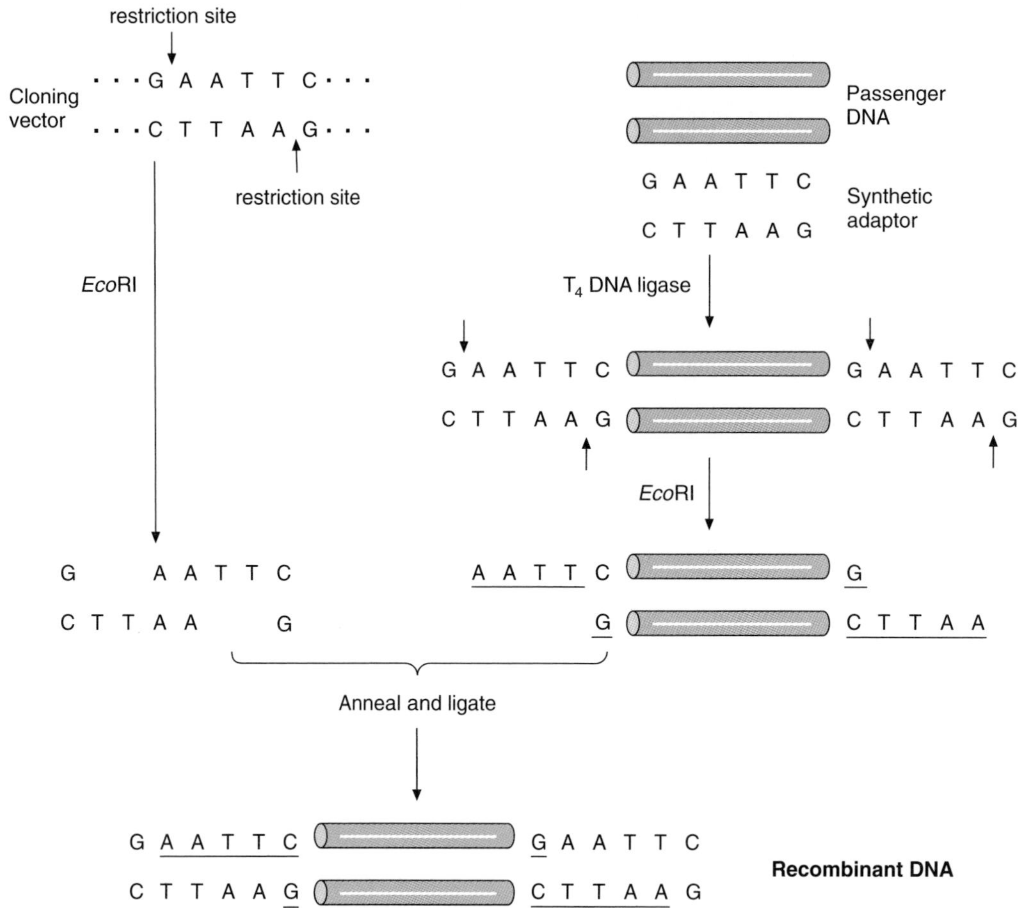


FIGURE 11.3
The use of a synthetic adaptor restriction sequence at either end of passenger DNA so the vector and passenger DNAs now have the same restriction recognition sites.

There are even artificially produced plasmids in use. One such plasmid is designated **pBR322.** This molecule has about 4360 base pairs and carries genes for ampicillin and tetracycline resistance along with an origin of replication sequence. Antibiotic resistance is an important screening advantage that allows one to distinguish cells that carry the plasmid from those that do not. When antibiotic-sensitive cells are plated onto a medium containing one or both antibiotics, only those cells with pBR322 survive and produce colonies. This special plasmid has another advantage over the naturally occurring plasmids; it can accumulate in larger than usual numbers of copies in *E. coli*.

In general, several characteristics make plasmids suitable carriers. Though these properties were first noted in plasmids, they, in fact, are also desirable in virus vectors (Sofer 1991):

1. The carrier should be small. It serves only as a carrier and a replicating molecule. The smaller the necessary vector DNA (e.g., origin of replication sequence is necessary) the more passenger DNA can be inserted into the molecule. Smaller vectors are also easier to get into bacterial cells and to purify; larger plasmids tend to be more fragile.
2. The nucleotide sequence of the vector should be known to facilitate its use (i.e., cutting out some unnecessary genes and inserting the desired ones).
3. It should replicate in high numbers in the host cell, assuming the inserted gene is not harmful to the host, in which case fewer plasmids per cell would be desirable. High copy numbers also make purification easier. Plasmids that replicate too many copies are said to be under relaxed replication control; that is, plasmid replication is not tightly controlled by the host cell. Other types of plasmids are under stringent host control and replicate infrequently or only when the host cell chromosome replicates.
4. The plasmid should carry a **marker** (a gene) whose product makes identification of the cells carrying the plasmid easy. A marker might be resistance to a particular antibiotic (e.g., markers carried by pBR322).

5. If possible, the vector should carry a second marker (resistance to a second antibiotic) that is inactivated upon insertion of the passenger gene. This second marker would avoid the problem of detecting a plasmid that is cut but does not pick up the passenger gene; inactivation of the second marker indicates that the passenger DNA has been inserted into the vector. So, for example, a cell that has picked up a plasmid carrying the desired gene would not grow on a medium containing the second antibiotic since the vector's gene for resistance to that antibiotic was inactivated upon insertion of the passenger gene.
6. The vector should have several unique restriction sites within the marker genes. This property will avoid the possibility of destroying the plasmid by fragmentation, as would happen if there were multiple common restriction sites within and outside the marker. Having more than one unique restriction site also makes the plasmid more flexible in use since a number of different restriction enzymes could be used to open it.

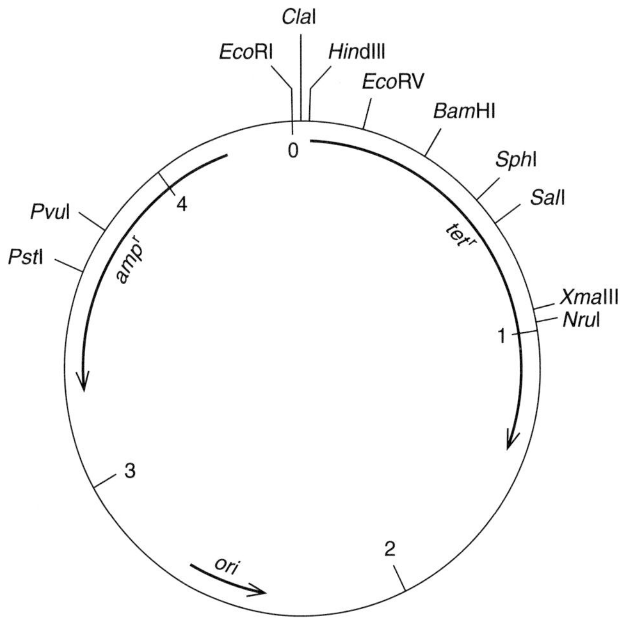


FIGURE 11.4
Plasmid pBR322: genes and restriction endonuclease recognition sites.

## Screening a Large Population of Cells

Because neither insertion of the passenger DNA into a vector nor the uptake of the vector is a certainty, there must be a procedure to isolate the cells carrying the desired gene from the thousands or millions of cells not carrying the gene. From the outside, as viewed with a microscope, all cells would look alike. Biochemically, however, there would be a difference because of the new gene, or more precisely, its product or the new gene's effect on a vector gene as we saw in item 5, a second marker, above. Screening techniques vary from replica plating of bacteria (Chapter 9) to the use of mutant mammalian cell lines deficient in some enzyme activity, such as the lack of thymine kinase activity. More recently, many more markers have been developed for mammalian cell lines. For example, a bacterial gene marker conferring resistance to a neomycin-related antibiotic called G418 that blocks protein synthesis thereby killing mammalian cells can be inserted into vectors for use with mammalian cell lines. The marker gene's product is an enzyme that destroys G418. So any cell that accepts the marker gene along with the passenger DNA will replicate in a medium containing G418, but those cells not carrying the marker will perish.

Antibiotic resistance is also used in work with plant cells. Markers such as resistance to the antibiotic kanamycin, which inhibits plant growth, are in widespread use. New plant traits are also being used as markers. Those individual cells that have accepted the new gene and the marker will produce traits different from those of the parent plant (e.g., color and foliage variations).

# PRODUCTION OF RECOMBINANT DNA

For our example of the production of recombinant DNA, we shall use as the vector the plasmid pBR322. This artificial plasmid contains 11 unique restriction sites and the genes for ampicillin and tetracycline resistance (Figure 11.4).

The restriction enzyme *Pst*I (from the bacterium *Providencia stuartii*) is used to cut DNA at the 5′ AG 3′ doublet at the palindrome recognition site as shown below:

5′ CTGCA`G 3′
3′ G`ACGTCC 5′

This site lies within the ampicillin resistance gene, so the gene is inactivated when the cutting takes place and the passenger DNA is inserted. The tetracycline gene is left intact, an important factor in screening.

We do not know which cells are carrying the recombinant DNA and we are interested in the ampicillin-sensitive and tetracycline-resistant cell, so the replica plating technique (see Figure 9.5) is used.

After cells and recombinant DNA are mixed and transformation is allowed to take place, cell samples are plated onto medium lacking both antibiotics. When colonies appear on the plate, it is replica-plated onto one set of plates that contain ampicillin and a second set of plates that contain tetracycline. Colonies carrying the recombinant DNA can be recognized because they will grow on tetracycline plates but not on ampicillin plates. The stages of recombinant DNA production are outlined in Figure 11.5.

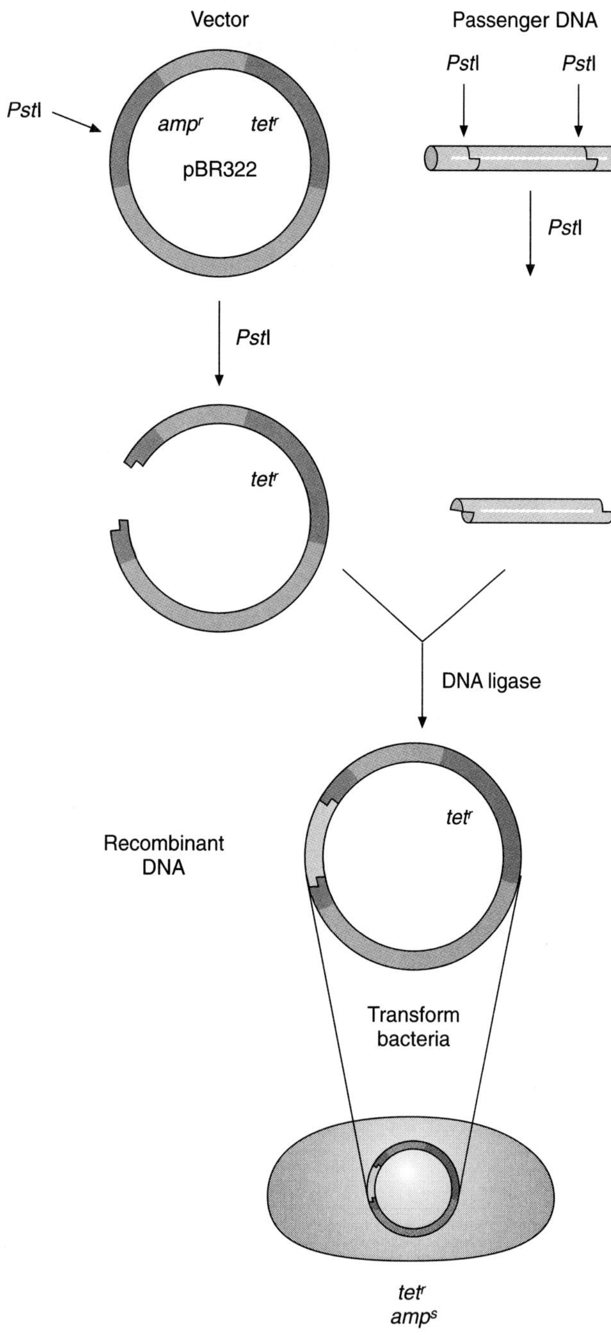


FIGURE 11.5
Brief overview of recombinant DNA production. The ampicillin gene is disrupted so the host cells become ampicillin-sensitive and tetracycline-resistant. Screening is designed to detect those cells.

This procedure for producing recombinant DNA and transforming cells with the recombinant DNA is often referred to as **gene cloning.** A few copies of a gene are amplified into millions upon millions of copies as the transformed bacteria reproduce; the plasmid DNA reproduces itself in each cell, and copies are distributed to daughter cells.

# THE POLYMERASE CHAIN REACTION

A major difficulty in gene manipulation has been that any given gene makes up an exceedingly small part of an organism's total genome. It is estimated that a human being has between 50,000 and 100,000 genes. Early techniques used to find a particular gene and clone it were very time-consuming for routine use: oligonucleotides often hybridize with DNA sequences other than their complements; the use of restriction endonucleases to cleave DNA also required the use of electrophoretic separation of the fragments to achieve some degree of purification of the desired gene; and gene cloning using plasmids and dideoxy sequencing also required large amounts of time.

In contrast, the **polymerase chain reaction** technique (referred to now simply as **PCR**) developed in the mid-1980s by Kary B. Mullis and colleagues has overcome these difficulties. In the space of hours, this cell-free reaction can make copies of a target DNA sequence in quantities that would take days or weeks using the earlier amplification techniques such as gene cloning. For example, with 25 cycles of the PCR, the target sequence can be amplified about 8 million times (Table 11.2), and within a few hours billions of copies can be made.

An important feature of the PCR is that the target DNA need not be isolated from the rest of its genome. But, once completed, the PCR yields sequences that can easily be separated from the rest of the DNA by gel electrophoresis. Also important is that the starting material containing the desired sequence can be as little as the amount of DNA contained in a single cell. Essentially, all that is required to begin the PCR procedure is to know or determine the nucleotide sequences flanking either side of the double-stranded target region. Primers of about a dozen nucleotides complementary to the 3′ ends of the flanking sequences are made. (Remember, DNA is synthesized in the 5′ to 3′ direction, and DNA polymerase requires a primer.)

The PCR uses the normal enzyme of DNA synthesis—the DNA polymerase—but the enzyme is derived from an unusual source. The PCR polymerase is isolated from the bacterium *Thermus aquaticus*, which lives in water at temperatures of about 75°C, a temperature that would denature most proteins. This enzyme, called Taq polymerase has an optimum temperature of 72°C and is reasonably stable at 94°C. (Keep these temperatures in mind as we describe the PCR.) The polymerase chain reaction is a cyclical reaction; at the completion of the reaction the system automatically returns to the first step and repeats itself. The cycling continues until a substrate is

*table 11.2*

PCR Amplification of DNA Fragment

| Cycle number | Number of double-stranded target molecules |
|---|---|
| 1 | 0 |
| 2 | 0 |
| 3 | 2 |
| 4 | 4 |
| 5 | 8 |
| 6 | 16 |
| 7 | 32 |
| 8 | 64 |
| 9 | 128 |
| 10 | 256 |
| 11 | 512 |
| 12 | 1024 |
| 13 | 2048 |
| 14 | 4096 |
| 15 | 8192 |
| 16 | 16,384 |
| 17 | 32,768 |
| 18 | 65,536 |
| 19 | 131,072 |
| 20 | 262,144 |
| 21 | 524,288 |
| 22 | 1,048,576 |
| 23 | 2,097,152 |
| 24 | 4,194,304 |
| 25 | 8,388,608 |
| 26 | 16,777,216 |
| 27 | 33,544,432 |
| 28 | 67,108,864 |
| 29 | 134,217,728 |
| 30 | 268,435,456 |
| 31 | 536,870,912 |
| 32 | 1,073,741,824 |

completely used up or the reaction is stopped. The steps of the PCR (Figure 11.6) are as follows (Watson et al. 1992):

1. The target DNA sequence, two primers, DNA polymerase, and a mixture of the triphosphate deoxynucleotides are mixed in a total volume of about 100 μl.
2. The mixture is heated to 94°C for about 5 minutes to separate the strands of the double-stranded DNA. The single strands become templates for the primers.
3. The temperature is lowered to between 30° and 65°C to allow the primers to bind to the templates. The temperature and times used in this step, the annealing step, vary. They depend on the sequence to be amplified. Some sequences require higher temperatures and longer times than others.

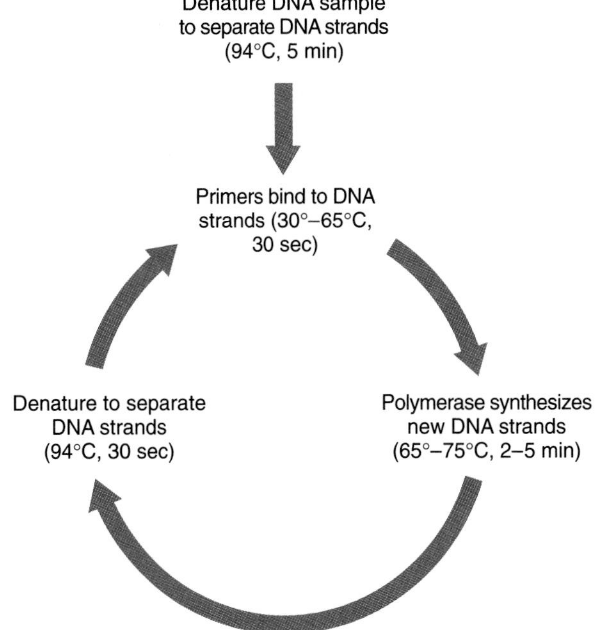


Figure 11.6
The polymerase chain reaction (PCR) cycle. At the conclusion of each cycle there are twice as many DNA molecules as there were at the start of the cycle. The end products of one cycle become the starting materials for the next cycle.

4. Once the DNA single strands and primers are bound, the temperature is raised to 72°C to allow synthesis of new strands by Taq polymerase starting at each primer. This step requires 2–5 minutes.
5. The temperature is raised to 94°C once again to separate the strands of the newly synthesized double-stranded DNA. And step 3 begins again.

Note, that at the completion of each cycle, the end products of the reaction become the substrates or templates for the next cycle. As more and more cycles are completed, more and more end product or substrate is generated so that each cycle doubles the amount of amplified DNA (Table 11.2), which is then the template for the next round of DNA synthesis.

## PROKARYOTIC CLONING: BACTERIA

Various cloning vectors have been designed to be used in bacteria. Each of these vectors was created to meet a particular demand. The vectors were themselves genetically engineered to create new tools for further genetic engineering experiments. The new vectors were formed from bacteriophage. These viruses are naturally occurring vectors.

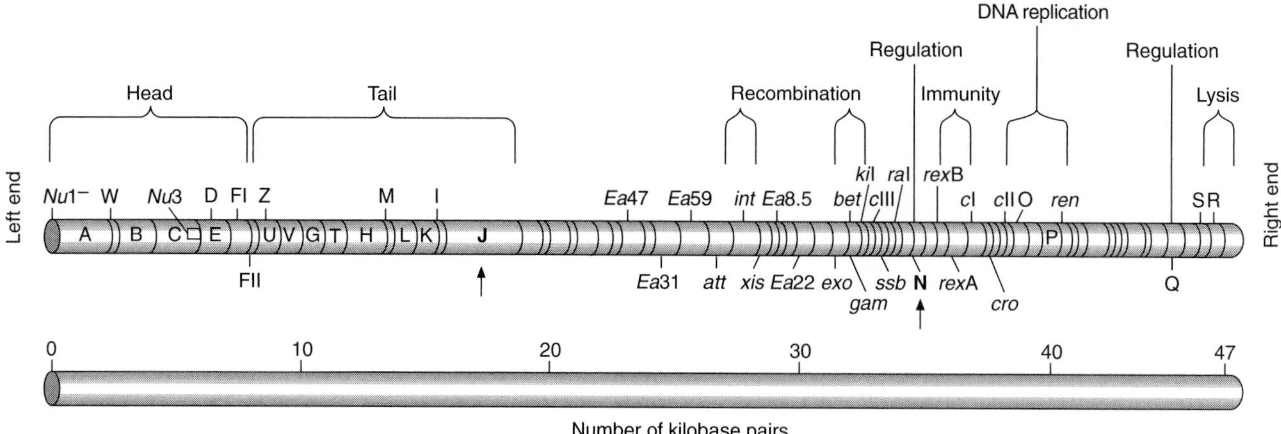


FIGURE 11.7
Lambda phage genome. The arrows (↑) at genes J and N bracket the genes removed when this phage is used as a cloning vector.

They are engineered to maintain this ability and to be able to carry larger fragments of DNA from any source.

## Lambda Phage

From lambda phage, large pieces of DNA that code for proteins needed to establish and maintain the lysogenic state (Chapter 3) and other genes not essential for lambda phage's function as a carrier can be removed and replaced with "stuffer" DNA. Stuffer DNA is what is eventually removed and replaced with passenger DNA.

Most of the DNA to be removed lies in the middle of lambda's genome, between gene J and gene N. A number of derivatives of lambda have been produced, each with different mutations to allow for different experiments. These derivatives have been given the name **Charon** in honor of the boatman of Greek mythology who ferried the dead across the river Styx to Hades. (Charon is pronounced "Sharon.") Figure 11.7 shows lambda's genome and the functions of some genes.

There are advantages to using lambda rather than a plasmid in cloning procedures: the virus can accept relatively large amounts of foreign DNA, up to 20 kbp; much is known about the organism's genetics and requirements; and, most important, it is one of the very few organisms that can be reconstituted in vitro. This in vitro "packaging" allows for easy insertion of recombinant DNA into a vehicle that naturally infects the bacterium *E. coli*.

There is, however, a very strict requirement concerning the size of the DNA inserted into the region from which DNA has been removed. Normally, the size of the DNA assembled into the head is about 48.5 kbp. This size must be maintained if the virus is to be an effective vector. If the total DNA, after recombination, is too long or too short, the phage's effectiveness as a carrier and deliverer of DNA is markedly decreased. So the amount of stuffer DNA in lambda can vary, but the total amount of recombinant DNA must be on the order 48.5 kbp. This size makes possible the insertion of genes of various sizes into lambda.

## Cosmids

Another vector designed to carry large amounts (40–50 kbp) of DNA was constructed by using the cohesive ends (sticky ends) of lambda. Cohesive ends are found naturally in lambda just after it infects a host cell. They are necessary for the circularization of lambda's DNA in preparation for its replication (Chapter 3). The cohesive ends are combined with a plasmid origin of replication. The term *cosmid* comes from *cos* ("cohesive") and *mid* ("plasmid").

## Bacteriophage M13

Bacteriophage M13 infects *E. coli*. M13 has a 6.5-kb single strand of DNA as its genome. M13's main advantage as a cloning vehicle is that, when released from cells, its progeny are single-stranded. And this single strand would be complementary to only one of the two strands of the cloned DNA. This property makes M13 an excellent choice when DNA sequence analysis must be done because there would be no need to separate the strands of dsDNA and then isolate the gene or coding strand. The phage DNA can be used as a template. And, perhaps more important, M13 DNA can be used in site-directed mutations (more on this topic later in the chapter).

## Phagemids

Last, combining the origin of replication nucleotide sequence of M13 and sequences of the artificial plasmid pBR322 creates another type of cloning vector: **phagemids.**

These vectors can be propagated as double-stranded plasmids. But introducing "helper phage" into the cell containing the phagemid allows single-stranded phagemid DNA to be produced, packaged, and released from the cell as M13 would be. The helper phage supply the genes of M13 required for replication that were cut away from M13 to obtain the origin of replication to make the phagemid.

One advantage of these vectors is their small size: about 3000 bp. This size makes possible the insertion of larger passenger DNA than M13 can carry. Remember, there are size limits to passenger DNA since the total amount of DNA the completed virus can pack is limited by the virus's size. Another advantage of phagemids is the ease with which double- or single-stranded progeny can be produced.

# EUKARYOTIC CLONING: YEAST, PLANTS, AND MAMMALIAN CELLS

## Yeast

Yeast were the first of the eukaryotes to be genetically engineered. Two of the most favored yeast used are *Saccharomyces cerevisiae* (baker's yeast) and *Schisosaccharomyces pombe*.

There are several advantages to using yeast as a cloning host. For eukaryotes, their generation times are short, about 2 hours. Though this is much longer than prokaryotic doubling time, yeast genomes are small, about 10 times larger than *E. coli's* genome. A haploid yeast cell has about $1.4 \times 10^7$ bp distributed among 16 chromosomes. Using haploid cells allows recessive mutations to be produced and maintained. Then, whenever needed, diploid cells can be produced by mating. This last advantage cannot be duplicated in mammalian cells, making the study of recessive mutations extremely difficult.

Yeast are also valuable as cloning hosts because, as eukaryotes, their posttranslational protein modification systems are like those of other eukaryotes—e.g., protein folding and glycosylation (sugar additions). The proteins yeasts make in cloning experiments can be exported by adding to the protein gene a nucleotide sequence coding for the export signal amino acid sequence. This makes for a relatively easy protein purification procedure since the exported protein is now in the medium separated from other yeast proteins.

And, last, some yeast genes are homologous to those of human beings, so a study of the yeast gene sheds light on the human gene. For example, yeast have a gene that is very similar to the normal H-*ras* gene of human beings. A single point mutation in this gene converts the proto-oncogene to an oncogene. A number of such altered H-*ras* genes are found in human tumors, including bladder carcinomas. Yeasts deprived of both alleles of their *ras* gene do not grow. Insertion of the H-*ras* gene corrects the defect, establishing some similarity between yeast and human genes. The hope is that study of both the yeast and human genes in yeast cells will shed some light on the functions of human genes.

The vectors used in yeast cells are plasmids called **shuttle vectors**—so named because they can replicate in *E. coli* and yeast. These vectors can be genetically engineered and easily cloned to high copy numbers in the bacterium, and they are easily purified. Then, inserted into yeast (haploid or diploid cells), the genes can be studied for their effect in a eukaryotic cell. Of course, these vectors must have origins of replication and selectable markers (e.g., antibiotic sensitivity or resistance) expressed in both organisms. These vectors can be grouped as follows:

1. Yeast integrating plasmid (YIp): these plasmids have bacterial origins of replication and antibiotic-resistance sites, a yeast marker gene for selection of transformed yeast, and unique restriction sites for insertion of new genes. The plasmids are stabilized within the cell by integration into the host cell's genome. The vector must integrate or be lost as the host cell replicates. Since the plasmid does not contain a yeast origin of replication site that the yeast DNA polymerases would recognize, it cannot be replicated in yeast cells unless it is integrated.
2. Yeast replicating plasmids (YRp): these are free plasmids in yeast cells. In addition to the assortment of genes found in YIps, replicating vectors have other genes that allow for the multiplication and permanence of the vectors within yeast cells. Some of these vectors have origins of replication sequences derived from a plasmid known as the 2μ circle that occurs naturally in yeast. The plasmids can replicate to high copy numbers in yeast cells.

   Other YRps have built into them a yeast "autonomously replicating sequence," or ARS element. When one of these elements, such as a yeast centromere sequence (CEN sequence), is added to the plasmid, it makes the vector extremely stable in yeast cells, and their replication is highly regulated so that their copy number equals that of the chromosomes in the cell, (i.e., haploid or diploid number).
3. Yeast artificial chromosome (YAC): these entities contain not only the usual bacterial sequences but also yeast markers, origin of replication sequences, a centromere, and also two sequences on either end that act as telomeres. The telomeres allow the YACs to replicate as linear chromosomal-like molecules. Most important, these molecules can carry hundreds of thousands of base pairs, something like 10 times what a cosmid can carry. And once inside a yeast cell, the YAC looks and behaves like a host cell chromosome.

One difficulty in using YACs is that before they can be inserted the cells must first be treated in such a way as to remove much of the cell wall material. The cell assumes a spherical shape because of the loss and is referred to as a spheroplast. When exposed to YACs, many of the spheroplasts will accept the artificial chromosomes. Of course, screening of the cells must still be done to detect those cells that have accepted a YAC—none of these techniques is 100% efficient in inserting vectors.

Yeast artificial chromosomes have been used to study the human disease fragile X syndrome (Chapter 10). The site of fragile X has been mapped to location Xq27.3. The YACs were used to produce DNA probes to seek out the gene, and from one of these probes a single-stranded DNA molecule complementary to the gene was cloned. The clones contained complementary copies of a gene called fragile X mental retardation-1 or FMR-1. This gene is expressed in the brain and has an exon coding for a continuous run of 30 arginine residues; arginine is the basic amino acid that is also found in histones. One suggestion is that along with increased methylation at the fragile X site, a high concentration of arginine, which might cause protein binding to the acidic DNA, may contribute to the disease process.

Because they are capable of carrying large amounts of passenger DNA, artificial chromosomes are now being used to construct what are called gene libraries of mammalian cells, including human cells (Chapter 12). Gene libraries are collections of vectors, such as YACs. Within individual vectors are fragments of DNA from a particular species of organism. These fragments are the genes of that species. Gene libraries will be discussed in somewhat more detail later in this chapter.

# Plants

## *Ti Plasmids*

Genetic manipulation of plants has a long history. It has been used to produce, for example, greater crop yields or resistance to weed killers so that herbicides can be used to make cultivation easier. The older methods of plant breeding, however, are time-consuming, and the outcomes are not always desirable; most often both known and unknown genes are transferred.

The plants of interest to human beings—wheat, corn, and rice—are monocotyledons (their seeds have a single leaf), and they are difficult to transfect with the vectors currently available. Also, because of the large and inconsistent numbers of chromosomes found among various groups, such as the grasses, plants produced from single cells grown in tissue culture, including those that have been genetically engineered, exhibit what is called somaclonal variation; they are not genetically stable—chromosomal number can vary from cell to cell. The dicotyledons (their seeds have two leaves), however, have been used successfully in gene manipulation experiments.

One of the most widely used vectors in plant genetic engineering is the **Ti (tumor-inducing) plasmid** found in the bacterium *Agrobacterium tumefaciens*. In dicotyledons, this plasmid produces tumor cells known as crown galls. The plasmid can be altered so that it can carry passenger DNA into plants without converting the cells into tumors.

Within the Ti plasmid are genes located on a segment called **T-DNA** (transfer-DNA). These genes code for enzymes needed for the synthesis of unusual amino acids, opines, and others that cause unregulated cell growth—tumors.

When a plant is wounded and infected by the bacterium, factors released by the plant stimulate the transcription of genes located on the Ti plasmid. The proteins produced by this transcription are needed to nick both ends of the T-DNA segment, releasing a single-stranded DNA molecule. Then, by a mechanism still incompletely understood, the single-stranded T-DNA enters the plant cell and integrates itself into one of the host cell's chromosomes. The break in the Ti plasmid is then repaired by DNA synthesis. Figure 11.8 illustrates this process.

The incorporation of the T-DNA into the plant chromosome induces the plant cell to synthesize and release opines on which the bacterium feeds. This entire process closely resembles bacterial mating, in which a portion of a DNA molecule is passed from a donor to a recipient cell, whereupon the passed DNA is incorporated into the host cell's DNA. Now a bacterium is "mating" with a plant cell. But the integration of the T-DNA results in tumor formation. T-DNA can be modified by removing the tumor genes and substituting foreign DNA to be transferred to the plant cell chromosomes.

The transfer of passenger genes requires genes located on the Ti plasmid, but not in the T-DNA segment, and those located on the T-DNA are needed for integration. Scientists have taken advantage of this separation of functions and have created what is called a binary system made up of two plasmids. One plasmid carries Ti plasmid genes whose proteins are needed to release single-stranded T-DNA, and the second plasmid carries the foreign genes plus T-DNA genes whose end product proteins are required to cut host cell DNA and to integrate the passenger DNA.

## *Plant Viruses*

Viruses are already adapted to spread throughout a plant body. They are now being tried as vectors in gene manipulation experiments. This would be a much easier delivery system than the Ti plasmid binary system. And the viruses might be used to infect the monocotyledons.

Much work is in progress to determine the potential usefulness of the vectors. Some problems have surfaced. One is the amount of passenger DNA that can be inserted into viral genomes (remember the size limitations because of viral particle size). Another problem is the tendency of these vectors to shed or release the passenger DNA.

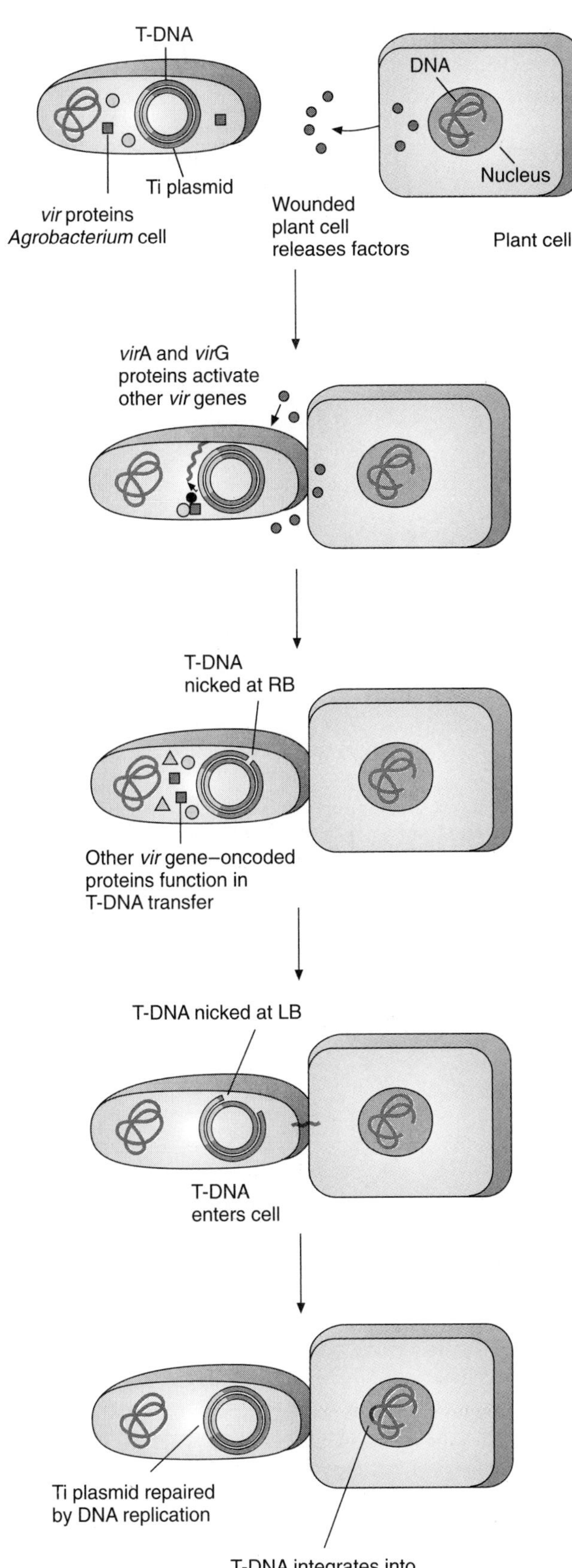


FIGURE 11.8
Stages in the transfer of T-DNA from the bacterium *Agrobacterium tumefaciens* to a plant cell chromosomal DNA molecule.

## *Physical Methods*

A number of physical methods for introducing foreign DNA into monocotyledons, in particular, are under investigation. The successful transformation of corn and rice has been achieved by electroporation. In this procedure, protoplasts are prepared (cell walls are removed by treatment with the enzyme cellulase) and mixed with a high concentration of plasmids containing passenger DNA. The mixture is subjected to an electrical pulse of 200–600 V/cm. The protoplasts are then grown in tissue culture for up to several weeks and then screened for the desired cells. Efficiencies of transformation on the order of 0.1% to 1% have been achieved. That is, of the population of cells subjected to this treatment, from 0.1% to 1% accept the foreign DNA.

Another technique used, especially in attempts to transform the organelles of photosynthesis, the chloroplasts, involves the bombardment of the organelles with DNA-coated beads of gold or tungsten. This procedure is necessary because the numbers of chloroplasts in a plant cell and their small size make the usual vector techniques unsuitable. This method relies on hitting leaves with the DNA-coated pellets, cutting the leaves into tiny fragments, and culturing the fragments, screening for those cells carrying the desired genes. The idea is to be able to study the genetic basis of photosynthesis. But this bombardment method is estimated to be at least 100 times less efficient than the transformation of the plant cell itself.

# Mammalian Cells

Because most cells taken from mammalian tissue cannot be established as immortal cell lines (cells that will grow and divide indefinitely) in tissue culture, permanent cell lines have had to be established from tumor cells originating in particular tissues. From these cells, we have learned about the specialized physiology of cell types such as blood, liver, muscle, and brain. And, because of these established cell lines, gene transfer experiments are possible. These experiments are predicated on the principle that the tumor cells function and are regulated, for the most part, as their normal cell counterparts.

What should be kept in mind as we look at the transfer of genes into mammalian cells is that transfers into the somatic cells of an animal will not affect progeny. To affect progeny requires the transfer of genes to germ line cells—egg and sperm. So, for example, the curing of a disease such as cystic fibrosis in an individual by genetically altering the genome of the individual's lung cells does not prevent the person from passing on the CF gene to offspring. To prevent the inheritance of a condition will require genetically engineering egg or sperm cells.

## *Coprecipitation of DNA*

The study of transformation in bacteria led to the discovery that mammalian cells could be made to take up foreign DNA if the DNA were coprecipitated with calcium phosphate. In fact, it was by using this method that scientists demonstrated that purified tumor virus DNA could transform normal cells into cancer cells. This was powerful evidence that the cancer state can be coded in nucleotide sequences of DNA.

Unfortunately, the stable incorporation of transferred DNA into mammalian cells is not a routine event in transfection experiments. Successful integration into host DNA is on the order of one cell in a thousand to one in a million. The transferred DNA is maintained in the host cell for a short period of time, perhaps days, and is progressively lost unless integrated. Unintegrated DNA is not replicated as the cell replicates its own chromosomes. As a result of this very low transfection rate, screening and selection procedures are critical.

There is very little an investigator can do to control the fate of DNA taken up by mammalian cells. Although once foreign DNA has been integrated into host DNA the arrangement is stable, the site of integration is a random affair. As a result, the expression and regulation of the activity of transferred genes can vary widely and depend on which genes lie next to the inserted ones.

Transferred genes inserted next to housekeeping genes may be more active than those integrated into less active segments of DNA. The transferred genes, apparently, are brought under the regulatory control of neighboring genes, perhaps because integration is between structural genes and their functional genes, the control elements—promoter or operator sequences.

## Other Techniques

Besides coprecipitation, a number of other techniques have been used with varying degrees of success. For example, exceedingly small glass needles can be used to inject DNA directly into the target cell's nucleus. Needless to say, this technique is slow and tedious, with perhaps as few as 100–300 cells injected per experiment. But, 100% of the cells receive the new DNA. A computer-assisted instrument has helped to increase, by at least a factor of 10, the number of cells that can be microinjected in a single day.

As with plant cells, electroporation has been used on some mammalian cells. Cells that are normally grown in suspensions, such as lymphocytes, are better suited for this technique than are cells that must be grown as monolayers in petri dishes.

Artificial lipid vesicles called liposomes are also being used as delivery systems. In this case, the liposomes fuse with the cell membrane and dump their DNA cargo into the target cell (Figure 11.9).

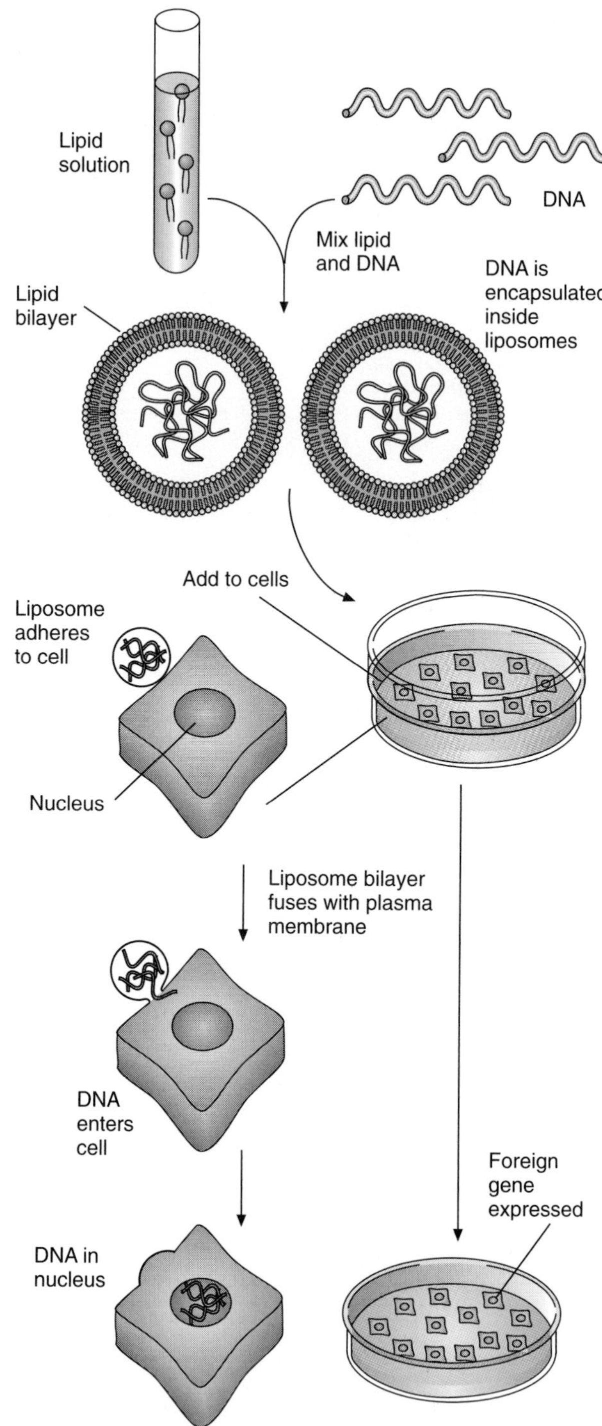


FIGURE 11.9
The use of liposomes to deliver DNA to mammalian cells. The liposomes are synthetic lipid bilayers which spontaneously fuse with the cell membrane, another lipid bilayer.

## Viral Vectors

Both DNA and RNA retroviruses have been used successfully in transfection experiments, yielding high efficiencies of gene transfer. But only the mammalian retroviruses routinely give rise to stable integration of the passenger DNA. Remember, many viruses, such as the bacteriophage, are normally lytic. They kill the host cell after replicating to high numbers, producing plaques in lawns of host cells grown in petri dishes.

Retroviruses, on the other hand, have a life cycle that includes a provirus stage. The provirus integrates into the host cell's DNA. These viruses have been found capable of infecting almost every type of mammalian cell.

Genetically engineered retroviruses capable of transferring genes, are prepared by inserting a cloned proviral DNA (lacking its own packing protein genes and the reverse transcriptase gene) carrying passenger DNA into special cells called packaging cells. These cells contain a defective, integrated provirus in their genomes. The integrated proviruses lack a number of the genes necessary to make and package their own RNA. The defective integrated provirus does, however, have all the genes needed for producing both packing proteins and the enzyme reverse transcriptase.

Upon infecting these packaging cells, the proviral DNA carrying the passenger DNA is transcribed into RNA, including the passenger DNA. This viral RNA and the reverse transcriptase are then packed into viral proteins and released from the cell. These particles can then be used to infect any other cell. The RNA is transcribed into DNA by the reverse transcriptase. The DNA becomes stably integrated into the new host cell's DNA as a provirus, and the foreign gene's end product is produced by the host cell.

### *Mouse Transfection*

It has been demonstrated in mice that the microinjection of foreign DNA (human interferon and rabbit beta-globulin genes were used) into one-cell embryos results in the stable integration of the DNA into both somatic and germ line cell genomes. The genes were subsequently transmitted to offspring in the same manner as normal mouse genes.

Foreign genes are not only expressed in the new host animal; they are expressed in a tissue-specific manner, suggesting they are being controlled in the way endogenous genes are. For example, if the gene for human insulin is transferred to mice, it may integrate in a random fashion at a location different from its endogenous counterpart, but transcription of the transferred gene is induced by the same biochemical signals as is the host gene and the insulin is produced only in the pancreas.

A note of caution, however; transgenes, when injected into one-cell embryos, can, on occasion, destroy the control or functioning of normal genes and thereby cause developmental abnormalities. The problem may be a matter of where in an operon the transgene integrates—some locations are allowed, others are not.

## GENOMIC LIBRARIES

Just as we use buildings, libraries, as repositories of information (books), so too, geneticists use "gene libraries," or vectors, to store genetic information. Geneticists, however, are forced to use numerous libraries to store their information. No one type of vector is sufficiently large to accommodate all the genetic information needed to be stored.

The various vectors used in cloning and gene manipulation experiments can also be used as a sort of storage facility for genetic information. A collection of random DNA fragments can be inserted into vectors, each of which is cloned in a suitable host. The resulting progeny (plasmids, cosmids, viruses, and so on) can then be stored for further reference or need.

The vector collection for any given species must, of course, be large enough to include all the genes of the species. In the case of the human genome, the collection itself would be enormous.

The libraries are generally constructed using lambda bacteriophage or cosmid vectors and in vitro packaging. These are good vectors to use for gene libraries because of the large amount of passenger DNA they can carry. For example, if an average passenger DNA segment in lambda is about 17 kbp and one is constructing a library of the human genome, which has about 3 billion base pairs per haploid genome, the entire genome would require $(3 \times 10^9)/(1.7 \times 10^4) = 1.8 \times 10^5$ individual vectors. To ensure that the chance of finding any particular segment will be greater than 99% requires about $10^6$ vectors. Because cosmids carry larger amounts of DNA (about 45 kbp), fewer of these vectors would be needed.

Once the vector recombinants are constructed, they are mixed with an excess of bacterial cells to be sure that only one vector infects a given cell. The bacterial cells can be stored indefinitely.

The problem, of course, is to find a particular fragment among the millions of cells. The hunt is carried out by making a radioactive **probe** that will hybridize (complementary strands of nucleic acids will bind one another) with the vector carrying the fragment needed.

The probe is made by plating samples of the bacterial culture on an appropriate medium and incubating it overnight to allow colonies to develop. A nitrocellulose filter is then touched to the colonies, thereby transferring some of the cells from each colony to the filter, as in the replica plating technique. The original plate is set aside.

The cells on the filter are then lysed with an alkali in such a way that the DNA from each colony sample sticks to the filter in the same position as the colony on the original plate from which the filter samples were taken. The radioactive probe is then added and allowed time to react with its complementary strand if it is present. The filter is then washed to remove any probe and unlabeled DNA that have not hybridized. The hybrid molecule is detected by autoradiography. (At this point, if hybridization has not occurred, the procedure is repeated with another sample from the bacterial culture.)

Cells carrying the correct fragment are then isolated from the original plate and grown in sufficient numbers to provide the desired amount of DNA fragment. If need be, the fragment can be increased in amount by the PCR.

Nonradioactive probes are also available. For instance antibodies against the gene protein end products or a biotin-avidin probe can be used. Biotin is bound to a nonradioactive DNA probe, which is then used in the hybridization step just described. The avidin, to which a fluorescent dye or enzyme has been attached, is added to a solution bathing the filter. Biotin and avidin form a very tight complex. So any hybridized DNA would also bind avidin, and this complex of DNA-biotin-avidin is detected by fluorescence or enzyme activity.

# MAPS, RESTRICTION MAPS, AND SEQUENCING

To acquire a detailed understanding of the genetics of an organism, to be able to manipulate genes, and to be able to compare genes—normal vs. abnormal—require a knowledge of the location of genes on chromosomes.

## Maps

Since Thomas Hunt Morgan's early studies (1910) of eye color and sex (linked on the X chromosome) in *Drosophila* and Alfred H. Sturtevant's first genetic map of the fruit fly genome in 1913, much has been learned about a multitude of organisms, including human beings.

By 1966, about 1500 human genes had been identified and mapped. Their locations were published in Victor A. McKusick's book *Mendelian Inheritance in Man*. Today the listing is well into the thousands. The rapidity with which the information is expanding now necessitates the use of computerized data banks.

Mapping now utilizes family studies, somatic cell genetics, cytogenetics, biochemistry, molecular genetics, and other fields of study. It is a procedure whereby the relative positions of genes on chromosomes can be identified. The idea behind the technique is that genes on a given chromosome do not assort independently; they are linked. However, pairs of alleles can exchange positions. The exchange, or **crossing-over,** occurs during meiosis when homologous chromosomes pair up and, after duplication, sister chromatids exchange portions of their chromosomes. The crossing-over produces recombinations of genes.

The exchanges happen at characteristic frequencies. Crossing-over locations are random and the cross-over frequency of a pair of linked genes varies directly with their physical separation on the chromosome. The nearer the genes, the less likely there will be a crossing-over event that causes recombination between them. Conversely, the farther apart the genes, the more likely it is that crossing-over will occur between them. Figure 11.10 is a representation of a crossing-over event.

The cross-over frequencies are used to determine the relative positions of genes on a chromosome. The genes are then separated by map units, which are a measure of the genetic distance between two linked genes. One **map unit** equals a recombination frequency of 1% (sometimes this is called 1 **centimorgan**). See Figure 11.11 for an example of chromosome 2 of *Drosophila*.

## Restriction Maps

Mapping chromosomes using cross-over frequencies has a major drawback: within any particular location there may be dozens upon dozens of other genes. These maps are like the maps of states that locate cities relative to one another but not blocks or individual streets or houses on a street. Molecular genetics is now concerned with locating the houses (genes) and determining the structure of the houses (the nucleotide sequences of the genes).

What allows us to begin a search for a gene and its structure are the restriction enzymes and a technique called gel **electrophoresis.** The creation of **restriction maps** by this technique relies on the fact that DNA is a charged molecule—negatively charged owing to the phosphate groups along the backbone (Chapter 1)—and so will migrate in an electrical field.

The endonucleases are used to cut prepared double-stranded DNA into precisely defined segments based on the locations of recognition sequences. The fragments are then separated by gel electrophoresis. The smaller fragments (cut out of DNA when restriction sites are close together) move faster and farther through the gel than do the larger fragments. So the fragments are separated on the basis of size and their net charge. Different endonucleases will give different banding patterns in gels because of the different number and location of their recognition sites (Figure 11.12a).

By cutting DNA with two or more endonucleases and separating the fragments by electrophoresis, a restriction map can be constructed as shown in Figure 11.12b. Each fragment can then be sequenced. We can use this technique to locate a particular nucleotide sequence on a chromosome and to compare how related fragments are similar and how they differ with regard to nucleotide sequence. After staining, the fragments are read much as the bar code on merchandise is read by scanners.

Because individuals of a species have differences in their nucleotide sequences on the same chromosomes, on average about every 200–500 bp, the numbers and locations of restriction endonuclease recognition sites ought to vary from one individual to another. In other words, genetic **polymorphism** ("many forms") should translate into fragment length polymorphism. And, indeed, it does. DNA from different individuals, even of the same species, show what is termed **restriction-fragment length polymorphism,** or **RFLP** (Figure 11.13).

Not only does RFLP show differences between normal individuals of a given species, but it can also be used to detect differences between normal individuals and

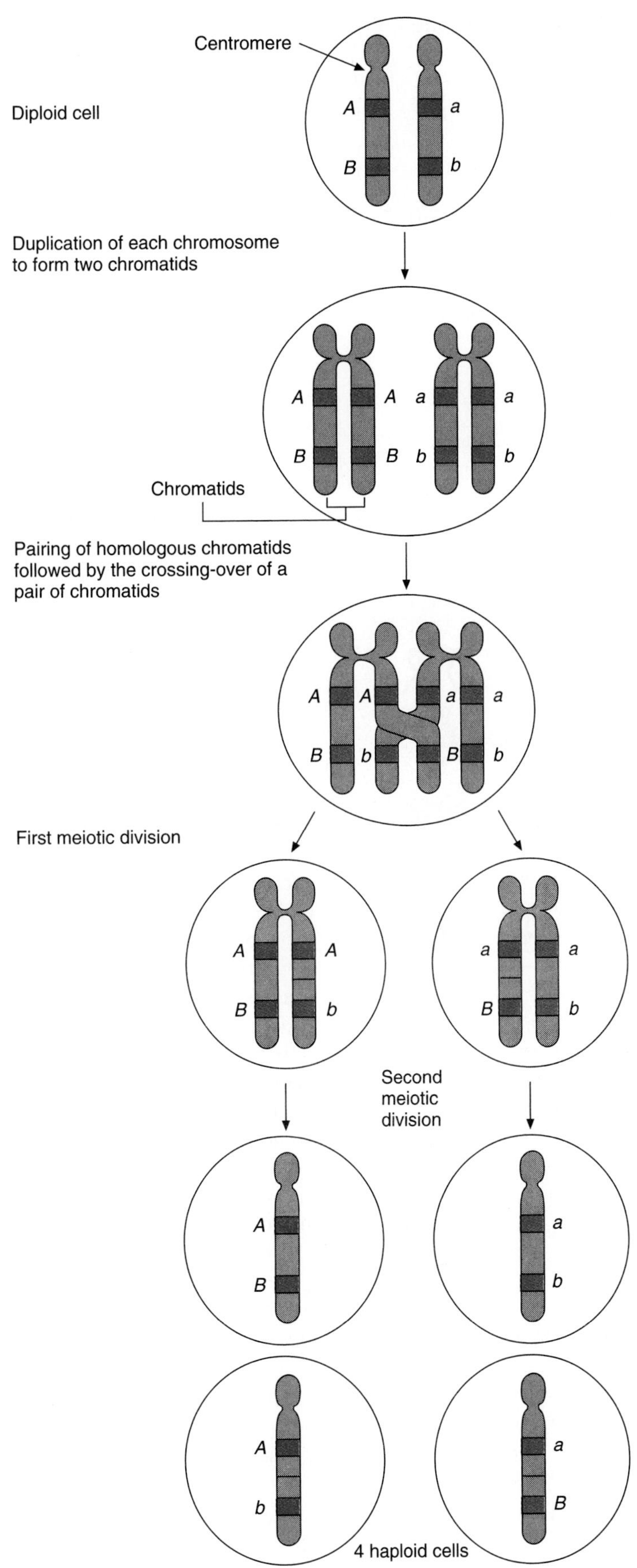


FIGURE 11.10
A representation of a cross-over event during meiosis. Recombination may occur wherever crossing-over takes place.

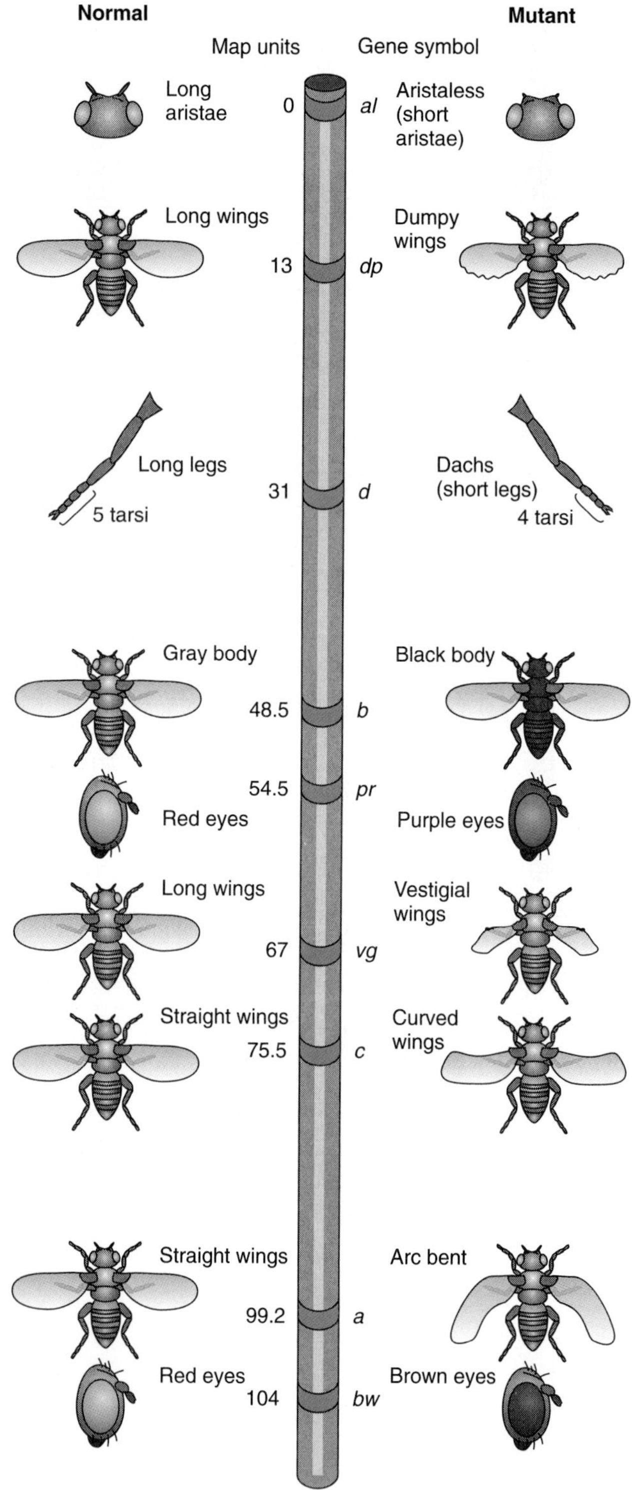


FIGURE 11.11
Drosophila chromosome 2. Some genes responsible for anatomical features are shown. Both normal and mutant characteristics are shown for each gene.

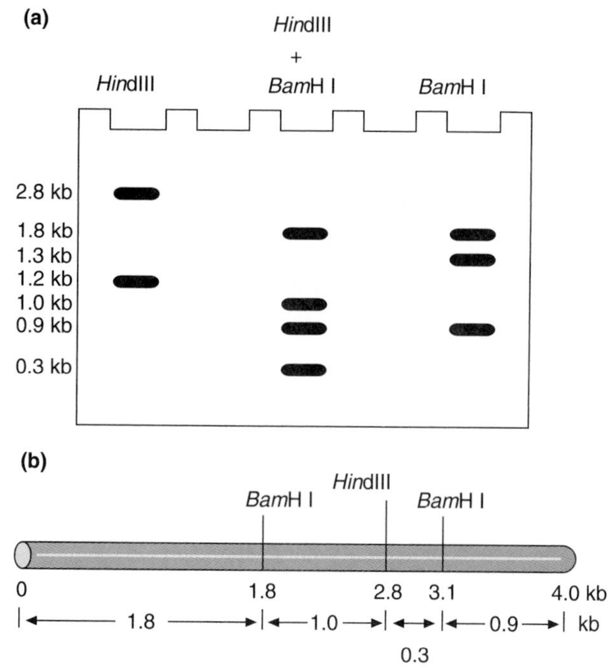


FIGURE 11.12
(a) Restriction fragments produced by Hind III and Bam HI and the fragment patterns produced by gel electrophoresis.
(b) Reading from left to right: a restriction map constructed from the fragment patterns.

those suffering from genetic disorders and it can be used to locate those differences.

Fragment length polymorphisms have been used to pinpoint the location of the genes for Huntington's and Alzheimer's diseases, cystic fibrosis, and many others. A word of caution, however: it should not be assumed that the RFLP procedure is the only method for locating genes. Most often, this technique is used in combination with other procedures (family studies, cytogenetics, probes, etc.).

For example, the gene for Duchenne muscular dystrophy was located without using the RFLP technique. Family studies were used to show that the disease strikes males more often than females, but female sufferers were invaluable in the search for the gene. Since cytogenetic evidence revealed that all females with the disease had a translocation of the Xp21 band, a search using all known X chromosome probes was done to find one that would hybridize with DNA from male patients. Probes of the Xp21 region showed a deletion in this area in all patients but rarely in healthy individuals.

## Sequencing

Once a fragment of interest has been identified and isolated using a technique called chromosome walking (more on this later in the chapter), it can be cloned using the PCR and sequenced to determine the exact order of the nucleotides of the fragment or gene.

Two methods of sequencing have evolved, and both are widely used. Each of the techniques was first introduced in the late 1970s. One is the Maxam and Gilbert procedure, and the other is the Sanger-Coulson method. (Sanger also pioneered amino acid sequencing of proteins.)

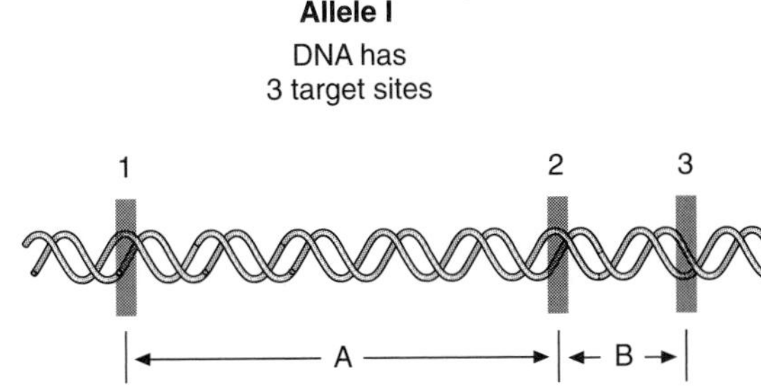


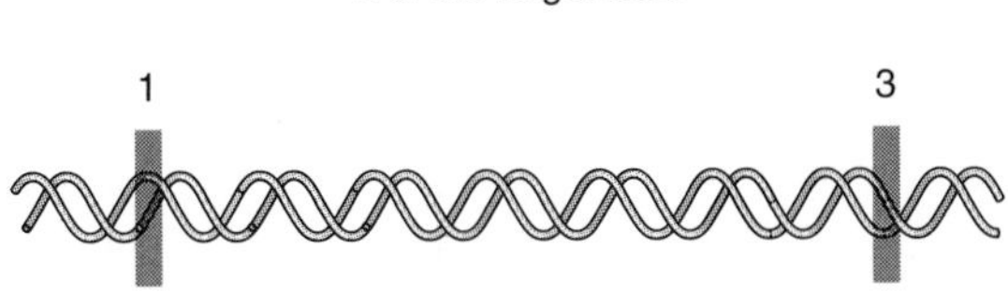


FIGURE 11.13
Changes in nucleotide sequences may alter the number and size of restriction fragments. These sequence changes produce restriction-fragment length polymorphisms.

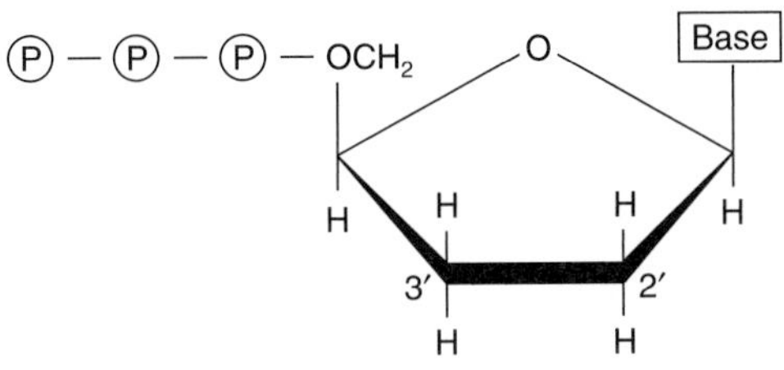


FIGURE 11.14
2′, 3′ dideoxynucleoside triphosphate. Note the absence of OH groups at the 2′ and 3′ positions.

The first, often called the chemical method, is the more complex of the two. It involves radioactive phosphate, a number of chemical cleavages of double-stranded–derived single-stranded DNA, fragment separation by gel electrophoresis, and autoradiography.

The simpler Sanger-Coulson method is sometimes referred to as the plus-minus, the primed synthesis, or the chain termination method. This method also utilizes radioactivity and gel electrophoresis. (Radioactivity is often used to increase the sensitivity of a procedure in which sample amounts are exceedingly small or difficult to detect.) Let's take a look at the Sanger-Coulson procedure.

The DNA fragment of interest is cloned into a vector, M13 or a phagemid can be used, to provide single-stranded DNA (ssDNA). A primer of about 20 nucleotides is added to the ssDNA's 3′ end so that the primer is oriented in a 5′ to 3′ direction. The ssDNA acts as the template for *E. coli* DNA polymerase I, from which the exonuclease activity portion has been removed.

In each of four test tubes, the reaction mixture is composed of ssDNA template plus primer, DNA polymerase I, the triphosphates of the four deoxribonucleotides, one of which is radioactive ($^{32}P$), and a small amount of a 2′,3′-dideoxyribonucleoside triphosphate (ddNTP) of one of the bases (Figure 11.14). The ddNTP lacks a 3′ OH group. As noted in Chapter 2, this site is necessary for the addition of the next nucleotide to the growing chain. Each test tube has a different dideoxyribonucleotide added.

As DNA synthesis proceeds, the incorporation of the dideoxyribonucleoside will occur at random and chain synthesis will stop at that ddNTP because another nucleotide cannot be added to its 3′ position (because it lacks an OH group).

Keep in mind, the contents of the four reaction vessels differ only in the dideoxyribonucleotide added. So, in each vessel, a series of shortened chains is produced, with each chain ending in the particular dideoxyribonucleotide added to the vessel at the start of the experiment. The fragments generated in each vessel are separated by gel electrophoresis according to size, and an autoradiogram is developed from the gel—the $^{32}P$ exposes the film, producing the radiogram. Figure 11.15 represents the gel patterns resulting from this procedure.

The gels are read from the bottom to the top (the fastest-moving fragments are of shortest length). Also, when read from the bottom to the top, the fragments correspond to the 5′ to 3′ direction of DNA synthesis, and, since the dideoxribonucleotides are added at random, the smallest fragments represent those portions closest to the primer.

This chain termination procedure has been automated and computerized, and a different fluorescing dye covalently linked to each different dideoxyribonucleotide replaces the radioactive phosphate. The reaction can now be carried out in a single vessel, and the gels can be read automatically by a computer-controlled detection system sensitive to the different wavelengths of the light emitted by the different fluorescent dyes.

This method allows an investigator to identify nearly 10,000 bases *per day* versus the 50,000 or so bases *per year* using the manual procedure. Assuming the human genome's 3 billion base pairs per haploid cell, it would take a single machine 1000 years to sequence the 23 chromosomes. But, of course, many such machines in many different laboratories will be used as this project proceeds (see Chapter 12).

One of the first and best examples of large-scale sequencing is the sequencing of the entire yeast chromosome III. This was done by the European Yeast Genome Sequencing Network, which consisted of 35 European laboratories. A total of 390,000 base pairs was sequenced with an error rate of about 1 in 1000 bp.

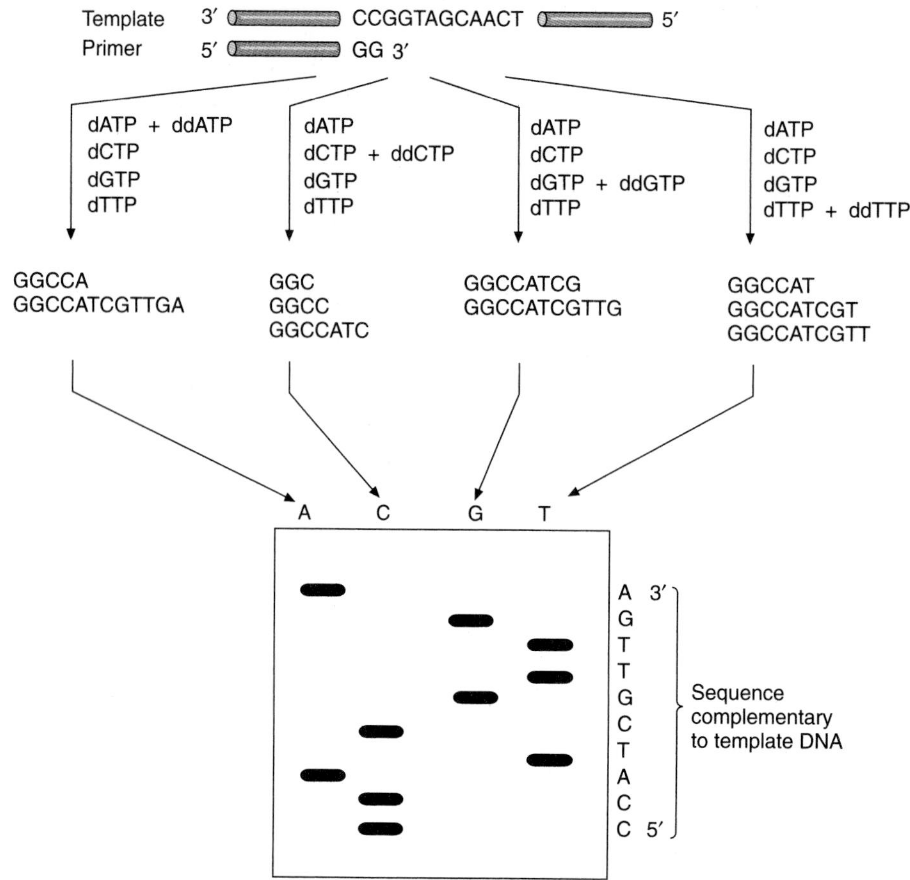


FIGURE 11.15
DNA sequencing using the chain terminating method of F. Sanger. (ddATP represents dideoxyadenosine triphosphate, etc). Fragments are read from bottom (5′) to top (3′).

## SOME OTHER TECHNIQUES

A number of other molecular techniques have been developed as a means to circumvent technical problems in procedures, to improve the efficiency of earlier methods, or to investigate newer questions in molecular genetics.

## cDNA

A procedure that relies on RNA to produce the gene was developed to circumvent the difficult technical problems of identifying, isolating, and purifying a particular gene before using it in manipulation experiments. In this technique, one of the species of RNA is isolated and used as the template. The enzyme reverse transcriptase is then used to produce a copy in DNA of the RNA nucleotide sequence. The copy can then be made into a double-stranded molecule by the enzyme DNA polymerase. (It should be noted that, if the RNA has been processed to remove introns, the copy DNA will not be identical to the gene from which the RNA was transcribed.) This molecule is referred to as copy or **complementary DNA, cDNA.** A collection of these DNAs cloned into a vector can be a library.

Often, cDNAs are produced from specific cell types in which not all the genes are active. Therefore, cDNA libraries of specific cell types are smaller than genomic libraries. Only those genes that are active would be copied into mRNA. And screening for a particular gene is easier if the protein end product is known. Rather than screening the entire genome library searching for the gene, one need screen only the cDNA library produced from the specific cell. So, for example, to get the insulin gene, one screens the cDNA library of pancreas cells.

There are other considerations that make cDNA libraries popular. The ease with which some mRNA can be isolated and purified simplifies the process. Some other considerations are (Sofer 1991):

1. Specialized mammalian cells often produce large quantities of a particular mRNA. Making cDNA is then much easier than attempting to get at the gene by digestion of the entire genome.
2. The mRNAs of eukaryotic cells have a polyadenylated tail, making it a relatively simple matter to isolate

them. Complementary molecules of poly Ts or Us can be used as molecular "hooks" to latch onto the 3′ tails. In this way, mRNA can be separated from the more abundant rRNAs and tRNAs.

3. And, because introns are not present in the mature mRNAs, the cDNA copied from these molecules will, in general, be much smaller than the corresponding gene. Gene size is an important consideration, because as size increases it becomes more difficult to handle and clone the entire gene in a single vector. (Remember that the size of the vector limits the size of the passenger DNA that each type of vector can carry.)

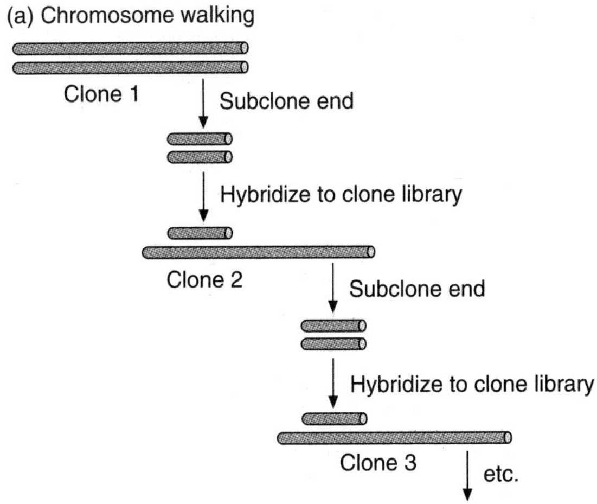


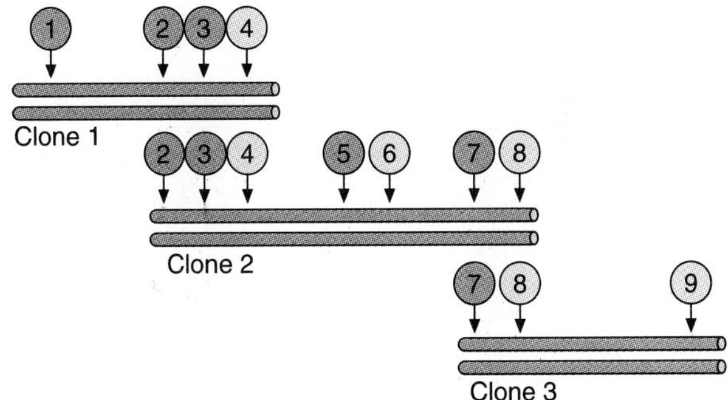


FIGURE 11.16
Chromosome walking. A clone of interest is isolated from a library. A subfragment of one end is subcloned, and this subclone is used to identify the next fragment clone. The process is repeated as often as necessary to build a set of overlapping clones. The direction of extension is determined by making a restriction map of each fragment and in this way large genes are reconstructed.

## Footprinting

Footprinting is a technique for identifying DNA-binding sites of proteins that are necessary for gene induction, repression, or regulation. In this procedure, labeled segments of DNA plus the protein of interest are bound, and then the DNA is digested with a restriction enzyme. A control-labeled segment of DNA (without the protein) is also digested. Both are then subjected to gel electrophoresis. The banding patterns of the segments are compared. The control DNA should have more bands representing fragments produced by the endonuclease than the protein of interest has. The protein-bound DNA has a number of protected sites and so has fewer bands. The missing bands identify the length of the protein binding site and its position relative to the ends of the DNA segment from which it was obtained.

## Chromosome Walking

The way in which large segments of DNA are identified is to sequence overlapping sections or fragments. This technique becomes necessary because very large genes may not fit into a single vector, so a number of vectors are necessary to store the genetic information of interest. Different segments of the gene end up in different clones of the genomic library. The procedure is also used to identify a segment of DNA for which no probe is available but the segment of interest is linked to a fragment or gene that has already been identified and cloned. Restriction maps are made of overlapping segments.

The procedure can be briefly outlined as follows:

A probe is used to locate by hybridization the known segment or gene in a genomic library. All fragments containing the gene are selected and sequenced. The fragments are then aligned (on the basis of their migration positions following gel electrophoresis), and segments farthest from the marker or known gene in both directions are subcloned to produce new probes. These probes are used to re-screen the genomic library for another collection of fragments. These fragments are sequenced and aligned and again new probes are made as before. The process is repeated as many times as required until the desired segment is encountered and sequenced (Figure 11.16).

## Triple-Helix DNA

Triple-helix DNA molecules are being thought of as a new type of molecular scissors for cutting DNA. Triplex DNA can be produced that is capable of cutting only a limited number of sites of DNA. In contrast, the restriction endonucleases most often cut DNA at multiple sites, making separation and analysis difficult. Triplex DNA can also be used to inhibit gene transcription.

In the triplex method, all four deoxyribonucleotides are used to create a short stretch of single-stranded DNA (20 or so nucleotides) complementary to a known sequence of a gene (e.g., promoter or operator regions). The

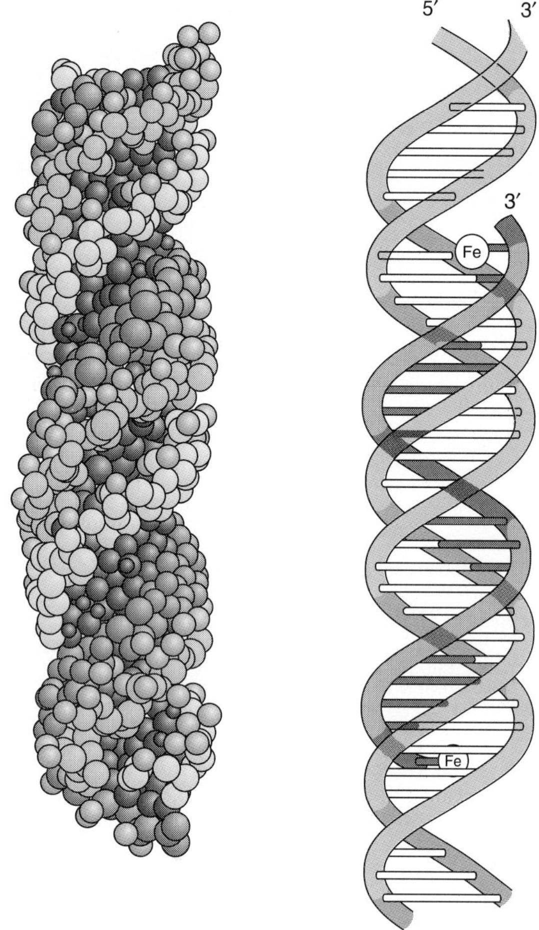


FIGURE 11.17
Schematic of a triple helix complex.

bases of this complementary DNA recognize their corresponding base pairs (i.e., thymine recognizes the A=T base pairs). And, by addition of an oxidizing chemical such as ethylenediaminetetraacetic acid (EDTA) complexed with iron, the third strand can act as a highly selective cutting tool. At the site of the oxidizing agent, the DNA is nicked on one strand. The nick prevents transcription, or, if transcription is already under way, the nick prevents its completion (Figure 11.17).

Experiments with yeast chromosome III indicated that the third-strand DNA correctly targeted a specific sequence in a chromosome that has almost 14 million base pairs. There is also evidence that the technique works on the human genome with its billions of base pairs.

## Antisense RNA

Antisense RNA molecules have a nucleotide sequence that is complementary to a particular mRNA. The antisense sequences can be used to block transcription very selectively by binding to complementary nucleotide sequences of genes, or they can be used to selectively inhibit translation by combining with specific mRNAs. These procedures can also be used to mimic mutations. The use of this technology is especially appropriate for the study of mutations in vertebrates.

Vertebrates have long generation times and gestation periods; many species have small numbers of offspring; and the phenotypic expression of a mutation may not reveal its exact location within a particular biochemical pathway. If, for example, an organism is incapable of utilizing a particular nutrient, is it because of a transport problem or because of the block at an enzymatic step in a pathway? Various antisense RNAs can be used to create "artificial mutants" that allow investigators to study the consequences of the mutation. For example, antisense RNA has been used to re-create in mice a condition called "shiver," which results from a defective nervous system protein—myelin basic protein. This procedure may provide a way of producing mice models of human diseases.

Antisense RNA procedures have been used to create new "mutants": unusual pigmentation in flowers; slower-ripening tomatoes, making transportation of the fruit easier; and disease-resistant species of tobacco. These experiments demonstrate the range of effects antisense RNA technology is capable of producing. This technology may also be used to produce compounds capable of blocking viral genes and oncogenes whose transcription contributes to the development of cancer. In the laboratory, the parasite *Trypanosoma brucei* was killed by an antisense RNA directed against a specific sequence of ribonucleotides found on most trypanosome mRNAs. The procedure works because double-stranded RNAs are degraded in vivo.

The first naturally occurring antisense RNA was found in *E. coli.* The gene encoding the cAMP receptor protein (CRP/CAP; Chapter 7) is inhibited not by a regulatory protein but by an antisense RNA molecule.

Antisense RNA can be synthesized in vitro using a bacteriophage RNA polymerase. The RNA's introduction into cells, however, is still a difficult procedure. Both microinjection and what are called expression vectors have been tried, but technical difficulties are still the order of the day. Expression vectors are carriers into which DNA coding for an antisense RNA molecule is spliced next to a promoter region. Upon entering a cell, the vector is transcribed, and antisense RNA is produced. In this way, vectors containing DNA sequences for different antisense RNAs can be placed in cells, and particular species or types of RNAs can be made nonfunctional upon combining with the antisense RNA. In this way, mutant phenotypes can be mimicked.

## Site-Directed Mutagenesis

In 1993, Michael Smith was awarded the Nobel Prize for his development of the sophisticated **site-directed mutagenesis** technique, which makes possible the mutation of a specific gene at a specific nucleotide. The technique relies on the fact if a (–) strand of DNA with an altered nucleotide sequence is hybridized with its complementary (+) strand, during rolling-circle replication (Chapter 3) the faulty (–) strand will act as the template from which (+) strands containing the "wrong" nucleotide will be produced.

This procedure (Figure 11.18) requires a gene of known nucleotide sequence (Chapter 12, sequencing genes) cloned into a phage, such as M13, the single-stranded filamentous DNA phage that replicates via RF forms (Chapter 3).

An example of how this procedure is done is as follows. A single-stranded copy of the DNA containing a gene of known sequence is isolated from a vector, such as M13, that infects *E. coli*. This is the (+) strand. The isolated DNA fragment is hybridized with a short synthetic segment of its complement, the (–) strand, in which a codon has had the new nucleotide inserted. The synthetic segment can easily be made using an automated DNA synthesizer.

Next, DNA polymerase is added. The complement is extended by the polymerase to complete its base pairing with the isolated strand. The new nucleotide of the (–) strand does not base pair since it is not the complement of the nucleotide on the (+) strand. The duplex is now used to infect *E. coli*. Via rolling-circle replication, M13's method of replication, the (–) altered strand acts as the template for (+) synthesis. The new (+) strands will contain the base that pairs with the new nucleotide inserted into the (–) earlier. In this way the original codon of the (+) strand is altered.

These newly synthesized (+) strands can now be harvested either from the bacterial cell or from the newly produced viral particles. It should be noted that the wild-type M13 DNA does not have any good regions for inserting foreign DNA. Promoters and linkers must be added to the molecule if it is to be used as a vector. But an advantage of using this phage is that, unlike other bacteriophage, M13 does not kill the host cell upon release, so the cell can become a continuous source of phage and altered genes.

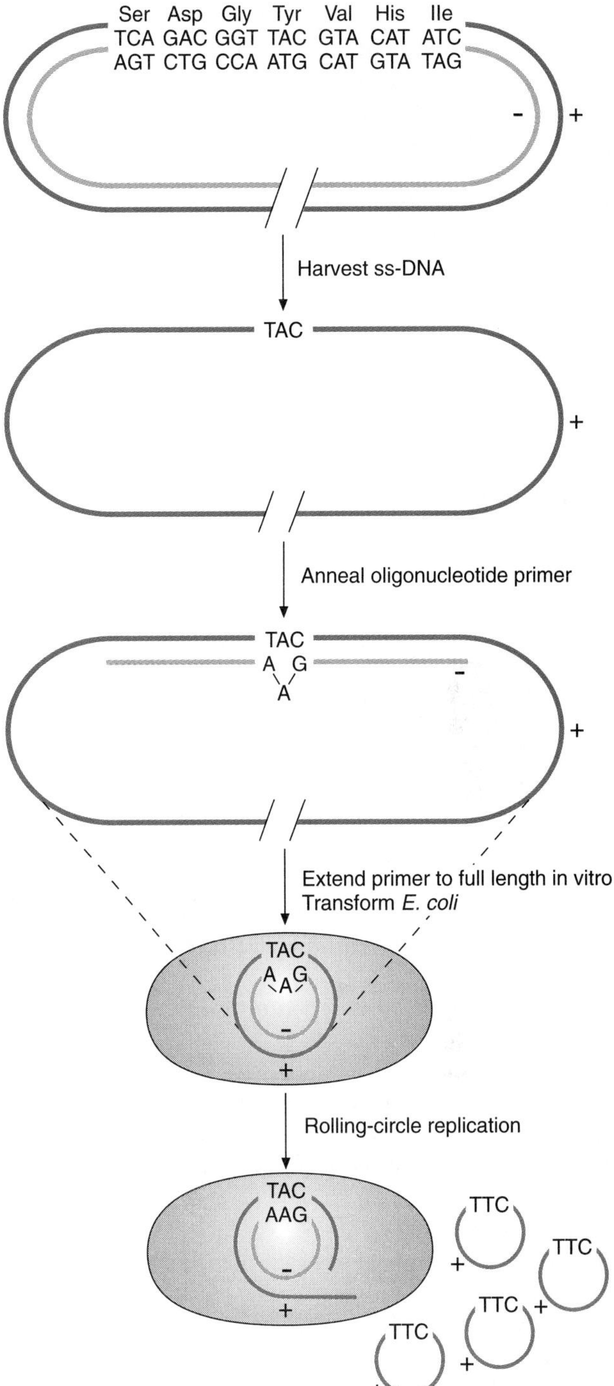


FIGURE 11.18
An example of site-directed mutagenesis using the phage M13 to clone the mutated codon. The codon for tyrosine (TAC) is changed to the codon for phenylalanine (TTC). A dsDNA is used to infect *E. coli*. Plus strands are harvested. Automated DNA synthesis is employed to produce a sequence of (–) strand nucleotides containing the complementary codon to TTC. The (–) strand sequence will hybridize with the (+) strand and will be extended in vitro. This altered dsDNA is used to infect *E. coli*. The (–) altered strand will be used as the template to produce mutated ssDNA, the (+) strand which now contains the codon TTC in place of the original codon TAC.

## Gene Targeting

Site-directed mutagenesis was developed with the idea of being able to selectively change a nucleotide sequence and then see the effect of that change on the function of the gene's product. This procedure, however, does not lend itself to the selective inactivation of genes, since it would most likely require a large number of changes or a prior knowledge of which changes might inactivate a particular gene.

This problem was overcome in the 1980s. K. R. Thomas and M. R. Capecchi reported in 1988 a procedure called gene targeting that would selectively inactivate genes (the terms *knockout* or *blocked-out* are also used to indicate this kind of gene inactivation). Capecchi's procedure involved altering a gene in vitro so that it contains, within one of its coding sequences, an exon, the neomycin-resistant gene (*neo*$^r$) and a second marker, the herpes *tk* gene, at the gene's end. The altered gene is then introduced into cells isolated from a mouse embryo (embryonic stem cells). During the process of **homologous recombination,** chromosomes pair and exchange segments. This leads to the incorporation of the altered gene, minus the *tk* portion, into the normal gene in some cases. Cells in which this exchange has occurred can now be identified. These cells, with the normal gene now carrying the neomycin-resistant gene, will grow in a medium containing the neomycin analog G418; other cells will die. Cells in which random incorporation of the *neo*$^r$ and the *tk* genes has occurred will be killed by the antibiotic ganciclovir. So only those cells carrying the incorporated altered gene (*neo*$^r$) survive and can be injected into embryos. After breeding the animals who had received the stem cells, those whose germ cells (egg or sperm) carry the inactivated gene derived from the stem cells are selected for further breeding since now the inactivated gene can be passed on to their progeny. However, many such mice may fail to develop if the inactivated gene is crucial for development.

This procedure opened the door to the study of the roles specific genes and their products play in biochemical pathways and in development. But since cells with altered genes are injected into embryos, the alteration of specific cell types was not possible because genes were knocked out in embryonic stem cells.

In 1994, Klaus Rajewsky, Hua Gu, and their colleagues reported the first successful use of gene targeting to selectively knock out a gene in a particular cell type in mice. These scientists were interested in the importance of the enzyme DNA polymerase beta in developing T and B cells, the critical cells of the immune system. The investigators succeeded in inactivating the DNA polymerase beta gene in mice T cells. This tissue-specific knockout procedure should now be applicable to other cell types. The procedure allows the investigator to inactivate a gene only in specific cells and during any stage of development.

Rajewsky and Gu developed a variation on Capecchi's technique. They used enzymes derived from bacterial viruses and yeast to accomplish the feat of inactivating a gene in a specific cell type.

The enzyme Cre recombinase from bacteriophage P1 (it infects *E. coli*) separates phage genomes, which become joined after infection. Cre does this by lining up short sequences of DNA called *lox*P. The DNA between these sites is then removed, leaving one *lox*P sequence behind. In the mid-1980s, B. Sauer noted that the Cre enzyme would also work in higher organisms. He introduced into cultured cells, engineered to contain an active *cre* gene, a gene flanked by *lox*P sequences. In short order, the *lox*P-flanked gene was removed from the cultured cells' genomes.

Following this demonstration, Rajewsky's group replaced a part of the normal DNA polymerase beta gene with a cloned copy of the segment that was flanked by *lox*P sequences. These mice were normal; they were able to make the polymerase beta protein despite the presence of the *lox*P sequences in the gene. Rajewsky then mated his mice to transgenic mice in which the *cre* gene is expressed only in developing T cells. The T cells of the offspring survived even though DNA polymerase beta was not made. Cre recombinase had succeeded in removing a sizable portion of the polymerase gene. The conclusion is that the polymerase is not absolutely vital during the entire T-cell developmental stage.

More important, however, is the fact that the technique of cell-type specific inactivation does work. With this knowledge and procedure scientists now have a method to study the fate and development of particular cells in the developing organism. And investigators are already busy studying brain development in mice. For example, they have found that, when the so-called engrailed genes are knocked out parts of the mid- and hindbrain do not develop.

## SUMMARY

Gene manipulation, or genetic engineering, requires the insertion of passenger DNA into a vector. The vector must be introduced into a host cell where, with a few exceptions (liposomes), it must be capable of reproducing both itself and its passenger DNA. And cells carrying these vectors must be isolated and grown in pure culture. This has all been made possible with the discovery of restriction endonucleases and DNA ligases, enzymes that selectively cut DNA and reform phosphodiester bonds, respectively.

Another extremely important contribution to genetic engineering was the development of the polymerase chain reaction (PCR). With this reaction, very small amounts of DNA can be increased so that gene manipulation is made much easier.

Over the years, a number of special and even artificial cloning vectors have been developed from plasmids and viruses for use in bacteria, yeast, and mammalian cell lines. And other techniques have been developed for use with difficult plant cells.

The new vectors have some important advantages over natural vectors. For example, larger amounts of DNA can be carried as passenger DNA, single-stranded DNAs can be produced, and the behavior of some vectors closely resembles the behavior of natural chromosomes, as is the case with yeast artificial chromosomes (YACs), making the study of the behavior of genes more accessible.

The cloning of various genes in bacteria, yeast, plants, and mammalian cells has as one goal the production of proteins of value to humankind—such as insulin and human growth hormone. Another goal is the production of organisms with certain desirable traits—such as plants with higher fruit or grain yields or plants that are resistant to herbicides, which would make weeding very much easier and more efficient.

The transfection of mice embryos has opened the door to the study of the response of mammalian genes to a new environment and of the host animal's response to foreign genes: Do these new genes stably integrate? Where and how do they integrate? How are the new genes controlled by the host cell?

The vectors developed for cloning experiments have also found a use as storage facilities for entire genomes. The vectors can serve as genomic libraries. Techniques have been developed to create such libraries and to retrieve the information stored therein.

The techniques of classical and molecular genetics are being used in attempts now under way to map the entire genomes of viruses, bacteria, nematodes, yeasts, insects, and even of the human being. Scientists are now capable of determining complete nucleotide sequences of genes. This will not only allow for the localization of genes on specific segments of chromosomes but will also give a detailed structural picture of genes.

To facilitate these investigations, researchers are continually developing new techniques. cDNA methods allow for the construction of complementary DNA from mRNA—in other words, by knowing the product of a gene we can construct the gene. Footprinting allows for identifying DNA binding sites. Chromosome walking allows for sequencing large segments of DNA. Triplex DNA is a highly selective cutting technique. And antisense RNA is a very powerful technique for selectively blocking transcription and mimicking mutations (creating artificial mutants).

The development of site-directed mutagenesis and gene targeting now makes possible the mutation of specific nucleotides and the inactivation of particular genes in individual cell types. This will open the door to the study of the role of particular gene segments in protein function and the study of the development of cell types and entire multicellular organisms.

## Study Questions

1. Describe in some detail the necessary elements of recombinant DNA production.
2. What is the importance of palindromes in restriction enzyme function?
3. What are the three methods of ligation, and what are the differences between them?
4. How might each of the different methods of ligation be used?
5. What are some characteristics of a good vector?
6. Why is screening a necessary procedure in recombinant DNA work?
7. Give a protocol of how a bacterium carrying a recombinant DNA molecule might be identified and isolated.
8. Describe a procedure for producing recombinant DNA.
9. Explain the logic behind the polymerase chain reaction.
10. Describe the steps of the PCR, and explain what is going on at each step.
11. Describe at least two vectors used in prokaryotic cloning. What are the advantages and disadvantages of each?
12. What are the differences among the yeast vectors, and what are the advantages of each?
13. Indicate some of the problems encountered in attempts to transfect plant cells and how the problems are being attacked.
14. The transfection of mammalian cells has also run into a few problems. What are the problems and some attempts at solutions?
15. Explain the differences between maps, restriction maps, and sequencing. What kind of information is obtained from each procedure?
16. How are restriction maps constructed?
17. Explain the chain termination method of DNA sequencing.
18. Describe at least two techniques other than those referred to in questions 1–17 above that are commonly used in molecular genetics.

## Readings and References


Barinaga, M. 1994. Knockout mice: Round two. *Science* 265:26–28.

Berg, P. 1981. Dissection and recombination of genes and chromosomes. *Science* 213:296–303.

Capecchi, M. 1989. Altering the genome by homologous recombination. *Science* 244:1288–92.

Capecchi, M. 1994. Targeted gene replacement. *Sci. Am.* 265 March:52–59.

Eguchi, Y., T. Itoh, and J. Tomizawa. 1991. Antisense RNA. *Annu. Rev. Biochem.* 60:631–52.

Fincham, J. R. S., and J. R. Ravetz. 1991. *Genetically engineered organisms: Benifits and risks.* Toronto: University of Toronto Press.

Gasser, C. S., and R. T. Fraley. 1992. Transgenic crops. *Sci. Am.* 266 June:62–69.

Gu, H., J. D. Maeth, P. C. Orban, H. Mossmann, and K. Rajewsky. 1994. Deletion of a DNA polymerase B gene segment in T cells using cell type–specific gene targeting. *Science* 265:103–06.

Hunkapiller, T., R. J. Kaiser, B. F. Koop, and L. Hood. 1991. Large-scale and automated DNA sequence determination. *Science* 254:59–67.

Lewin, B. 1997. *Genes VI.* New York: Oxford University Press.

McKusick, U. A. 1990. *Mendelian Inheritance in Man: Catalogs of Autosomal Dominant, Autosomal Recessive, and X-linked Phenotypes*, 9th ed. Baltimore: Johns Hopkins University Press.

Moffat, A. S. 1991. Triplex DNA finally comes of age. *Science* 252:1374–75.

Morgan, T. H. 1910. Sex limited inheritance in *Drosophila*. *Science* 32:120–22.

Mullis, K. B. 1990. The unusual origin of the polymerase chain reaction. *Sci. Am.* 262 (April):56–65.

Noren, C. J., S. J. Anthony-Cahill, M. C. Griffith, and P. G. Schultz. 1989. *Science* 244:182–88.

Rosenberg, S. A. 1990. Adoptive immunotherapy for cancer. *Sci. Am.* May:62–69.

Sofer, W. H. 1991. *Introduction to Genetic Engineering*. Boston: Butterworth-Heinemann.

Sturtevant, A. H. 1913. The linear arrangement of six sex-linked factors in *Drosophila*, as shown by their mode of association. *J. Exp. Zool.* 14:43–59.

Watson, J. D., M. Gilman, J. Witkowski, and M. Zoller. 1992. *Recombinant DNA*. 2d ed. New York: W. H. Freeman.

Weaver, R. F., and P. W. Hedrick. 1997. *Genetics*. 3rd ed. Dubuque, Iowa: Wm. C. Brown.

Weintraub, H. M. 1990. Antisense RNA and DNA. *Sci. Am.* 262 (January):40–46.

CHAPTER TWELVE

# THE PROMISES AND THE PERILS OF THE NEW GENETICS

## OUTLINE



## CHAPTER OBJECTIVES

*This chapter will discuss:*

- The operations of the Human Genome Project
- The nature of infectious and noninfectious diseases and the use of genetic engineering to cure or prevent them
- The genetic disease cystic fibrosis and gene therapy–based treatment
- Laboratory procedures based on molecular genetics techniques
- Some uses of molecular genetics procedures in various fields of study
- The risks and dangers some people see in the applications of molecular genetics procedures

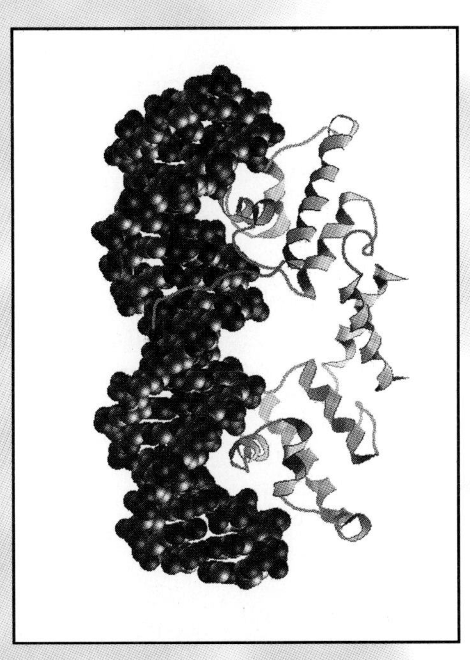

## INTRODUCTION

In the relatively short time since the publication of the Watson and Crick paper concerning the structure of DNA (1953), the field of molecular genetics has touched and profoundly affected the entire science of biology, its myriad subspecialties, and even fields outside of biology. The tools, techniques, and procedures of molecular genetics have found their way into diverse areas such as agriculture, archeology, embryology, human physiology, and paleontology. No doubt, however, the areas of biology most affected have been those most closely related to the human condition. The power of molecular genetics has been brought to bear on the many problems associated with the human physical and mental states.

Most people agree that the field of molecular genetics holds great promise. Sooner rather than later, what had once been thought of as intractable health problems will begin to yield to the power of the new technologies. This has become an article of faith among scientists and laypersons alike. Both cures and preventions of disease are now discussed with hope and optimism rather than with despair. Even so, some diseases, such as AIDS and malaria, have proved to be extremely intractable to molecular biology.

Are there, however, some pitfalls along the paths now being taken? How, for example, is information about an individual's genetic makeup to be used? And, ought we now to be identifying carriers of particular diseases, especially those for which there is no cure?

These, of course, are not, strictly speaking, questions of biology. They are not even questions for which biologists or molecular biologists have, necessarily, a special insight. They are philosophical, ethical, moral, and legal questions. They are, nonetheless, questions that society must confront. The answers may have profound consequences. We may, for example, be led to the ultimate genetic experiment—the management and control of human conception and ultimately of human evolution. We have, in effect, already managed our evolution somewhat; for example, we have raised taboos against certain marriages, and modern medicine extends the lives of the genetically disadvantaged to the point that they can produce offspring with the same defects. By and large, however, with few exceptions, we have not directly or specifically influenced the evolution of human beings on a massive scale (attempts at eugenics were urged in the early 1900s, and, of course, there was the Nazi program to eliminate the mentally and physically defective). Molecular genetics may soon provide the knowledge and the tools for such a direct assault on the direction of human evolution. The Human Genome Project and the techniques and procedures of gene manipulation may provide the basis for control of our evolution. That may become the ultimate peril.

## THE HUMAN GENOME PROJECT

Since the publication in the 1500s of the Flemish anatomist and surgeon Andreas Vesalius's work *De Humani Corporis Fabrica* (*The Construction of the Human Body*), investigators have been discovering, identifying, classifying, and mapping the human body. The goal has always been to get a better understanding of the body, an understanding of both its construction and its functions. Now, scientists are pushing beyond construction and function to blueprints. They are asking What is the master plan of the human body? In other words, what are the sets of instructions that control the development, construction, and functioning of the body? And, they ask, how are these instructions themselves controlled and relayed within a cell and between and among cells as the body develops and operates?

From the point of view of molecular genetics, the pursuit began with E. B. Wilson's 1911 report linking colorblindness in human beings to the X chromosome. For the next 60–70 years progress in identifying and mapping human genes was relatively slow. In fact, it wasn't until 1968 that the first gene linked to a chromosome other than X was reported. Roger Donahue and colleagues mapped the gene for the Duffy blood group to autosome 1.

But since the late 1980s, there has been a veritable explosion of information concerning the locations of human genes. At the moment, new genes are being found and localized on various chromosomes at the rate of about a dozen a week. In fact, some researchers have suggested that by 2010 all of the human genes will have been located. Positioning genes somewhere on the 46 chromosomes is not, however, the same as knowing their structure—their nucleotide sequences. That is the ultimate goal—to be able to read the instructions.

You can get some idea of the enormity of what has come to be called the Human Genome Project by comparing it with other sequencing projects of genomes much smaller than the 3 billion base pairs of the human. The genomes of the bacterium *Escherichia coli* and the yeast *Saccharomyces cereviseae*, each with some 4 million base pairs, have just recently been completely sequenced. There has been much progress, however, in the sequencing of the 100 million base pairs of the haploid genome of the nematode *Caenorhabditis elegans*. Table 12.1 lists some of the organisms now being intensely studied.

Laboratories in the United States, Europe, and Japan are working on their own human genome projects. In the United States, the sequencing of the entire 46 human chromosomes is estimated to cost between $3 billion and $5 billion over a span of about 15 years. Molecular genetics has become "big science," rivaling physics.

Each segment of DNA must be cloned and then sequenced. The procedures are labor-intensive and expensive

**table 12.1**

ORGANISMS BEING STUDIED FOR GENE SEQUENCE

| ORGANISM | CLASSIFICATION | NUMBER OF BASE PAIRS |
|---|---|---|
| *Escherichia coli* | Bacterium | 4,000,000 |
| *Saccharomyces* sp. | Yeast | 12,000,000 |
| *Arabidopsis thaliana* | Plant | 100,000,000 |
| *Caenorhabditis elegans* | Nematode | 100,000,000 |
| *Drosophila melanogaster* | Insect | 165,000,000 |
| *Mus musculus* | Mammal | 3,000,000,000 |
| *Homo sapiens* | Mammal | 3,000,000,000 |

at present. It is, therefore, imperative that automated techniques and systems be devised that can handle large-scale sequencing of large fragments at reasonable cost.

For example, standard agarose gel electrophoresis is useless in trying to separate fragments of 30 kbp or larger. Good separation is achieved only with fragments of 10 kbp or smaller. But with the development of a separation technique called pulsed field gel electrophoresis (PFGE), in which the electrical field is periodically reversed, DNA molecules some 200–3000 kbp long can be well separated. Remarkably, why this technique works is not clearly understood. What is known is that the size of the molecules that can be separated depends on the length of time the electrical field is applied in one direction before it is changed. The longer the charge is applied in one direction, the larger size range of molecules that can be separated.

Table 12.2 gives information currently available concerning what has been done on the human genome. The information in Table 12.2 was obtained using, among others, many of the techniques and procedures described in Chapter 11. So, for example, chromosome walking, PCR, genetic linkage mapping, YACs, cosmids, cDNA, PFGE, and more were used. The table includes the results of studies in the United States, Europe, and Japan.

The rate at which the genome is being sequenced is increasing ever so rapidly. To date, only a few percent of the 3 billion base pairs within a haploid human cell have been sequenced.

## CURABLE DISEASES

Since before the dawn of science, humans have longed to find cures for the multitude of human afflictions (not to mention animal and plant diseases). Diseases can be conveniently divided into two groups: infectious and noninfectious.

The defining characteristic of infectious diseases is that they are brought about by other organisms—bacteria, viruses, protozoa, and so on. These diseases may or may not be transmissible (contagious or communicable). Infectious diseases include AIDS, pneumonia, the common cold, tetanus, and botulism. AIDS, pneumonia, and the common cold are transmissible; tetanus and botulism are not. Transmissible diseases are the result of the activities of the infecting organism—the organism is the problem.

Noninfectious diseases are disorders resulting from the activities of a microorganism's products—an exotoxin or an endotoxin is the problem. Noninfectious diseases also include all those abnormalities not brought about by other organisms: for example, heart disease, strokes, Alzheimer's disease, cystic fibrosis, Huntington's disease, obesity, mental illnesses. Noninfectious diseases are not transmissible.

There are cases that do not fit neatly into one or the other category. A case in point is diphtheria. The organism that causes this disease, *Corynebacterium diphtheria,* is dangerous both as an organism and because of the potent neurotoxin, an exotoxin, it produces. And the debate about whether some cancers are due to an infection process has still not been resolved.

Molecular genetics, in particular genetic engineering, is now at the forefront in the search for cures and preventive measures for both infectious and noninfectious diseases. Cures, of course, involve the reestablishment of a healthy condition, whereas preventive measures involve establishing safeguards against disease—vaccines or abstinence to protect against sexually transmitted disease, for example. So, we can talk about curing noninfectious diseases such as Alzheimer's disease and cystic fibrosis and preventing such infectious diseases as AIDS, malaria, pneumonia, and polio.

In attempting cures for the noninfectious diseases, we must deal with the individual's fundamental makeup, his or her genetic constitution. A genetic error has lead to a physiological problem. In the case of infectious diseases, we make use of an individual's own defense system to ward off attack by another organism; we stimulate the individual's immune system or the disorder is treated with antibiotics or drugs, when possible.

Each of these approaches involves distinct problems. In the one case, a defective gene or genes must be located and repaired or replaced. In the other case, an antigen that

table 12.2

THE HUMAN GENOME

| CHROMOSOME NUMBER | ESTIMATED NUMBER OF GENES | NUMBER OF MAPPED GENES | NUMBER OF DISEASE-RELATED GENES |
|---|---|---|---|
| 1 | 1450 | 247 | 55 |
| 2 | 3950 | 138 | 24 |
| 3 | 3200 | 85 | 24 |
| 4 | 3050 | 99 | 25 |
| 5 | 2900 | 92 | 23 |
| 6 | 2750 | 131 | 27 |
| 7 | 2700 | 132 | 25 |
| 8 | 2250 | 64 | 22 |
| 9 | 2200 | 77 | 26 |
| 10 | 2200 | 79 | 15 |
| 11 | 2200 | 154 | 46 |
| 12 | 2050 | 136 | 23 |
| 13 | 1800 | 31 | 13 |
| 14 | 1750 | 66 | 19 |
| 15 | 1650 | 65 | 17 |
| 16 | 1400 | 83 | 17 |
| 17 | 1350 | 135 | 26 |
| 18 | 1250 | 28 | 9 |
| 19 | 1150 | 132 | 24 |
| 20 | 1050 | 45 | 14 |
| 21 | 900 | 39 | 8 |
| 22 | 950 | 71 | 17 |
| X | 2350 | 225 | 111 |
| Y | 1000 | 18 | 1 |

will not damage the host but will stimulate the immune system must be constructed, or an antibiotic or drugs that will eliminate the cause of the disease must be found.

## Infectious Diseases

Where infectious diseases such as AIDS and malaria are concerned, what is needed is either some invariant, or unchanging, part of the organism (e.g., an invariant protein) or an organism that has been killed or is no longer capable of producing the disease. Use of the invariant part is preferable, but both approaches can be used to produce a vaccine without fear of harming the host. For a variety of reasons, major unanticipated stumbling blocks have been encountered using these approaches.

A vaccine for AIDS is not expected to be ready for human trials until 2010, but a malaria vaccine may be ready for human trials in the United States within the next several years. These vaccines and many others (hepatitis B) are the result of genetic engineering.

An ideal vaccine would be produced by isolating and cloning in a bacterium or yeast a gene for a protein whose structure does not vary within the organism over time and whose structure is a strong stimulant of the immune system. The vaccine protein must be one that is essential to the organism's ability to infect a host so that any interference with the organism's natural protein (e.g., by antibodies) prevents infection.

## Noninfectious Diseases

The gene for cystic fibrosis has been found, isolated, cloned, altered, and inserted into diseased cells in vitro and a "cure" has been achieved—in vitro. But can the altered gene be delivered to a patient's lung cells, and will it cure cystic fibrosis? Always keep in mind that the patient may be cured by this approach, but his sperm or her eggs will still carry the defective gene, so the disease can still be passed on to offspring. Avoiding this possibility would require genetically engineering the reproductive cells, a formidable task at the moment, even if it were allowed by government regulatory agencies. Some diseases for which genetic treatments are feasible are shown in Table 12.3.

For the moment, the first diseases of a noninfectious nature for which gene therapy holds the best chances of a successful cure would be those caused by a single-gene defect. Table 12.4 lists some potential candidate diseases for a genetic engineering cure.

*table 12.3*

**Possible Cures or Preventions via Gene Therapy**

| Disease | Cure | Prevention |
|---|---|---|
| ADA | + | |
| AIDS | | + |
| Alzheimer's disease | + | |
| Cancer | + | + |
| Diabetes Type II | + | |
| Huntington's disease | + | |
| Malaria | | + |
| Neurofibromatosis | + | |
| Retinitis pigmentosa | + | |

*Note:* A plus (+) indicates that gene therapy is known to be feasible.

Within specialized areas of medicine, the techniques and procedures of molecular genetics such as DNA probes, restriction-fragment length polymorphism studies, and linkage analysis are leading to the localization and characterization of the genes responsible for an array of disorders. Table 12.5 lists some neurological disorders for which locations are known and some characterizations of the gene defect.

## CYSTIC FIBROSIS

Let us take a look at one single-gene disorder, its genetic cause, and an approach to its cure—gene therapy. Most single-gene disorders occur only rarely. But cystic fibrosis (CF) is more common. It is the most prevalent single-gene disorder among Caucasians, striking approximately 1 in

*table 12.4*

**Single-Gene-Defect Diseases, Candidates for Gene Therapy**

| Disorder | Incidence | Normal product of defective gene | Target cells | Status |
|---|---|---|---|---|
| Cystic fibrosis | 1 in 2,500 Caucasians | Substance important for keeping air tubes in lungs free of mucus | Lung cells | Aerosol delivery of gene directly to lungs is a theoretical possibility |
| Duchenne muscular dystrophy | 1 in 10,000 males | Dystrophin (structural component of muscle) | Muscle cells (particularly embryonic ones that develop into muscle fibers) | Work is preliminary. Nondystrophin genes injected into muscle have directed synthesis of the encoded proteins |
| Familial hypercholesterolemia | 1 in 500 | Liver receptor for low-density lipoprotein (LDL) | Liver cells | Animal studies are in early stages |
| Hemoglobinopathies (thalassemias) | 1 in 600 in certain ethnic groups | Constituents of hemoglobin | Bone marrow cells (which give rise to circulating blood) | Globin production in animals receiving gene needs to be improved |
| Hemophilia A | 1 in 10,000 males | Blood-clotting factor VIII | Liver cells or fibroblasts | Good chance for clinical trials (with fibroblasts) |
| Hemophilia B | 1 in 30,000 males | Blood-clotting factor IX | | |
| Inherited emphysema | 1 in 3,500 | Alpha$_1$-antitrypsin (liver product that protects lungs from enzymatic degradation) | Lung or liver cells | Work is very preliminary |
| Lysosomal storage diseases | 1 in 1,500 acquires some form | Enzymes that degrade complex molecules in intracellular compartments known as lysosomes | Vary, depending on disorder | Most diseases would require delivery of gene into brain cells (a difficult task) as well as into other cell types |
| Severe combined immunodeficiency (SCID) | Rare | Adenosine deaminase (ADA) in about a quarter of SCID patients | Bone marrow cells or T lymphocytes | Clinical trial of lymphocyte therapy for ADA deficiency is under way |

*table 12.5*

SOME NEUROLOGICAL DISORDERS AND GENE DEFECTS

| GENETIC CLASSIFICATION AND DISEASE | CHROMOSOME | GENE DEFECT |
|---|---|---|
| Autosomal dominant | | |
| Bilateral acoustic neurofibromatosis | 22 | Unknown |
| Charcot Marie-Tooth disease (type 1) | 1q2 | Unknown |
| Familial Alzheimer's disease | 21q21 | Unknown |
| Familial amyloidotic polyneuropathy | 18q11.2–q12.1 | Single-base pair substitution in mRNA for transthyretin |
| Huntington's disease | 4p16 | Unknown |
| Manic-depressive illness | 11p, Xp | Unknown |
| Myotonic dystrophy | 19 centromere | Unknown |
| Spinocerebellar atrophy | 6 | Unknown |
| Von Recklinghausen's neurofibromatosis | 17 | Unknown |
| Autosomal recessive | | |
| Gaucher's disease | 4q21 | Amino acid substitution in glucocerebrosidase |
| $G_{M1}$ gangliosidosis | 3p1 | Partially characterized |
| $G_{M2}$ gangliosidosis | | |
| Sandhoff disease (type 2) | 5q13 | Mutation in gene encoding β chain of hexosaminidase |
| Tay-Sachs disease (type 1) | 15q22–q25 | Mutation in gene encoding α chain of hexosaminidase |
| Wilson's disease | 13q14.11 | Unknown |
| Recessive with germinal chromosomal deletion | | |
| Central neurofibromatosis | 22q11–q13 | Unknown |
| Meningioma | 22q12.3-qter | Unknown |
| Retinoblastoma | 13q14 | Partially characterized |
| X-linked recessive | | |
| Adrenoleukodystrophy | Xq27–q28 | Unknown |
| Duchenne muscular dystrophy | Xp21 | Deletions in 5% to 10% |
| Lesch-Nyhan syndrome | Xq27 | HPRT deficiency, many variations |
| Pelizaeus-Merzbacher disease | Xq21–q22 | Defect in myelin proteolipid protein |

Source: Data from J. B. Martin, "Molecular Genetics: Applications to the Clinical Neurosciences" in *Science*, November 6, 1987, American Association for the Advancement of Science, Washington, DC.

2500 newborns. Among the normal Caucasian population, 1 in 25 persons is a heterozygote. Cystic fibrosis is less common among other ethnic groups, but it is of some significance among Southern Europeans, Ashkenazi Jews, and American blacks. The disease represents an autosomal recessive disorder. Patients must inherit both alleles for the disease to manifest itself. The product of the normal gene is a protein found in chloride channels of lung and other cells. These channels are vital to maintaining membrane potentials essential for the activity of nerve and muscle cells.

The search for the CF gene was difficult because no gene product had been identified when the search began. The usual approach to gene searches is to have a protein or RNA end product and search for its gene. Even with this disadvantage, the search proved successful when the gene was finally located in 1989. This work, done in separate laboratories, was led by geneticists Lap-Chee Tsui, Francis Collins, John Riordan, and others.

The gene for the disease was located at 7q31.2–q31.3. Figure 12.1 shows sequencing gel patterns of the normal sequence in the gene and the mutated sequence, which shows a 3 bp deletion (CTT) in the tenth exon of the gene. This deletion causes the amino acid phenylalanine to be deleted from the CF chloride channel protein.

The channels are regulated by the phosphorylation of a protein kinase that is cAMP-dependent (see Chapter 7 for enzyme phosphorylation and cAMP activity) and re-

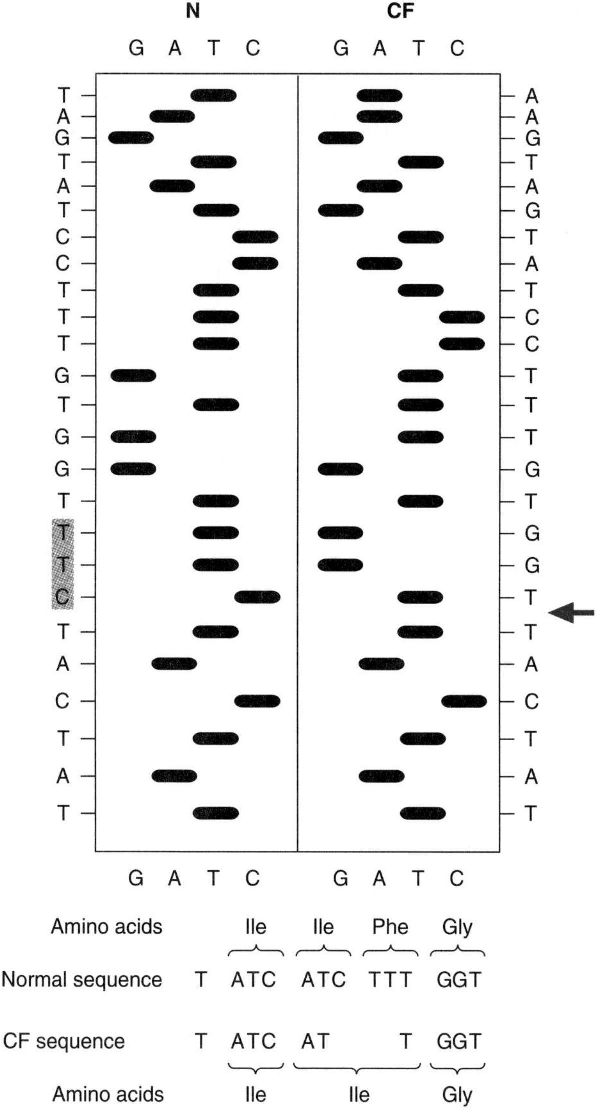


FIGURE 12.1
The sequencing gel patterns for normal (N) and cystic fibrosis (CF) genes and the 3bp deletion. Note that as a result of the deletion, the amino acid phenylalanine is missing from the amino acid sequence of the protein.

quires ATP as the energy source for the transport of the chloride ion. In patients with CF, the normal release of the chloride ions from cells as the cAMP level rises does not occur. Protein kinase activated by cAMP fails to open the channels to release the ions.

So a defective gene results in the defective regulation of the chloride channels, which in turn results in the accumulation of a thick, sticky mucus in the lungs and frequent infections especially by species of the bacterial genus *Pseudomonas*. Other complications involve the gastrointestinal tract, pancreatic malfunction, liver disease, and infertility (nearly 100% in males and frequently in females). Life expectancy is about 25–30 years; the disease is inevitably fatal.

Using various techniques over a period of 4 years, the researchers identified and localized the gene. Linkage analysis was first used to locate the position of the gene on chromosome 7. It then became the first human disease gene to be cloned solely on the basis of its position on the chromosome. Linkage analysis was done using RFLPs as markers and probes that hybridized with those markers to establish the location of the markers surrounding the gene. The recombination frequency between one marker (*met*) and the CF gene was 0.013, and that between a second marker (D7S8) and the CF gene was 0.009. These frequencies indicated that these markers were very close to the CF gene—about 1–2 million base pairs away. From this point on, using first a modification of chromosome walking called chromosome jumping, the researchers pinpointed the gene's location and determined its sequence.

Chromosome jumping is used to search DNA lengths on the order of 100 kbp. Cellular DNA is digested into these lengths with a restriction enzyme, and pulsed field gel electrophoresis separates the large pieces. The 100 kb pieces are circularized by the addition of a known marker gene, and the circle is then reopened. But it is reopened at some point other than the place where the known marker was added. In effect, we now have a marker flanked by two large pieces of DNA. This opened DNA is cloned into a vector such as lambda phage to create a library of 100 kb DNAs. Using probes to a known site suspected of being near the gene of interest, such as the *met* locus, a search of the 100 kb library is made looking for *met*. This technique allows investigators to scan or jump across larger segments of DNA than is possible by chromosome walking. Once *met* is found, chromosome walking is used on smaller digested pieces of DNA to locate the gene of interest.

But how does the investigator know when the gene is found? The procedure used in the case of the CF gene was to assume that the gene is important enough to have been conserved in other species of animals and to search for the conserved sequence. To do this, investigators hybridized radioactive probes of the suspected regions of the human DNA to fragment DNAs from species such as the cow and mouse. As it turned out, four sequences from other species hybridized with the human radioactive probes. Three of these regions were eliminated: one because it mapped close to *met* but too far from another marker, D7S8; a second because it did not contain an open reading frame (a triplet sequence coding for an amino acid sequence); and the third, because no mRNA corresponding to the sequence that hybridized was found, suggesting that this region was not transcribed so it could not be a gene.

The fourth region turned out to contain a segment made up of some 250,000 bp. An exhaustive screening of cDNA libraries with a probe from this fourth region eventually matched an mRNA from the sweat gland of a normal individual. The mRNA, when translated, yields a protein of 1480 amino acids. Several pieces of evidence were then gathered to verify that this gene was, in fact, the CF gene.

1. The gene was transcribed in all tissues affected by cystic fibrosis.
2. The nucleotide sequence of the gene had a region that codes for a sequence of amino acids that is a transmembrane domain (this is an important clue, since the protein of CF is involved in membrane transport of chloride ions).
3. Further sequencing of the gene suggested that the protein product of the gene was not a channel protein per se but was probably a regulatory protein (so the protein has been named the cystic fibrosis transmembrane conductance regulator, or CFTR).
4. The 3 bp deletion shown in Figure 12.1 is found in 70% of CF patients (it is presumed that the remainder of the CF patients have a mutation elsewhere in the gene).
5. The CFTR is a member of a superfamily of proteins that are found in bacteria, insects, and mammals; deletion of the gene in animals yields the disease.

In cell culture, genetically engineered retroviruses were able to deliver the CFTR gene successfully. Another virus, adeno-associated virus (AAV) carrying the gene has successfully integrated into chromosome 19. Clinical trials on human beings are now under way. A drawback of these systems is that the limited packaging of the passenger DNA of these viruses makes engineering them difficult because little if any extraneous nucleotides can be accommodated. Liposomes containing the normal gene are also under study as potential delivery systems. This technique has shown some promise in mice.

A number of questions that need to be answered regardless of the approach used have been identified by one of the chief investigators of the CF gene (Collins 1992):

1. What cells must be treated?
2. What fraction of the responsible cell types must be corrected to achieve a clinical benefit?
3. Is an overabundance of CFTR toxic?
4. How long will expression of the transferred gene persist?
5. Will the immune system react to and possibly destroy the vector, since it is foreign to the CF patient as are the cells now carrying the new CFTR protein?
6. Can we be sure we are transferring only the desired gene and that we are not inadvertently transferring unwanted genes to germ line cells?

## SOME LABORATORY PROCEDURES

The techniques and procedures of molecular genetics are finding their way from the research laboratories to the clinical laboratories.

### Viruses

For example, viral nucleic acid detection methods that rely on the hybridization of chemical- or enzyme-labeled probes are now in use. The various procedures use single probes with multiple labels, multiple probes each with a single label, or multiprobes and multilabels (Figure 12.2).

These techniques are referred to as signal amplification methods because they are designed to increase the concentration of label on the target nucleic acid thereby increasing the chance of detecting the nucleic acid. Other viral detection methods depend on adding a known primer to a likely viral nucleic acid sample and then following the PCR protocol to increase the concentration of the nucleic acid. Once this has been done, procedures are followed (such as those just described) to detect the viral nucleic acid. For this procedure, there are currently available commercially produced reagents, primers, and controls.

### Bacteria

Nosocomial infections in the United States are mostly of bacterial origin, and a number of molecular techniques are used to identify the causative agents of particular infections. For example, since many bacterial species and strains contain plasmids often of different types and of known number with known molecular weights, an analysis of the plasmids, number, and DNA molecular weight can help in identifying organisms. In some cases, as with the genus *Enterobacter*, species may contain a sufficient number of plasmids to make strain differentiation possible. Plasmid analysis has been used to type other bacteria such as *Staphylococcus aureus*, *S. epidermidis*, and *Enterococcus* sp. These organisms are all important causes of hospital infections.

DNA probes have also been developed for use in DNA-DNA hybridization procedures for use in identifying organisms. This has become an important epidemiological tool in studies concerning the spread of antibiotic-resistance genes among bacteria.

## FROM ANTHROPOLOGY TO ZYGOTES

Molecular biology, in particular molecular genetics, has found its way into a wide variety of scientific fields. Most of these fields are, not surprisingly, specialties of biology, but many are not. Many of the uses to which the principles and practices of molecular biology are put may be trivial in some senses, but many, if not most, will have important and lasting influences on our lives in one way or another. And, of course, some of these attempted applications will succeed, but many will fail.

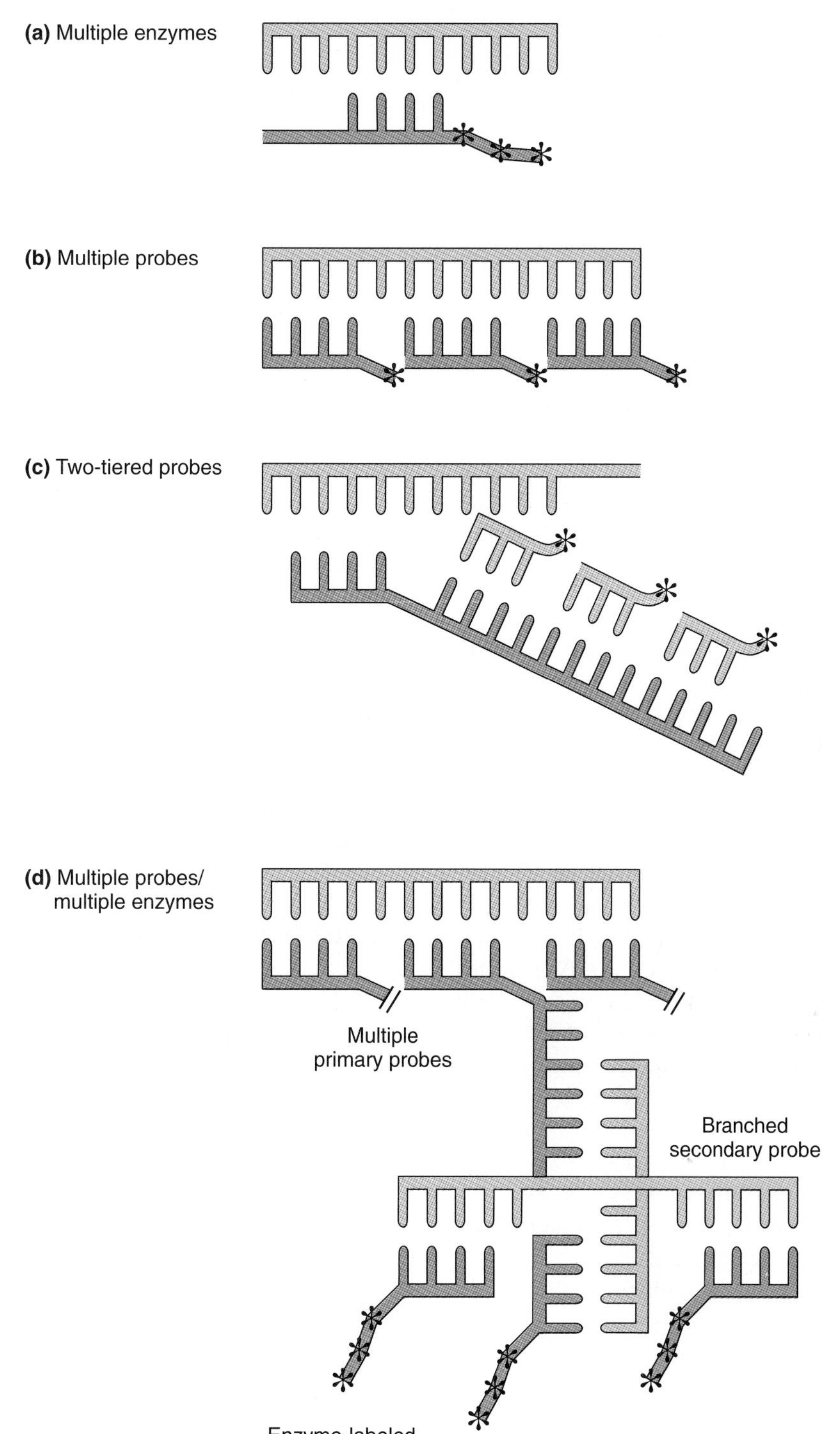


FIGURE 12.2
Some methods of signal amplification used to increase the probability of detecting viral DNA.

We will take a brief look at some of the areas now being influenced by molecular biology and molecular genetics. This is not intended to be an all-encompassing survey but rather an attempt to demonstrate the wide variety of fields in which molecular biology/molecular genetics plays a part. What should strike the reader is the fertility of the scientists' imagination in putting techniques and procedures to novel use.

## Anthropology

A number of studies have suggested the branchings of the human family tree. The tree is constructed using population genetics analysis to trace the worldwide distribution of hundreds of marker genes such as the genes controlling the Rh factor, an antigen found on blood and other human cells.

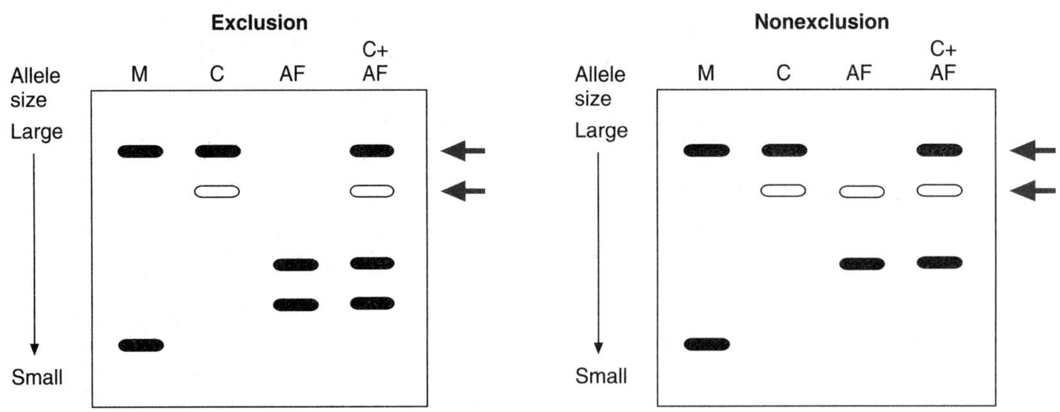


FIGURE 12.3
Disputed parentage: A simplified example showing the exclusion of the alleged father and the nonexclusion or possibility that the alleged father is the father. M = mother, C = child, AF = alleged father. Note the absence of a common band between C and AF in the exclusion case and the appearance of such a band in the nonexclusion case.

Molecular data concerning sequence analysis (RFLP) was used in separate studies and was found to be in excellent agreement with the classical data used to construct the family tree. An African origin for humankind has been postulated from these data.

In a still-controversial study (Cann et al. 1987), the worldwide distribution of the genes encoded in mitochondrial DNA (13 protein, 22 tRNA, and 2 rRNA genes, totaling some 16,569 bp) have been studied in an attempt to locate the origin of the human species. This study also places the origin in Africa, at least our maternal origin. Very nearly all mitochondria found in a human being derive from those supplied by the egg (the sperm rarely contributes mitochondria), so these organelles are used in studies of maternal inheritance. The idea behind the so-called mitochondrial clock is based on the number of mutations that have accumulated over time. The older the mitochondria, it is supposed, the more mutations should have accumulated.

Much of the controversy over this approach stems from the assumption made by the investigators that mitochondrial genes, over the course of millennia, mutate at a constant rate. To establish the rate, investigators compared human mitochondrial DNA with that of another group, such as chimpanzees, known to have diverged from the human species at a given time. After calibrating the mitochondrial DNA clock in this way, investigators analyzed mitochondrial DNAs for a number of different human populations around the world. It was concluded that an ancestral human population was more likely to have been located in the area where the African samples were collected than in any of the other areas from which mitochondrial DNA samples were collected. According to these investigators, our common mother lived in Africa about 200,000–300,000 years ago.

Other studies have sought the origins of the people who colonized the western Pacific Ocean, from Hawaii to Tahiti to Australia. One prominent theory is that those who navigated the vast ocean distances originated from Southeast Asia some 3600 years ago. But a 9 bp sequence deletion and a so-called cluster II in a noncoding region of mitochondrial DNA (mtDNA) suggest that Polynesia may have been settled simultaneously by Southeast Asians and Melanesians.

## Paternity Testing

RFLP procedures based on the variable number tandem repeat (VNTR) segments in different DNAs and the use of the PCR to amplify small amounts of DNA are being used to help resolve disputed parentage. Offspring of particular parents should share a number of common bands when samples of treated DNA from the parents and offspring are compared after electrophoresis. This procedure takes advantage of the fact that, though individuals have different banding patterns (except for identical twins), some sets of bands are subject to Mendelian inheritance. Use of multilocus probes for various DNA segments increases tremendously the statistical probability of excluding or confirming parentage. For example, single-locus probes may generate one or two bands, but multilocus probes can produce 10 or more bands. See Figure 12.3 for a simplified version of an RFLP concerning disputed parentage.

## Archeology

In the field of archeology, the new subspecialty of molecular archeology is helping to explore Egyptian history using the PCR to clone the DNA of 4000-year-old mummies and to evaluate the source of bone fragments some 50,000 years old. Also being studied are the genetic relationships of prehistoric peoples. For instance, genetic bottlenecks are being investigated. Genetic bottlenecks occur because pestilence, famine, war, or the like leaves only a small number of individuals to repopulate an area; the population decline leaves fewer varieties of genes to be distributed among the survivors and their offspring.

Other scientists are looking at the inheritance of particular phenotypes such as skin pigmentation and physical size. These studies, it is hoped, will help to shed some light on how genetic disorders are inherited in particular populations, on how long these disorders have occurred among human beings, and on where the origins of the diseases might lie.

Investigators are also interested in ancient plant life for, among other reasons, classification and for the purpose of reconstructing the early history of our planet. In fact, recently, DNA was extracted from an 18-million-year-old leaf that turned out to be a magnolia specimen. What has astounded and encouraged scientists is that DNA can survive that long. Of course, how long DNA survives may be more a function of chance and circumstance than of any inherent property of DNA.

The techniques of molecular genetics are even being used to solve mysteries of long standing. At the beginning of the Russian revolution of 1917, the Romanovs, the Russian royal family, were arrested and soon disappeared. They were executed and buried in a secret place. Rumors of where this place might be have circulated since then. One such rumor placed the burial site at or near the Russian city of Ekaterinburg, where the remains of several individuals were discovered in 1991. Using bits of hair and bone from the remains and samples of blood from relatives of the Romanovs, including the Duke of Edinburgh, husband of Queen Elizabeth II, British and Russian investigators using DNA fingerprinting have concluded, with a 99% probability of certainty, that the remains found are those of the royal family. Interestingly, the remains did not include those of Alexi, the heir to the throne, or Anastasia, the youngest of the girls.

## Ethology

Even in the field of animal behavior, molecular genetics is making inroads. Scientists are attempting to isolate and clone the genes encoding the polypeptides that appear to elicit particular behavioral patterns when injected into other animals that do not normally show the behavior. For example, scientists are studying the egg-laying behavior of the mollusk *Aplysia*, a shell-less marine snail.

Other scientists are studying the effect of the so-called *per* (for *period*) gene located on the X chromosome of *Drosophila* species; males of different species have different songs. Germ-line gene-transfer experiments are being used to investigate control of singing and other locomotor activities found in the *per* locus. And, because various behaviors are species-specific and under *per* control, it has been suggested that this locus may play an important role in speciation.

Behaviors, ranging from mating rituals, to survival instincts, and even to artificially induced alcohol consumption in mice, are at least in part under genetic control. Behavioral geneticists have also concluded that such human afflictions as Huntington's disease, Alzheimer's disease, Tourette's syndrome, and aggressive behavior have some genetic component. Others believe that an interplay between genes and the environment result in certain behavior patterns such as personality and intelligence.

Behavioral geneticists, themselves, however are aware that better methodologies for identifying behavioral genes are needed. They also must find ways to study the extremely complex mechanisms by which genes interact to produce particular behaviors. It is highly unlikely that such complicated and elaborate outcomes are the result of single genes.

## Law

Perhaps no use of molecular genetics technology has been more publicized than the use of RFLP in forensic science. In this area, the techniques are used primarily to identify individuals in both civil and criminal cases that may involve disputed parentage, murder, rape, assault, and so on.

Alec Jeffreys, working in England in 1985, was studying the myoglobin gene and discovered that it contained a 33 bp repeat in one of the introns. What fascinated Jeffreys and his colleagues was the finding that the repeat was found in a number of other places in human nuclear DNA. Using the restriction endonuclease *Hae*III—because it cut the 33 bp repeat not internally but rather at its flanks—Jeffreys found that individuals had different numbers of these repeats at different loci along the genome. This discovery is the basis for a procedure now referred to as **DNA fingerprinting.** What is called evidentiary material such as blood, semen, bone, and soft tissue is collected and is the source of the DNA to be analyzed. The idea is to collect as much material as possible to maximize the amount of DNA that can be isolated, though the sample may be exceedingly small (e.g., blood or semen samples). Then the sample must be processed in such a way that its DNA content is not damaged or destroyed. Sample DNA as small as 5 ng (billionths of a gram) can be analyzed.

The entire procedure for the analysis of RFLP patterns involves gel electrophoresis, transfer of DNA fragments to a nitrocellulose membrane, the labeling of DNA probes, and the hybridization of the labeled probes to the sample DNA (Figure 12.4).

This procedure relies on the fact that scattered along the genome are a variable number of identical sequences (repeats) joined head to tail. These repeats are called hypervariable loci or variable number tandem repeats, for short, VNTR loci. When DNA flanking these VNTR loci is cut using restriction enzymes, the fragments of VNTRs produced will vary in length depending on the number of restriction sites in the particular locus. Typically, the

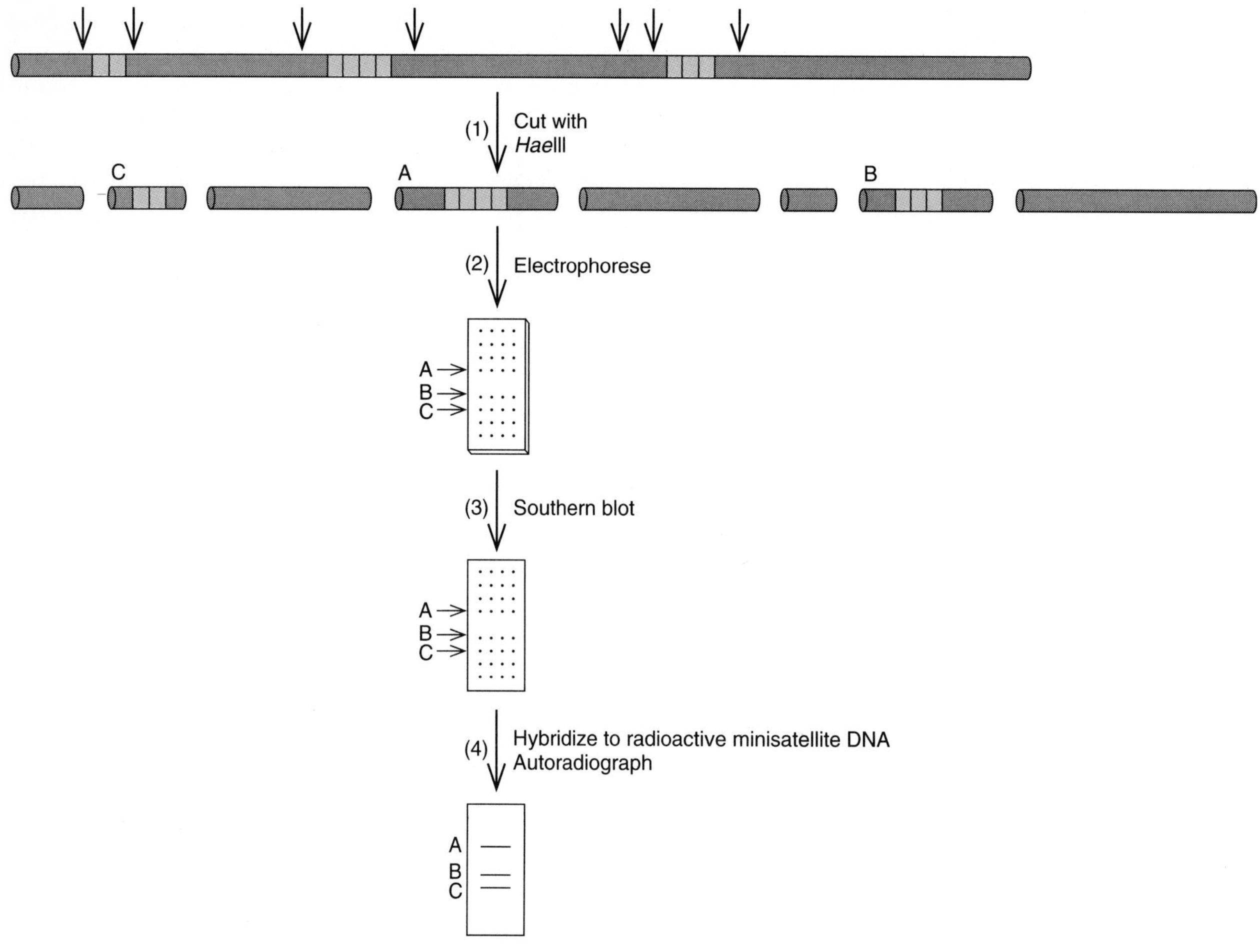


FIGURE 12.4
DNA fingerprinting procedure. The DNA is cut with restriction enzymes, generating fragments that are separated by gel electrophoresis. Fragments are then transferred to a nitrocellulose filter (southern blot) and hybridized to radioactive probes. An autoradiogram is then prepared for analysis.

number of repeats ranges from 3 to 20. Labeled DNA probes of the VNTRs are used to detect the fragments containing the VNTRs. The piece sizes are determined by gel electrophoresis. So detection of the fragments is dependent on the restriction enzyme and the DNA probe used.

In contrast to this technique, VNTR polymorphism is based on the variation in the number of VNTR sequence repeats between restriction sites. That is, fragment sizes are determined by the number of repeat units contained in each fragment. It has been found that the number of repeated sequences varies not only between individuals but also among individual genetic loci on separate chromosomes. And, whereas RFLP may produce only a few different-sized DNA fragments, VNTR-based polymorphism may give more than 100 fragment sizes. Figure 12.5 compares the two procedures.

The patterns of bands produced in a DNA fingerprinting procedure turn out to be unique for any given individual except for identical twins (unless such things as mutations and crossing-over occurred). When large numbers of VNTR loci from individuals are analyzed in this way, the chance that any two individuals would have the same electrophoretic pattern of fragments is extremely low.

The first highly publicized use of this technique was by Jeffreys. In 1983 a 15-year-old girl was sexually assaulted and killed in Leicestershire, England. Semen was recovered from the victim. Three years later another 15-year-old girl was also attacked and slain. Semen was again recovered from the victim. The two semen samples were DNA fingerprinted, and they matched.

A month after the second attack, a 17-year-old boy with a history of sexual problems was arrested. Blood samples taken from the boy were also DNA fingerprinted by Jeffreys. His conclusion was that the boy was not responsible for the attacks. Police then sent notices to residents of the surrounding towns notifying them that a blood sample would be taken from all males and analyzed.

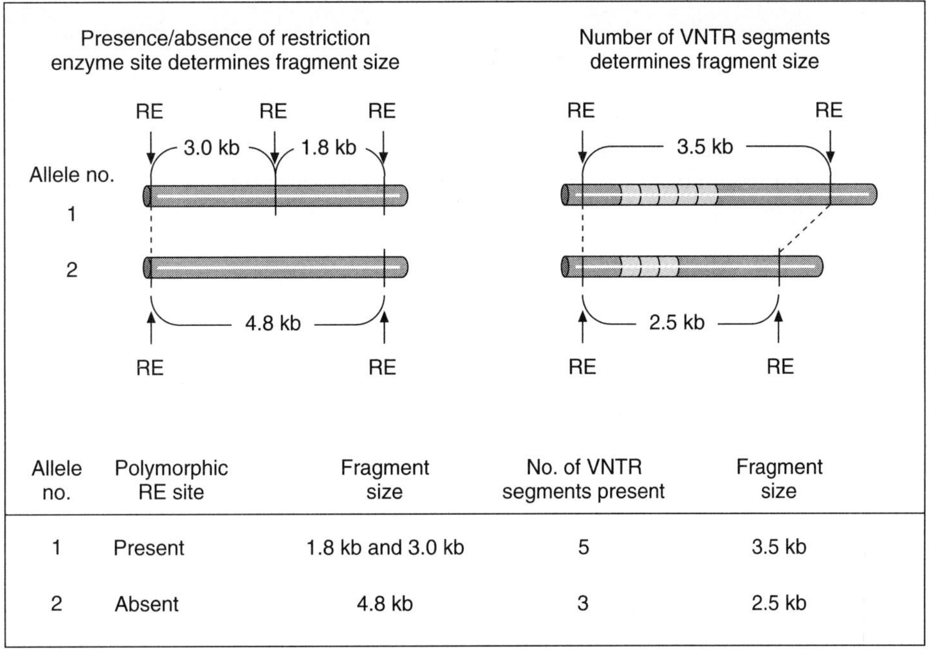


| Allele no. | Polymorphic RE site | Fragment size | No. of VNTR segments present | Fragment size |
|---|---|---|---|---|
| 1 | Present | 1.8 kb and 3.0 kb | 5 | 3.5 kb |
| 2 | Absent | 4.8 kb | 3 | 2.5 kb |

FIGURE 12.5
(a) Restriction enzyme–based polymorphism. Differences in fragment size are due to the presence or absence of a restriction enzyme. (b) VNTR-based polymorphism. Differences in fragment size are based on the number of VNTR sequences between flanking restriction enzyme sites. RE = restriction enzyme; VNTR segments = variable number tandem repeats of different lengths.

One individual, called "Colin," upon receiving his notice convinced a co-worker to give his blood under Colin's name because, he, Colin, feared his criminal record for flashing would make him a suspect. Colin later admitted the deception to other co-workers, one of whom informed the police. Colin was arrested and confessed to both crimes.

The use of these techniques in criminal cases requires extremely tight controls and careful attention to the protocol details because of the seriousness of the consequences. After some controversy involving controls and procedures used in DNA fingerprinting, analysis, and interpretation of results, convictions in a number of U.S. criminal cases were reversed. Recently, however, after a new set of controls and procedures was adopted, DNA fingerprinting is once more acceptable in many U.S. courts at the state and federal levels. And, in at least one case, it was used to reverse a wrongful conviction. A man convicted and sent to prison in New York State was released on the basis of DNA fingerprinting; he had served 11 years for a rape that DNA fingerprinting proved he had not committed.

# Medicine

Human proteins having great medicinal value have been produced using cloned genes. This method of manufacturing the proteins is necessary because the naturally occurring molecules are not readily available, and they are difficult to obtain and expensive. These proteins include insulin (the first recombinant product licensed for human use), human growth hormone, a hepatitis B virus subunit vaccine, and tissue plasminogen activator (tPA, given to heart attack patients: this protein cleaves plasminogen, releasing a protease called plasmin that, in turn, degrades fibrin, the clot-forming protein).

Insulin and human growth hormone are produced by genetically engineered bacteria; the hepatitis vaccine is made by an altered yeast; and the tPA is manufactured by a mammalian cell line. The type of cell used to produce a given protein is determined primarily by the chemical and physical complexity of the protein. Insulin, for example, is a simple protein and is easily made by bacteria. Other human proteins are much more complex and not as easily made, if at all, by bacteria or yeast, so mammalian cell lines need to be used.

## *Insulin*

As a general example of recombinant protein production, let us look at the manufacture of human insulin. It is produced in human beings as a prehormone single-chain polypeptide that is then processed to remove a number of internal amino acids, yielding a two-chain molecule of some 51 amino acids held together by sulfide

table 12.6

SOME GENETICALLY ENGINEERED PLANT SPECIES

| | | | | |
|---|---|---|---|---|
| Alfalfa | Corn | Kiwi | Poplar | Sugarbeet |
| Apple | Cotton | Lettuce | Potato | Sugarcane |
| Asparagus | Cranberry | Muskmelon | Raspberry | Sunflower |
| Broccoli | Cucumber | Oilseed rape | Rice | Sweet potato |
| Cabbage | Eggplant | Papaya | Rye | Tobacco |
| Carrot | Flax | Pea | Soybean | Tomato |
| Cauliflower | Grape | Pepper | Spruce | Walnut |
| Celery | Horseradish | Plum | Strawberry | Wheat |

bridges. Because of its small size, F. Sanger (in 1952) was able to determine its amino acid sequence. From this sequence, synthetic genes were constructed (remember colinearity from Chapter 8) and attached to the β-galactosidase gene in a vector, pBR322. This artificial plasmid also contains a promoter sequence, which is necessary to transcribe the gene, and an ampicillin-resistance gene to aid in screening.

The vector is inserted into *E. coli*. Production of insulin is then controllable by induction using lactose as the inducer (Chapter 7). The cells produce the insulin, which then accumulates in the cells from which it must be extracted.

Since the insulin and β-galactosidase genes are joined, their product is a fusion protein, β-galactosidase-insulin. The two genes, however, are joined by a methionine codon (AUG). Therefore, the proteins are joined by the amino acid methionine. This is done purposefully. The chemical cyanogen cleaves peptide bonds following a methionine. Furthermore, insulin does not contain internal methionines; β-galactosidase does. So treating the fusion protein with cyanogen releases insulin and degrades β-galactosidase. This reaction makes purification of the insulin relatively easy.

### *Some Other Proteins*

Scientists are also at work creating sheep, goats, and cows that have been genetically engineered to secrete human proteins into their milk. In effect, the mammary glands, as one worker put it, can be used as an impressive bioreactor. It is hoped this method will generate large quantities of the desired protein at a low and reasonable price. Though, at present, the yields in milk are too low for commercial production.

Among the proteins being tested as candidates for production in transgenic animals are tissue plasminogen activator, a clotting factor needed to treat hemophilia; erythropoietin, needed to reduce the side effects on bone-marrow blood-cell production caused by certain drug treatments for several diseases including AIDS; and alpha-1-antitrypsin, which may be of value in treating degenerative diseases including emphysema.

## Plants

To feed and clothe ever-increasing populations, plant geneticists have been breeding superior plants for years. But their older methods are time-consuming and uncertain. Since the early 1980s, when transgenic plants were first created, more than 50 such plants have been developed. Genes that confer resistance to herbicides, insects, and disease have been transferred to a number of plants. Some major problems have, however, surfaced in attempts to produce transgenic plants among critical feed plants such as maize. In this species, gene transfers have left the plants infertile for some reason. Where the major cereals are concerned, a difficult problem has been to regenerate whole plants from protoplasts, which must be made in order to force DNA into the cell (see the earlier discussion on this problem in Chapter 11, monocots vs. dicots). Table 12.6 is a partial listing of genetically engineered plant species.

In the spring of 1994, a milestone was reached when the first genetically altered food was put on the market. The Food and Drug Administration okayed the sale of genetically engineered tomatoes. The fruit was produced by Calgene Inc. and is engineered to ripen longer on the vine to make it more flavorful and to have a longer shelf-life than ordinary tomatoes. The fruit remains firm for picking and shipping. The tomato has been named Flavr Savr and sells at premium prices.

## Zygotes

Sexually reproducing organisms have germ cells, eggs and sperm, which pass on genetic information, including defective genes, to the next generation. The defective genes may or may not affect the offspring. The effect on offspring is determined by whether the allele is recessive or dominant or whether it is on the copy of the X chromosome inactivated during egg production. As noted in Chapter 11, it is now possible to insert genes into one-celled embryos. These genes can become stably integrated into the organism's somatic and germ line cells. Integration, obviously, must occur very early in development before the germ line cells are separated from the primitive somatic cells. This technique will not only serve as a step toward the long-sought goal of curing genetic diseases, but it will also become a powerful tool for determining the functions of specific gene products—proteins and RNAs.

In line with this idea, mutated genes have been successfully inserted into mouse embryos, and the genes are stably integrated and passed on to progeny. Altering genes in defined ways and then inserting them into embryos provides a way of studying the alteration by analyzing the phenotypic expression of the alteration. Then, it is hoped, a correlation can be made between the gene and its expression at the biochemical and physiological levels.

The technique used to make transgenic mice uses two types of cells: embryonic stem cells (ES cells) from one mouse and the embryo of another mouse. ES cells are obtained from cultured blastocysts (the stage of development of a fertilized mammalian egg when a central fluid-filled cavity is formed and the cell mass is ready for implantation into the uterine wall). Genes are inserted, and the cells are grown in tissue culture in such a way that differentiation is inhibited. The cells are screened for those cells that have accepted and integrated the genes. When injected into mouse embryos, ES cells participate in the development of all other types of cells. But, because there are now nontransgenic and transgenic cells present, the embryo will develop as a chimeric adult. Often the ES cells and the recipient blastocysts are taken from mice having different coat colors. In this way, the ES-derived cells can be distinguished from the recipient blastocyst-derived cells, and the effect of the transferred gene on the animal can be studied more easily.

ES cells are preferred over fertilized eggs because once the genes are injected into fertilized eggs no control over the genes' integration is possible. In other words, the genes can no longer be manipulated. As noted in Chapter 11, the sites at which injected genes will be integrated cannot be controlled by the investigator. By using ES cells, one can screen for those cells that have integrated the new gene in the desired location and then use these cells as the transfer vector into the embryos. DNA can be introduced into ES cells by three means: transfection of naked DNA, retroviral infection, and electroporation. The federal government does not now allow this type of research in human beings.

## INACTIVATING CELLULAR GENES

One way to study gene function is by selectively inactivating target genes. A procedure to do this has been developed (Chapter 11). In mammalian cells, introduced genes integrate into sequences unrelated to the inserted gene's in the vast majority of cases. This process is called heterologous recombination. There are rare occasions when the inserted gene recombines at precisely its identical sequence. This is referred to as homologous recombination. If this homologous recombination could be obtained often enough, the precise placing of inserted genes would be possible, and the study of specific genes would be easier. Procedures have been developed that selectively encourage the proliferation of those cells in which homologous recombination has occurred.

One procedure involves the insertion of a gene for a selectable marker, which lacks its own promoter sequence, into a cloned gene of interest. We then have a double gene. The marker might be resistance to an antibiotic, for example, and the insertion disrupts the cloned gene of interest, so it is nonfunctional.

If the cloned gene (and marker gene) is integrated into its precise homologous sequence, it interrupts and renders the host gene inactive. But, if the integration is close enough to the host gene's promoter sequence, the double gene can be activated by the cell in its normal fashion. The marker gene will be turned on, and those cells carrying the marker gene in the correct position can be identified in a screening procedure. The cells thus identified have the original gene of interest inactivated. Double genes that insert in other regions of the genome far from promoter sequences are not activated, so cells with that type of inserted gene will not survive in the medium containing the antibiotic.

## SOME ETHICAL ISSUES: "GOOD" AND "EVIL"

Lurking in the near future are some questions and problems that will result from the applications or proposed applications of some techniques and procedures of molecular genetics. Some issues such as the right to privacy are already upon us.

These applications will confront us with possibilities never before imagined. Some of these questions are of a practical nature: Is it worth engineering slowly ripening fruit to extend shelf-life? Other questions involve constitutional issues concerning the right of privacy: Who should have access to human genetic data, and how may that information be used? Still other questions are about important philosophical concerns: Should human beings tamper with their own and other organisms' evolutionary development?

The discussion of the ethical, moral, and legal issues should engage all members of society and not just the scientists. Though scientists may have special insights into problems and questions of biology, and genetics in particular, they do not necessarily have any special insights into ethical, moral, or legal dilemmas.

Genetic engineering holds many promises, as we have seen in this and the previous chapter, but there are risks associated with the manipulation of the very biological essence of organisms—namely, their genomes.

## The Asilomar Conference

At the very birth of the subspecialty of genetic engineering, the scientists themselves were aware that enormous power and risks lay in their hands. In July of 1974, an unprecedented appeal was made in a letter simultaneously printed in *Nature, Science,* and *The Proceedings of the National Academy of Sciences,* three of the most widely read scientific journals. The letter was signed by leading scientists. Among the signers were Paul Berg, a pioneer in recombinant DNA technology and a soon-to-be Nobel Prize winner, and the letter's author, James D. Watson, the co-discoverer of the structure of DNA and also a Nobel Prize winner.

The letter called for a voluntary moratorium on recombinant DNA work until the full implications and risks of such work could be assessed. Further, the scientists asked that a conference be convened to address these issues.

In February of 1975, a conference was held in Asilomar, California. It was attended by some 139 leading investigators. It was nearly unanimously agreed that some restrictions on DNA cloning were in order until the risks of gene manipulation could be identified and better understood. It was agreed that only genetically disabled bacteria incapable of living and reproducing outside the laboratory should be used in recombinant DNA work. Other recommendations concerning laboratory design and safety measures were also made.

Soon thereafter, the National Institutes of Health (NIH) adopted the guidelines, including a prohibition against using recombinant DNA techniques in the study of viral cancer genes.

The recommendations were to take effect in July 1976 and were to be administered by a special committee of the NIH, the Recombinant DNA Advisory Committee (RAC). In many of the discussions concerning the risks of gene manipulations, laymen and their political representatives were included (RAC now has laymembers). In England, a similar committee was established—the Genetic Manipulation Advisory Group (GMAG).

By 1979, it had become obvious that the actual risks of this work are slight, assuming the proper laboratory procedures are observed. And, in fact, the ban on working with viral cancer genes has been lifted. To this day, no serious problems have been encountered because of genetic engineering experiments.

## Ethical, Moral, and Legal Issues

There still are issues to be confronted, most especially, issues that include human beings. It is one thing to genetically engineer a tomato plant but quite another to manipulate the genes of human beings.

There is little concern that somatic cell gene therapy will pose vexing ethical questions. This sort of treatment for disease is seen as an extension of other therapies that also introduce foreign genes into patients—organ and bone marrow transplants for example. But a clear line is drawn at gene therapy aimed at egg, sperm, and embryos since these introduced genes would be passed to future generations.

There are people who balk at closing the door on what may be a valuable tool in attempts not just to cure but to eradicate a disease, such as cystic fibrosis or multiple sclerosis or neurofibromatosis. Other people fear, however, that as gene therapies become available and commonplace, persons carrying defective genes will be pressured to "take the cure." In effect, those people would be coerced into accepting a procedure they may object to for any number of reasons, including religious beliefs.

Further, where and how does the notion of informed consent fit in? And other scenarios come to mind. Society may perceive the cost of caring for handicapped persons as being unacceptable if a cure is available. So one may be forced to either accept gene therapy or be left to manage alone.

Some argue that the state has a compelling interest in protecting the lives of its citizens, in particular, the children. And, just as mandatory vaccination is seen as a good, so, too, will gene therapies as they become available; and they too will become mandatory.

Other concerns are the issues of mandatory genetic screening as the Human Genome Project makes more and more information available regarding normal and defective genes; the privacy of genetic information; and discrimination based on genetic defects. In other words, Who is entitled to genetic information, and How may the information be used?

Should genetic screening be made mandatory? Or, even if screening is voluntary, once it has been done must the individual disclose the information in, say, applying for a job or for insurance? What responsibility will physicians bear concerning disclosure of genetic information? Will such information be regarded as just another piece of health-related data like blood pressure or weight?

Employers may have a strong incentive in favor of genetic screening to reduce insurance costs and to protect themselves against lawsuits. They could refuse to allow individuals with particular defective genes to work at certain jobs, or they could even refuse to hire such individuals for those jobs in the first place. Women who carry particular genes might be refused some kinds of employment for fear of their giving birth to defective babies and then blaming it on their jobs and employers.

Many states have mandatory testing for sexually transmitted diseases before a marriage license can be granted. Will genetic testing also come to be required? Other questions center on how carriers of defective recessive alleles are to be treated. Should it, for example, be required that one's intended spouse be informed of these defects?

And once the detection of defective genes becomes routine, will there be greater and greater demands for abortions, creating still more controversy? Will society, here too, find the costs associated with the care of defective children to be a burden it is not willing to accept? Such an attitude, no doubt, would also increase the number of abortions, adding more fuel to this already inflamed issue.

Further, what is the ethical or moral position regarding the detection of genetic defects for which there are no cures? Our ability to find these genes is far ahead of our ability to cure the diseases they engender. For some diseases, such as some immunodeficiency disorders, a cure may be relatively easy, but for other diseases, Huntington's disease, for example, which are more systemic and involve many tissues and organs, a cure may be more difficult. And, last but not least, there is the concern that the awesome powers of molecular genetics might be used by some in an attempt to create the "perfect" human being.

Where should the line be drawn, if at all, in the uses to which gene manipulations can be put? Most people would accept the curing of disease as a noble end. But what about using genetic engineering to produce a particular eye color, or skin tone, or height?

It should be abundantly evident by now that there are more questions than answers about genetic engineering. It should also be abundantly clear that the questions must be answered. We leave them unanswered at our peril.

Where other organisms are concerned, the problem may not entail quite such difficult ethical, moral, or legal questions. But there still are problems that have been identified and remain to be resolved.

J. R. S. Fincham and J. R. Ravetz (1990) listed some of the problems associated with the release of genetically altered organisms into the environment:

1. The more generalist an organism is with regard to its feeding habits and habitat the more likely it is to be a problem.
2. Where no natural predators of a new organism exist, other creatures are likely to be preyed upon or displaced.
3. Genetically altered organisms may produce unintended consequences.
4. In a new environment, originally harmless organisms that have been genetically engineered may become significant problems.

For those reasons and others, the release of genetically engineered organisms into the environment has been strictly regulated by the government, and only a few experiments have, to date, been allowed to proceed from the laboratory to the stage of field trials.

As the experiments take place, we will, no doubt, gain a better understanding of how to manage these new creations, whether corn, tomatoes, microorganisms, or cows, and how they behave in the natural environment. Then we will be in a better position to protect ourselves and the environment against harmful consequences.

## Summary

Since the 1953 publication of the Watson and Crick paper on the structure of DNA, advances in molecular genetics have reached the stage where their uses are now almost commonplace in such diverse areas as archeology, medicine, botany, and zoology. But along with the promises of solving some long-standing intractable problems, there are some risks.

The techniques and procedures of molecular biology and genetics are making possible the sequencing of the entire human genome (the Human Genome Project); the curing of such disorders as cystic fibrosis; the development of vaccines for diseases such as hepatitis and AIDS; the laboratory procedures that will facilitate the identification of viruses, bacteria, and other microorganisms; and the manufacture of special proteins such as insulin. Scientists are also using these techniques and procedures to study the human family tree, Egyptian mummies, and animal behavior. In criminal and civil law, molecular techniques are helping to identify perpetrators of crimes such as rape and murder and to resolve cases of disputed parentage.

But a note of caution has been sounded by a number of people, both scientists and nonscientists. These people have been questioning what, if any, kinds of human gene transfers ought to be prohibited (e.g., transfers that affect egg and sperm and, therefore, human evolution), what kind of genetic information about an individual ought to be private, who has a right to genetic information, how might such information be used, and what might the consequences be of releasing genetically altered organisms into the environment?

In short, there are ethical, moral, and legal implications associated with the use of a good many molecular techniques and procedures and with genetic engineering applications. The resolution of the many doubts surrounding genetic engineering must include input not only from scientists but also from philosophers, sociologists, lawyers and judges, and laypersons, because how the techniques and procedures of molecular genetics and molecular biology are used will profoundly affect the future of humankind.

## Study Questions

1. Explain the goals of the Human Genome Project and indicate the procedures being used to meet those goals.
2. What are the differences between infectious and noninfectious diseases? Give several examples of each type of disease.

3. What are the approaches to curing or preventing both types of diseases?
4. Give a brief account of the genetic basis for cystic fibrosis (CF).
5. Explain the procedures used to locate the CF gene.
6. What is the evidence that the reported CF sequence is indeed the CF gene?
7. What are the questions needing answers regardless of which CF treatment approach is used?
8. Briefly describe how molecular genetics procedures are being used in fields unrelated to biology.
9. What is the basis behind the use of RFLP in identifying individuals?
10. Explain some of the promises and technical problems of transferring genes to embryos.
11. What are some of the ethical, moral, and legal issues associated with the uses of molecular genetics procedures, and how might some of these issues be resolved?

## Readings and References


Beardsley, T. 1994. Big-time biology. *Sci. Am.* 271 November:90–97.

Bowden, J. H. III, S. Shelton, and T. E. Mifflin. 1994. A guide to DNA fingerprinting. Part 1. *Clinical Chem. News* February:3, 13.

Cann, R. L., M. Stoneking, and A. C. Wilson. 1987. Mitochondrial DNA and human evolution. *Nature* 325:31–36.

Cavalli-Sforza. 1991. Genes, peoples and languages. *Sci. Am.* November:265(5):104–10.

Collins, F. S. 1992. Cystic fibrosis: Molecular biology and therapeutic implications. *Science* 256:774–79.

Farley, M. A., and J. J. Harrington, eds. 1991. *Forensic DNA Technology*. Chelsea, Mich.: Lewis.

Fincham, J. R. S., and J. R. Ravetz. 1991. *Genetically engineered organisms: Benefits and risks*. Toronto: University of Toronto Press.

*Genetic witness: Forensic uses of DNA tests*. 1990. Washington, D.C.: U.S. Office of Technology Assessment.

Hanson, E. D. 1983. *Recombinant DNA research and the human prospect*. Washington, D.C.: American Chemical Society.

Jeffreys, A. J., V. Wilson, and S. L. Thein. 1985. Hypervariable "minisatellite" regions in human DNA. *Nature* 314:67–73.

*Laboratory Medicine*. 1992. DNA Technology Issue. Chicago: American Society of Clinical Pathologists.

Lee, T. F. 1991. *The Human Genome Project: Cracking the genetic code of life*. New York: Plenum Press.

Levin, M., and H. Stauss, eds. 1990. *Risk assessment in genetic engineering*. New York: McGraw-Hill.

Marshall, E. 1994. A showdown over gene fragments. *Science* 266:208–10.

Moffat, A. S. 1995. Exploring transgenic plants as a new vaccine source. *Science* 268:658–60.

Nowak, R. 1994. Genetic testing set for takeoff. *Science* 265:464–67.

Rennie, J. 1993. DNA's new twists. *Sci. Am.* 268 March:122–32.

Ross, P. E. 1992. Eloquent remains. *Sci. Am.* 266 May:115–25.

Scheller, R. H., and R. Axel. 1984. How genes control an innate behavior. *Sci. Am.* 250 March:54–62.

Schmidt, K. 1994. Genetic engineering yields first pest-resistant seeds. *Science* 265:739.

*Science*. Special Issues.

June 5, 1987. Frontiers in recombinant DNA; June 16, 1989. The new harvest: Genetically engineered species; October 12, 1990. The human map; June 21, 1991. Frontiers in biotechnology; October 11, 1991. Genome issue: Maps and databases; November 22, 1991. Cancer issue; May 8, 1992. Molecules. Advances in genetic diseases; October 2, 1992. Genome issue; May 14, 1993. Frontiers in biotechnology: Biologically based therapies; June 17, 1994. Genes and behavior.

Shelton, S., J. H. Bowden III, and T. E. Mifflin. 1994. A guide to DNA fingerprinting. Part 2. *Clinical Chem. News* March:3, 14.

Staskawicz, B. J., F. M. Ansubel, B. J. Baker, J. G. Ellis, and J. D. G. Jones. 1995. Molecular genetics of plant disease resistance. *Science* 268:661–67.

Verma, I. D. 1990. Gene therapy. *Sci. Am.* 263 November:68–84.

Watson, J. D., M. Gilman, J. Witkowski, and M. Zoller. 1992. *Recombinant DNA*. 2d ed. New York: W. H. Freeman.

Wheeler, D. A., C. P. Kyriacou, M. L. Greenacre, Y. Qing, J. E. Rutila, M. Rosbash, and J. C. Hall. 1991. Molecular transfer of a species-specific behavior from *Drosophila simulans* to *Drosophila melanogaster*. *Science*. 251:1082–85.

# GLOSSARY

## A

**acceptor site** The position on the ribosome to which incoming charged tRNAs attach during the elongation step of translation; the A site.

**acceptor stem** The 3′ end of tRNA to which an amino acid is attached.

**adenine** A nitrogenous base of DNA and RNA; a purine; base pairs with thymine in DNA and with uracil in RNA.

**adenosine** The nucleoside form of adenine.

**adenosine diphosphate** A nucleotide form of adenine containing two phosphate groups.

**adenosine monophosphate** A nucleotide form of adenine.

**adenosine triphosphate** A nucleotide form of adenine containing three phosphate groups; often used as a source of energy.

**A-DNA** The conformation of double-stranded DNA gotten at relative humidity of about 75%.

***Agrobacterium tumefaciens*** A plant-invading bacterium containing the Ti plasmid that can be used as a vector to transform plants.

**alkaptonuria** A genetic disease characterized by a buildup of homogentistic acid due to the absence of the liver enzyme homogentistic oxidase.

**alkylation** The substitution of an alkyl group, usually methyl or ethyl, for an active hydrogen atom in an organic compound.

**allele** One form of a given gene having a different DNA sequence than another form.

**allolactose** The real inducer of the *lac* operon; a form of lactose.

**allosteric enzyme** An enzyme molecule capable of changing shape and function when stimulated by another molecule.

**Alu sequence** A 300-bp sequence that contains the recognition site (AGCT) for the restriction endonuclease *Alu*I; occurring about 300,000 times in the human genome.

**amber** The termination codon UAG.

**amber mutation** Conversion of an amino acid–specifying codon to the termination codon UAG.

**Ames test** A bioassay for possible mutagens and carcinogens developed by Bruce Ames in 1974 based on reverse mutation to histidine independence.

**amino acids** The nitrogen-containing subunits polymerized to form polypeptides. There are 20 commonly occurring amino acids.

**amino acyl site** *See* acceptor site.

**amino acyl tRNA synthetases** The enzymes used to activate and attach amino acids to tRNAs. There is a specific enzyme for each of the 20 commonly occurring amino acids.

**amino tautomer** The normal form of the bases adenine and cytosine found in nucleic acids.

**amino terminus** The beginning end of polypeptides containing a free amino group.

**amplification** The selective replication of a gene in greater number than usual.

**anabolism** The synthesis of complex molecules from simpler ones, usually requiring energy and specific enzymes; biosynthesis.

**annealing DNA** The process by which separated strands of DNA are united, often requiring heat.

**antennapedia complex (ANT-C)** A locus in *Drosophila* containing the genes controlling the differentiation of cells from the head to the anterior portion of the second thoracic segment.

**antibody** *See* immunoglobulins.

**anticoding strand** The template strand of DNA from which mRNA is transcribed.

**anticodon** The three unpaired bases of tRNA that read specific codons of mRNA.

**anticodon loop** The loop of tRNA, usually shown in diagrams at the bottom of the molecule, that contains the anticodon bases.

**antigens** Substances capable of stimulating an immune response: e.g., antibody synthesis.

**anti-oncogenes** A group of genes that negatively regulate normal cell division; mutation of these genes may lead to malignancy.

**antiparallel** The orientation of the strands of nucleic acids, particularly DNA, in which one strand runs in the 5′ to 3′ direction, and the other in the 3′ to 5′ direction.

**aporepressor** A repressor protein in its inactive form; an allosteric protein in the absence of its corepressor.

**apurinic site** A nucleotide location in DNA at which a purine base is removed.

**apyrimidinic site** A nucleotide location in DNA at which a pyrimidine base is removed.

**A site** *See* acceptor site.

**ATP** Adenosine triphosphate.

**ATPase** An enzyme that hydrolyzes ATP, releasing energy for biochemical reactions.

**attenuation** A mechanism for regulating the transcription of operons encoding amino acid–synthesizing enzymes; premature termination of transcription.

**attenuator** A regulatory sequence of DNA located upstream of structural genes where attenuation can occur.

**autoradiography** A photographic process by which the position of radioactive substances in biological materials can be determined.

**autosome** Any chromosome other than the sex chromosomes X and Y.

**auxotroph** Organisms mutated so that they grow on minimal medium only that has been supplemented with growth factors not required by the parental or wild-type organism.

## B

**backbone** The repeating alternating sugar-phosphate groups along the strands of DNA.

**back mutation** A reverse mutation in which an inactive or mutant form of a gene is restored to its normal function.

**bacteriophage** Viruses that infect bacterial cells.

**base** A cyclic, nitrogen-containing compound found in DNA and RNA; the purines and pyrimidines.

**base pair** Hydrogen-bonded bases, one purine and one pyrimidine, joining the strands of double-stranded DNA (or double-stranded RNA); abbreviated bp.

**B-DNA** The conformation of double-stranded DNA gotten at relative humidity of about 92%; the "ideal" Watson-Crick molecule.

**β-galactosidase** The enzyme encoded in the *lac* operon that converts lactose to allolactose and splits lactose into its component sugars—glucose and galactose.

**bidirectional DNA synthesis** The replication of DNA starting at a common site and proceeding in opposite directions at the same time.
**biodegradation** *See* catabolism.
**biosynthesis** *See* anabolism.
**biothorax complex (BX-C)** The locus in *Drosophila* containing the genes controlling the differentiation of cells of the thorax and abdomen.
**blunt ends** The flush ends of DNA produced by some restriction endonucleases.
**bp** *See* base pair.
**bromodeoxyuridine** A thymine analog that can be incorporated into DNA and that base pairs with guanine to eventually produce a mutation via a AT → GC transition.
**Burkitt's lymphoma** A reciprocal translocation in B lymphocytes involving the long arm of chromosome 8 and chromosome 14 usually but sometimes chromosomes 22 or 2. The break point on chromosome 8 is always near the *myc* oncogene (encoding a nuclear protein), while the break point on the other chromosome is always near immunoglobulin heavy chain gene (14) or light chain genes (22 or 2). The translocated *myc* gene becomes activated and the cancer results.
**burst size** The number of progeny phage produced by a single cell.

# C

**cAMP** Cyclic adenosine monophosphate in which a phosphodiester bond joins the 3′ and 5′ carbons; a regulatory molecule in many prokaryotes and eukaryotes.
**cap** The methylated guanosine bound to the 5′ end of mRNA, hnRNA or snRNA.
**CAP** Catabolite activator protein; an allosteric protein joined by cAMP to form a complex that activates catabolite-repressed operons; a positive regulator protein.
**carboxyl terminus** The end of a polypeptide containing a free carboxyl group; the C-terminus.
**carcinogen** A substance that causes cancer.
**carrier proteins** Proteins used to move other molecules within and between cells or between the bloodstream and the lymphatic system.
**catabolism** The biochemical processes by which substances are degraded into simpler units; e.g., glucose and lactose metabolism; energy release may be a by-product; biodegradation.
**catabolite activator protein** *See* CAP.
**catabolite repression** The repression of an operon involved in sugar metabolism (lactose, arabinose, etc.) by glucose or a breakdown product of glucose.
**catalytic site** Part of an allosteric enzyme molecule where substrate molecules are attached and converted to product.
**CCAAT box** A sequence found in eukaryotic promoters recognized by the enzyme RNA polymerase II needed for mRNA synthesis.
**cDNA** Complementary DNA or copy DNA; made using the enzyme reverse transcriptase and an RNA template.
**cellular immune response** The protection against foreign materials that involves specially altered lymphocytes.
**centimorgan** *See* map unit.
**centromere** The constricted region of a chromosome to which spindle fibers attach in preparation for mitosis.
**channel proteins** A group of proteins involved in the selective permeability of cell and organelle membranes.
**Chargaff's rule** The molar ratios in DNA of total purines to total pyrimidines and of adenine to thymine and of guanine to cytosine are not far from 1. Complementarity.
**charging** The attachment of amino acids to tRNAs.
**Charon phage** A group of cloning vectors derived from the bacteriophage lambda.
**chloroplast** The organelle of eukaryotic green plants in which photosynthesis takes place.
**chromatids** The individual chromosomes of a duplicated chromosome joined by a single centromere.
**chromatin** The material of eukaryotic chromosomes composed of DNA, histone and nonhistone protein, and RNA.
**chromosome** The physical structure containing the genes and composed of DNA in prokaryotes and chromatin in eukaryotes.
***cis*-acting** A term used to describe a genetic region that serves as an attachment site for DNA-binding proteins (e.g., enhancers, operators, and promoters) thereby affecting the activity of genes on the same chromosome.
**clones** Genetically identical organisms or cells derived from a single common ancestor or genetically engineered replicas of DNA sequences.
**closed promoter complex** A promoter complex containing RNA polymerase but whose base pairs are still intact.
**coding strand** The strand of DNA (or RNA) containing the nucleotide sequences of genes.
**codominance** The simultaneous phenotypic expression of two alleles in a heterozygote.
**codon** A three-nucleotide sequence in mRNA coding for a specific amino acid or for termination of translation; also called triplet.
**cohesive ends** *See* cos sites.
**colinearity** The relationship between nucleotide sequences in DNA (the gene) and mRNA and the amino acid sequence of a polypeptide.
**colony hybridization** A procedure for the identification of a bacterial clone containing a particular gene. A labeled probe is used to select the clone.
**complementarity** Nucleotides on one strand of DNA always appear opposite their complementary nucleotides on the other strand according to Chargaff's base pairs (A-T, T-A, G-C, C-G). Chargaff's rule.
**complementary strands** Two strands of nucleic acids that can base pair (DNA-DNA, DNA-RNA, RNA-RNA).
**concatemer** DNA molecule composed of multiple genome lengths.
**consensus sequence** An average nucleotide sequence most often found in a region of DNA of interest: e.g., Pribnow and Hogness boxes.
**conservation transposition** The removal of an intact transposon from donor DNA and insertion into target DNA. *Compare* replicative transposition.
**conservative replication** A form of DNA replication in which the parental strands are not found in the progeny molecules that are newly synthesized.
**constitutive** Something that is always present or active, such as an enzyme or a gene.
**contig** A group of cloned DNAs having contiguous or overlapping sequences.
**continuous synthesis** The synthesis of the DNA strand whose template is oriented 3′ → 5′ relative to the direction of movement of the replication fork; leading strand synthesis.
**core element** A DNA sequence in eukaryotic promoters recognized by RNA polymerase I.
**core enzyme** All the combined prokaryotic RNA polymerase protein subunits except the sigma polypeptide.

**corepressor** A substance that combines with an aporepressor to activate the repressor; e.g., tryptophan is the corepressor of the *trp* operon.

**cosmid** A cloning vector for large segments of DNA having cohesive ends of lambda phage and a plasmid origin of replication; can be packaged in lambda phage heads and replicate as a plasmid.

**cos sites** The cohesive ends of linear lambda phage DNA; sticky ends.

**Cot** In DNA renaturation experiments, a number related to the complexity of the strands being renatured—i.e., the more complex the nucleotide sequences (the more unique sequences they contain) the larger the Cot value. For example, nonrepetitive DNAs would have a greater Cot value than poly T–poly A renaturing strands.

**Cro** The product of lambda *cro* gene. It represses the repressor gene, allowing lytic replication to occur.

***cro* gene** The gene encoding the Cro repressor protein.

**crossing-over** Recombination; the physical exchange of homologous segments of chromosomes during meiosis.

**C-terminus** *See* carboxyl terminus.

**CTP** *See* cytosine triphosphate.

**C-value** In picograms ($10^{-12}$ grams), the amount of DNA in the haploid genome of a species.

**C-value paradox** The paradox that often there is no relationship between an organism's evolutionary complexity and its C-value.

**cyclic adenosine monophosphate** *See* cAMP.

**cytidine** The nucleoside containing the pyrimidine base cytosine.

**cytosine** A nitrogenous base of DNA and RNA; a pyrimidine; base pairs with guanine.

**cytosine triphosphate (CTP)** The nucleotide form of cytosine that contains three phosphates.

# D

**deamination (DNA)** The removal of an amino group from the bases adenine or cytosine and the substitution of a C=O group, converting them to hypoxanthine and uracil, respectively. Causes transition mutations.

**degenerative code** The genetic code in which an amino acid may be specified by more than one codon: e.g., leucine.

**deletion** A mutation in which one or more base pairs are removed or larger segments of chromosomes are removed.

**denaturation (DNA)** The separation of the complementary strands.

**denaturation (protein)** The loss of the normal conformation of a protein usually resulting in the loss of its biological activity.

**deoxyribonucleic acid** *See* DNA.

**deoxyribonucleotide** *See* nucleotide.

**deoxyribonucleotide triphosphates** The form of nucleotides required for DNA synthesis.

**deoxyribose** The sugar found in DNA in which the oxygen at the 2′ position is missing and is replaced by a hydrogen atom.

**depurination** Removal of a purine.

**derepress** *See* inducer.

**dideoxyribonucleotide** A nucleotide lacking oxygen at both the 2′ and 3′ positions; used to stop elongation during DNA sequencing procedures.

**differentiation** The phenotypic expression of the specialized characteristics of a particular cell type.

**dihydrouracil loop** The D loop of tRNA, which contains a number of modified uracil bases.

**dimer (protein)** A molecule consisting of two polypeptides, which may be the same (homodimer) or different (heterodimer).

**diploid** The number of chromosomes in the zygote and all other cells except sperm and egg; denoted as $2n$. *Compare* haploid.

**discontinuous synthesis** DNA synthesis in which the template strand is running 5′ → 3′ relative to the movement of the replication fork; lagging strand synthesis where Okazaki fragments are produced.

**dispersive replication** A mechanism of DNA replication by which new molecules are produced and are composed of parental segments and newly synthesized segments.

**D loop (tRNA)** *See* dihydrouracil loop.

**DNA** Deoxyribonucleic acid; a polymer of deoxyribonucleotides joined via phosphodiester bonds.

**DNA-directed DNA polymerases** The enzymes required for DNA synthesis, in particular, DNA polymerase I and III.

**DNA-directed RNA polymerases** The enzymes required for RNA synthesis, which uses DNA as the template; in eukaryotes there are RNA polymerase I, II, III, and in prokaryotes a single RNA polymerase is used.

**DNA fingerprinting** An identification technique that relies on simple tandem-repetitive sequences of different lengths but having a common core of 10–15 bases; the varying lengths identified using radioactive probes complementary to the core sequence mark particular individuals.

**DNA footprinting** A technique used to identify the DNA-binding sequences of proteins; the binding site is first protected by the protein and then the exposed DNA is degraded.

**DNA glycosylase** An enzyme used in repair mechanisms that breaks the glycosidic bond between a base and its deoxyribose.

**DNA gyrases** Enzymes that relax positively coiled DNA by introducing negative coils during DNA unwinding as replication occurs.

**DNA ligases** Enzymes that produce the phosphodiester bonds between adjacent 5′ and 3′ termini during DNA synthesis and repair.

**DNA melting** The breaking of the hydrogen bonds between base pairs.

**DNA photolyase** The enzyme that repairs UV-produced thymine dimers in DNA.

**DNA polymerase I** A DNA-directed DNA polymerase used in nick translation during DNA synthesis and for DNA repair.

**DNA polymerase III** Holoenzyme; a DNA-directed DNA polymerase; the enzyme that actually synthesizes DNA during replication.

**DNase** A DNA-degrading enzyme.

**domain** A particular section of a polypeptide that has a distinct function in a completed protein.

**double helix** The form of DNA found in all cellular forms of life; two complementary strands of DNA are twisted around one another.

**downstream** In the 3′ direction with regard to a reference sequence.

***Drosophila melanogaster*** One of the 900 species of fruit fly widely used in genetic studies.

**duplication** A chromosomal or gene aberration in which there are two chromosomes, chromosomal segments, or genes.

# E

***E. coli*** See *Escherichia coli*.

**editing** The process whereby newly inserted nucleotides are checked against their complementary bases to be sure a proper base pair is made during DNA replication.

**EF** *See* elongation factors.

**effector molecule** A molecule other than an allosteric enzyme's substrate that controls the allosteric enzyme.

**electrophoresis** A technique in which charged molecules are migrated in an electrical field; used to separate DNA, RNA, or protein molecules/segments for further study.

**elongation factors** Proteins necessary for the binding of charged tRNA to mRNA/ribosome or the translocation of the growing polypeptide chain during elongation stage of translation; EF-Tu, EF-Ts, EF-G.
**encode** To contain the genetic information (nucleotide sequence) for RNA or polypeptide synthesis.
**endonucleases** Enzymes that degrade DNA from within.
**enhancer** A regulatory sequence of nucleotides in DNA that strongly influences the rate of transcription.
**enol tautomer** The unusual tautomer of uracil, thymine, or guanine, which base pair abnormally thereby causing mutations.
**enucleated egg** An egg from which the nucleus is either removed or destroyed.
**enzymes** Protein molecules that increase the rate of a biochemical reaction but are not consumed in the process.
**eRF** *See* release factors.
**error-prone mechanism** A DNA repair mechanism that overcomes damage but makes many errors; e.g. in *E. coli*, opposite a thymine dimer, any bases are incorporated and replication can continue.
***Escherichia coli*** A species of gram-negative bacterium widely used for biochemical, genetic, and physiological study.
**euchromatin** The extended segments of chromosomes that contain most of the genes and that are, at least potentially, capable of being transcribed.
**eukaryote** Organisms whose cells have a nucleus surrounded by a membrane (a true nucleus); all organisms except viruses and bacteria.
**excision repair** A DNA repair mechanism by which damaged segments of DNA are cut out and replaced with newly synthesized segments.
**exons** The segments of DNA or RNA that contain information for polypeptide or RNA synthesis; exons are expressed. *Compare* introns.
**exonucleases** Enzymes that degrade DNA from the ends.

# F

**5′ end** The end of DNA or RNA having a free (though it may have a hydroxyl or phosphate group or a cap), unbonded 5′ carbon; it is not attached to another nucleotide.
**fMet** The special methionine, N-formyl-methionine, whose nitrogen group is prevented by a formyl group from reacting; the first amino acid incorporated into a polypeptide chain during translation.
**fMet-tRNA** The special tRNA that carries N-formyl-methionine.
**frame-shift mutation** A change in the reading frame of mRNA caused by an insertion or a deletion and resulting in a mutation.

# G

**galactosidase permease** The transport protein of *E. coli* encoded in the Y gene of the *lac* operon; actively transports lactose into the cell.
**gamete** The haploid sex cell: egg or sperm.
**GC box** Guanine-cytosine repeats of various lengths recognized by eukaryotic RNA polymerase II; located upstream of the consensus sequence TATA and the transcription start site.
**gene** A linear sequence of nucleotides coding for a polypeptide or for an RNA molecule; the unit of inheritance.
**gene cloning** The process by which many copies of a gene are made by insertion into a bacterium or virus.
**gene expression** The process by which a gene is decoded and its information is used to produce a polypeptide or RNA molecule.
**genetic code** The entire group of codons specifying amino acids or termination; the code contains a total of 64 codons.
**genetic marker** A specific sequence in a genome that can be used as a reference point in mapping procedures.
**genome** The entire complement of genetic information carried by an organism.
**genomic library** A group of vectors containing random DNA fragments from a given species and cloned in appropriate hosts.
**genotype** The complete genetic constitution of an individual.
**germ cell** Egg or sperm.
**germline mutation** Any change in the DNA sequence of sperm and/or egg.
**glycosidic bond** The bond joining the base and sugar in a nucleoside.
**G-protein** A group of membrane proteins activated by GTP that in turn activate enzymes on the inner surface of the membrane; the inner, activated enzymes produce "second messages"; in this way, external signals are transported and amplified inside the cell.
**group I introns** Self-splicing introns in which the splicing takes place at a guanosine or guanosine nucleotide.
**group II introns** Self-splicing introns in which splicing is begun by the formation of an intron "lariat."
**GTP** *See* guanosine triphosphate.
**guanine** A nitrogenous base of DNA and RNA; a purine; base pairs with cytosine.
**guanosine** The nucleoside form of guanine.
**guanosine triphosphate** A nucleotide form of guanine containing three phosphate groups; GTP.
**gyrase** *See* DNA gyrases.

# H

**hairpin** A structure of DNA or RNA produced by internal base pairing of inverted repeats; also called a stem-loop.
**half-life** The time required to reduce by half the population of molecules.
**haploid** The number of chromosomes in a sperm or an egg; denoted as $n$. *Compare* diploid.
**helicase** The enzyme that unwinds the double-helix in preparation for replication.
**helix-turn-helix** The structure of certain DNA-binding proteins which allows them to bind the major groove of DNA; many regulatory proteins have this conformation.
**hemoglobin** The oxygen-carrying protein of red blood cells.
**heterochromatin** Chromatin that is genetically inactive and maximally condensed during interphase; can be constitutive or facultative.
**heterogeneous nuclear RNA** hnRNA; the original transcript of genes prior to processing into mRNA; contains both exons and introns.
**histones** DNA-binding proteins that form a part of nucleosomes; rich in basic amino acids especially lysine and arginine.
**HIV** *See* Human immunodeficiency virus.
**hnRNA** *See* heterogeneous nuclear RNA.
**Hogness box** A eukaryotic DNA sequence (TATAAA) some 19–27 bp upstream of the transcription start site; binding site of RNA polymerase II.
**holoenzyme** *See* DNA polymerase III.
**homeobox** A DNA sequence of about 180 bp coding for a DNA-binding protein of some 60 amino acids; located near the 3′ end of some homeotic genes; in vertebrates, called Hox genes.
**homeodomain** The 60–amino acid segment of some DNA-binding proteins that allows the proteins to bind tightly and specifically to DNA.

**homeotic gene** A gene in which a mutation causes the translocation of one body part to another location.
**homologous recombination** Recombination of chromosomes or DNAs requiring extensive complementarity or base pairing between the molecules.
**hotspots** Sequences of nucleotides that mutate well above the average rate.
**housekeeping genes** Genes whose products are required by all cells of an organism for the normal maintenance of the cells.
**Hox genes** *See* homeobox genes.
**human immunodeficiency virus** HIV; the virus responsible for AIDS (acquired immunodeficiency syndrome).
**humoral immunity** The immune response that involves antibodies.
**hybrid cell** A cell formed by the fusion of two dissimilar cells.
**hybridization (of polynucleotides)** The joining of DNA and RNA strands to form a hybrid molecule.

# I

**ICR** Internal control region
**IF** *See* initiation factors.
**imaginal disks** Insect cells that give rise to various body parts or organs.
**imino tautomer** The unusual forms of adenine and cytosine whose base pairing with each other leads to mutations.
**immune system** The protection mechanism of vertebrates that produces antibodies and special T cells.
**immunoglobulins** Protein molecules that are produced by plasma cells in response to foreign substances and that are a form of specific protection; antibody molecules.
**imprinting (genetics)** A phenomenon in which an autosome inherited from the mother and one from the father give rise to different phenotypes.
**independent assortment** The Mendelian principle that states that genes on different chromosomes are randomly distributed among gametes.
**inducer** A substance that derepresses or releases negative control of an operon; required to initiate synthesis of a protein.
**induction** The initiation, by a prophage, of a lytic cycle; the synthesis by an organism of new enzymes.
**inheritance** How genes are transferred from one generation to the next.
**initiation codon** The codon (AUG) at which translation begins.
**initiation factors** Proteins needed to initiate translation during protein synthesis; IF-1, IF-2, IF-3.
**inosine** Hypoxanthine riboside; a rare base that base pairs with cytosine; a common intermediary in adenine and guanine biosynthesis.
**insertion** A mutation in which one or more base pairs are added or larger segments of chromosomes are added.
**insertion sequence (IS)** Transposable elements found in bacteria containing only inverted terminal repeats at the ends of the DNA strands and genes needed for transposition.
**internal control region (ICR)** Promoter DNA sequences found within tRNA genes and recognized by RNA polymerase III.
**interphase** During the cell cycle, the phase between $G_1$ and $G_2$ when DNA synthesis occurs and chromosomes are not visible.
**intervening sequences** *See* introns.
**introns** *Int*ervening sequences in DNA or RNA that do not contain information needed for either polypeptide or RNA synthesis. *Compare* exons.
**inversion** The rotation of chromosomal segments so that the inverted sequence is reversed compared with the sequences of the rest of the chromosome.
**inverted repeat** Two sequences of DNA oriented in opposite directions in the same double-stranded DNA molecule; read in the same direction—e.g. 5′ to 3′—the sequences are identical.

# K

**karyotype** A photographic representation of all the chromosomes of a cell, individual, or species; it shows the number, size, and morphology of the chromosome set.
**kbp** Kilobase pair; 1000 base pairs.
**keto tautomer** The normal or usual form of uracil, thymine, or guanine found in DNA or RNA.
**kilobase pair** *See* kbp.
**Klenow fragment** A segment of DNA polymerase I from which the 5′ → 3′ exonuclease activity has been removed by a protease but retaining the polymerase's polymerization and proofreading activities.

# L

***lac*A** The *E. coli* transacetylase gene found in the *lac* operon.
***lac*I** The gene that encodes the *E. coli* allosteric repressor protein of the *lac* operon.
***lac* operon** The *E. coli* set of genes allowing the bacterium to metabolize the milk sugar lactose.
***lac* repressor** The allosteric protein molecule that exerts negative control over the *lac* operon.
**lactose** The milk sugar; a disaccharide composed of glucose and galactose.
***lac*Y** The gene that encodes the transport protein permease of *E. coli;* a gene of the *lac* operon.
***lac*Z** The *E. coli* β-galactosidase gene in the *lac* operon.
**lagging strand** The strand of DNA synthesized using the template strand oriented in the 5′ → 3′ direction relative to the movement of the replication fork; the strand composed of Okazaki fragments and produced discontinuously in semidiscontinuous replication of DNA.
**lambda phage** A temperate bacteriophage of *E. coli* containing double-stranded DNA; used as a vector in gene manipulation.
**lariat** The structure formed during the splicing out of group II rRNA introns.
**leader** The 5′-end, untranslated mRNA sequence or the 5′ end sequence of the attenuation mRNA of the *trp* operon.
**leading strand** The continuously synthesized strand of DNA replicating semidiscontinuously; it is the strand whose template is oriented 3′ → 5′ relative to the movement of the replication fork.
**leucine zipper** A domain or segment of some DNA-binding proteins; regular spacing of leucines allows formation of dimers of these molecules, which then bind DNA.
**ligases** *See* DNA ligases.
**linkage** The physical association of genes on a chromosome.
**locus** The physical position of a gene or genes on a chromosome.
**lysogenic** A state in which a bacterium harbors a prophage (is capable of being lysed).

# M

**major groove** The larger depression in the DNA double helix that results from the twisting of the two strands.
**map unit** A unit expressing the relative distance between genes; a centimorgan (one hundredth of a Morgan) is equal to a cross-over frequency of 1%.
**marker** A gene or group of genes that serves as a reference to a known location on a chromosome.
**megabase pair** One million base pairs.
**meiosis** The processes by which the chromosome number is halved after cell division during gamete formation.

**Mendelian genetics** The field of genetics having to do with the transmission of genes and traits from one generation to the next.
**message** A sequence of ribonucleotides carrying information for the synthesis of a polypeptide; see mRNA.
**messenger RNA** mRNA; the sequence of ribonucleotides carrying the information for the sequence of amino acids of a polypeptide; mRNA is derived from DNA or another RNA molecule in the case of some viruses.
**metal finger** The fingerlike domain of DNA-binding proteins containing zinc bound to cysteines and histidines; they insert into DNA's major groove.
**methylation (of DNA)** The addition of a methyl group ($-CH_3$) to DNA; may serve to inactivate genes and to identify parental DNA from newly synthesized DNA in bacteria.
**minor groove** The smaller depression in the DNA double helix that results from the space separating the two strands along the backbone.
**minus ten (−10) region** *See* Pribnow box.
**minus thirty-five (−35) region** A nucleotide sequence (TTGACA) found 35 bp upstream of the prokaryotic transcription start site.
**mismatch repair** The correction of improperly paired bases.
**missense mutation** A nucleotide change in a codon, resulting in the incorporation of a different amino acid into a polypeptide.
**mitosis** The processes during cell division by which the duplicated chromosomes are sorted into the daughter cells so that each new cell has the same chromosomal makeup as the parent cell.
**molecular biology** The field of biology dealing with biological processes at the molecular level.
**molecular genetics** The field of genetics having to do with the structure and activities of genes at the molecular level.
**monocistronic transcription units** Genes that are activated as single entities.
**Morgan** *See also* map unit; a unit expressing the relative distance between genes; equals a cross-over frequency of 100%.
**morphological mutation** A mutation affecting the appearance of a cell or organism.
**mRNA** *See* messenger RNA.
**mutagen** A substance that causes mutations.
**mutation** Any change in the nucleotide sequence of the genome, whether DNA or RNA.
**mutation frequency** The proportion of mutants in a population.
**mutation rate** The number of mutation events per unit time or per cell generation.
***myc*** A cellular oncogene located on the long arm of human chromosome 8; it encodes a nuclear protein and is rearranged in Burkitt's lymphoma.

# N

**negative control** The form of control that represses an operon or gene and depends on the action of a repressor protein.
**negative strand (−)** The strand of DNA or RNA synthesized as the complement of a viral parental or positive strand.
***Neurospora crassa*** A common bread mold used by Beadle and Tatum in their genetic research.
**N-formyl-methionine** *See* fMet.
**nick** A single-stranded break in DNA.
**nick translation** Removal of RNA primer during replication.
**nitrocellulose** A chemically charged paper used to bind nucleic acids and proteins in hybridization procedures.
**nonsense codons** The termination codons UAG, UAA, UGA; they do not specify an amino acid.
**nonsense mutation** A mutation in which an amino acid–specifying codon is changed to one specifying termination; causes premature termination of protein synthesis.
**Northern blotting** A technique in which RNA molecules are transferred to a support medium.
**nucleic acid** A chain of deoxyribonucleotides (DNA) or ribonucleotides (RNA).
**nucleocapsid** A viral nucleic acid surrounded by its protein coat or capsid protein.
**nucleolar organizer** The site around which a nucleolus forms.
**nucleolus** A nuclear structure of eukaryotes associated with a segment of chromosome called the *nucleolus organizer region*, or nucleolar organizer, containing the rRNA genes, proteins, and enzymes for the synthesis of rRNA.
**nucleoside** A nitrogenous base bound to either deoxyribose or ribose.
**nucleosome** The structure of eukaryotic chromosomes consisting of about 150–200 bp of DNA wrapped around an octamer of histones.
**nucleotide** A base linked to a sugar (deoxyribose or ribose), which is in turn linked to at least one phosphate group; the fundamental unit of nucleic acids.
**nutritional mutant** A mutation that changes a prototroph or wild-type organism into an auxotroph; the mutant requires growth factors.

# O

**ochre** The termination codon UAA.
**ochre mutation** A change in a nucleotide sequence that produces the ochre termination codon.
**Okazaki fragment** The fragments of DNA produced during discontinuous synthesis and named after the discoverer, R. T. Okazaki.
**oligonucleotide** A short chain of about 20 nucleotides joined by phosphodiester bonds.
**oncogene** A gene that induces tumor growth or uncontrolled cell proliferation; cellular oncogenes are denoted by "c" and viral ones by "v": e.g., c-*src* and v-*sis*.
**oncogenesis** Tumor formation.
**oncogenic** Tumor-causing.
**one gene–one polypeptide hypothesis** The hypothesis that states that one gene codes for one polypeptide (not necessarily a complete functional protein); the hypothesis is now considered to be valid.
**opal** The termination codon UGA.
**opal mutation** A change in a nucleotide sequence that produces the opal termination codon.
**open promoter complex** The transcription complex consisting of a tightly bound RNA polymerase to its promoter, allowing for local melting of the double helix.
**open reading frame** A reading frame not including termination codons.
**operator** A regulatory sequence of nucleotides upstream of the transcription start site to which repressor proteins bind, exerting negative control over the operon.
**operon** A set of genes coordinately controlled by a single operator sequence.
**ORF** *See* open reading frame.
**origin of replication** A unique sequence of nucleotides acting as the start site for DNA synthesis.

**overlapping genes** Nucleotide sequences for some genes that are completely contained within the sequences of other genes.

# P

**palindrome** A sequence of nucleotides of DNA reading the same on both DNA strands in a particular direction, e.g., 5′ to 3′; an inverted repeat.

**p arm** Petite arm; the small arm of a chromosome.

**pBR322** One of the original plasmid cloning vectors; contains ampicillin- and tetracycline-resistant genes and a number of restriction endonuclease recognition sites.

**PCR** *See* polymerase chain reaction.

**peptide bond** The carbon-to-nitrogen bond linking amino acids in a polypeptide/protein molecule.

**peptidyl site** The site on the ribosome to which the growing polypeptide chain is transiently attached; the site is close to the 5′ end of mRNA; the P site.

**peptidyl transferase** The enzyme located on the large ribosomal subunit that catalyzes peptide bond formation.

**p53** A protein molecule having a variety of cellular functions such as transcription, cell cycle control, and as a tumor suppressor. Mutations often result in cancer.

**phage** *See* bacteriophage.

**phagemid** A plasmid cloning vector having a single-stranded origin of replication allowing for single-stranded DNA production; requires a helper virus.

**phenotype** The expression of the genotype; may include biochemical, morphological, and behavioral expressions.

**phenylketonuria** A genetic disease caused by a deficiency of the enzyme phenylalanine hydroxylase, which converts phenylalanine into tyrosine.

**φX174** A single-stranded *E. coli* phage that replicates via a conservative model.

**phosphodiester bond** The sugar-phosphate bond linking nucleotides in nucleic acids.

**photoreactivating enzyme** *See* DNA photolyase.

**photoreactivation** Light-stimulated repair of thymine dimers by the photoreactive enzyme.

**physical map** A chromosomal/DNA map based on such things as location of restriction sites or characteristics other than gene location.

**PKU** *See* phenylketonuria.

**plaque** The hole made by bacterial or other viruses in a layer of cells; caused by destruction or slow growth of the host cell.

**plasmid** Extrachromosomal double-stranded, circular DNA found in eukaryotes and prokaryotes (such as yeast) that replicate independently of the host cell DNA.

**point mutation** A change in one base in DNA or RNA.

**polyadenylation** The process of adding adenines to nucleic acids producing poly A tails.

**poly A polymerase** The enzyme that produces poly A tails.

**poly A tails** Polyadenylic acid; refers to the 100 or 200 adenines typically added to the 3′ end of eukaryotic mRNAs.

**polycistronic mRNA** An mRNA consisting of more than one message; contains information from more than one gene.

**polygenic mRNA** *See* polycistronic mRNA.

**polymer** A chain of molecules.

**polymerase chain reaction** The procedure for amplifying DNA; PCR.

**polymorphism** The condition in which a gene appears in more than one form in a population.

**polynucleotide** A polymer of nucleotides such as DNA or RNA.

**polypeptide** A molecule composed of a linear arrangement of amino acids joined by peptide bonds.

**polyribosome** An mRNA molecule to which a number of ribosomes are attached.

**polysome** A polyribosome.

**population genetics** That branch of genetics dealing with the variation in genes within and between populations.

**positive control** Control of the expression of an operon by an effector molecule such as cAMP.

**positive strand (+)** The genome of some simple viruses that has the same information as the viral mRNA.

**postreplication repair** The repair of UV-damaged DNA, e.g., thymine dimers, occurring after replication of the DNA; recombination repair.

**posttranscriptional control** Control of gene expression after mRNA synthesis by processing and modification.

**posttranscriptional modification** The changes to mRNA occurring after its synthesis: e.g., poly A tails.

**posttranslational modification** Changes to polypeptides/proteins occurring after synthesis: e.g., addition of sugars to produce glycoproteins.

**Pribnow box** The attachment site (TATAAT sequence) of the sigma polypeptide in prokaryotes located 10 bp upstream of the transcription start site; also called the –10 (minus ten) region.

**primary structure** The sequence of amino acids of a polypeptide/protein or the sequence of nucleotides of DNA or RNA.

**primary transcript** The initial product of a gene; unprocessed RNA.

**primase** The primosome enzyme that synthesizes the primer RNA during DNA replication in *E. coli*.

**primer** The RNA synthesis before actual DNA synthesis in *E. coli*; provides the attachment site for DNA polymerase III.

**primosome** The structure of *E. coli* composed of about 20 proteins within which primer RNA is made.

**probe (DNA)** A short piece of single-stranded DNA labeled with radioactivity and used to detect particular segments of DNA via hybridization of the probe and the unknown DNA.

**prokaryotes** The bacteria; organisms lacking a nuclear membrane.

**promoter** The DNA sequence located upstream of the transcription start site to which RNA polymerase, sigma, and positive control elements attach before transcription starts.

**proofreading** The procedure by which base pairs are checked during DNA synthesis to assure proper base pairing.

**prophage** A phage that has incorporated itself into a host cell's DNA and replicates as the host DNA replicates.

**protein** A polymer of amino acids joined by peptide bonds; polypeptide.

**protein sequencing** The determination of the order of amino acids in a protein molecule.

**proto-oncogene** The cellular analog of a viral oncogene; does not contribute to cancer unless overexpressed or mutated.

**prototroph** An organism capable of growth on defined minimal media; the wild-type strain.

**provirus** An animal virus that has incorporated itself into the chromosome of the host cell and is replicated as the chromosome replicates, e.g., HIV.

**pseudogene** A nonallelic copy of a normal gene mutated so that it is nonfunctional.

**P site** *See* peptidyl site.

**purine** A nitrogenous base; (usually) either adenine or guanine in DNA and RNA.

**pyrimidine** A nitrogenous base; (usually) either cytosine or thymine in DNA or cytosine or uracil in RNA.

# Q

**q arm** The long arm of a chromosome.
**quaternary structure** The structure of a complex protein composed of two or more polypeptides, e.g., hemoglobin.

# R

**ras** A family of human oncogenes encoding the protein p21.
**reading frame** The nucleotide sequence of mRNA beginning with the initiation codon and ending with a termination codon, with the codons in between specifying amino acids.
***recA*** The gene of *E. coli* that specifies the RecA protein needed for homologous recombination and the SOS repair system.
**receptor proteins** A class of proteins found on the cell's surface as transmembrane molecules whose function is to transmit messages to the cell's interior.
**recombinant DNA** The DNA composed of two or more DNAs from different sources; may occur naturally or be produced in vitro.
**recombination** The reassortment of genes.
**recombination repair** *See* postreplication repair.
**regulatory genes** Genes that encode polypeptides that form proteins whose function is to control the expression of structural genes. Also called regulatory sequences, regulation units, regulator regions, control units, control sequences, and control regions. *Compare* structural genes.
**regulatory proteins** The molecules that regulate gene activity or other cell activities.
**regulatory sequences** *See* regulatory genes.
**release factors** Polypeptides that read termination codons, causing the release of polypeptides from the ribosomes; RF-1, RF-2, RF-3 in prokaryotes and eRF in eukaryotes.
**repetitive DNA** Nucleotide sequences appearing more numerous times in haploid chromosomal DNA.
**replicases** Enzymes of the class RNA-directed RNA polymerase.
**replication** The means by which nucleic acids pass information on to the next generation.
**replication fork** The point where the two strands of DNA are separated and replication occurs.
**replicative form** The double-stranded nucleic acid molecules formed during one stage of some single-stranded viral DNA or RNA replication: e.g., $\phi$X174 replication.
**replicative transposition** The movement of a transposon from one DNA site to another during which replication of the transposon occurs so that one copy remains at the original site while another copy is moved. *Compare* conservative transposition.
**replicon** The place on a chromosome where RNA polymerase binds in preparation for primer synthesis prior to DNA synthesis; the site where DNA synthesis originates; in prokaryotes there is one site, and in eukaryotes there are many sites.
**replisome** The *E. coli* structure within which is found the complex of proteins, including the primosome, needed for DNA replication.
**Rep protein** The protein that during replication helps uncoil the parent DNA.
**repression** The inactivation of an operon.
**restriction endonuclease** Enzymes capable of cutting DNA internally but at specific nucleotide sequences or close to these sequences.
**restriction fragment** A piece of DNA produced by the action of restriction endonucleases.
**restriction-fragment length polymorphism** RFLP; the variation among individuals of the distribution of given restriction sites so that restriction fragments produced vary in length.
**restriction map** A map of a DNA segment showing the locations of particular restriction sites.
**restriction site** The particular nucleotide sequence recognized by a specific restriction endonuclease.
**retinoblastoma** A cancer of immature retina cells.
**retrovirus** An RNA virus that reproduces by first making a DNA complementary copy of its RNA genome.
**reverse transcriptase** RNA-directed DNA polymerase; the enzyme that allows retroviruses to make the DNA complement of the viral RNA genome.
**reverse transcription** The process by which complementary DNA is made using RNA as the template; reverse transcriptase is used in the process.
**reversion** A mutation that negates the effects of an earlier mutation in the same gene.
**RF** Replicative form, as seen in the replication of some single-stranded DNA and RNA viruses. *See also* release factors.
**RFLP** *See* restriction-fragment length polymorphism.
**rho** A polypeptide needed for some forms of transcription termination in *E. coli* and some of its phages.
**ribonucleic acid** *See* RNA.
**ribonucleotide triphosphates** The forms of the ribonucleotides needed for RNA synthesis.
**ribose** The five-carbon sugar found in RNA.
**ribosomal RNA** The species of RNA found in the large and small subunits of ribosomes; rRNA.
**ribosome** The RNA-protein particle on which mRNA is translated by tRNA and peptide bonds are formed between amino acids, producing polypeptides/proteins.
**ribozymes** A class of RNA molecules capable of enzyme functions.
**RNA** Ribonucleic acid; a polymer of ribonucleotides joined via phosphodiester bonds.
**RNA-directed DNA polymerase** The reverse transcriptase enzyme.
**RNA polymerase** The enzyme that transcribes DNA, producing mRNA, tRNA, or rRNA, or that replicates RNAs.
**RNA splicing** RNA processes during which introns are removed and the exons joined to one another.
**rolling-circle replication** A form of viral replication in which one strand of the DNA remains intact while the other strand is continually produced and cut at the appropriate site to produce a complete genome.
**rRNA** *See* ribosomal RNA.

# S

***Saccharomyces cerevisiae*** The eukaryote baker's yeast.
**sarcoma** A cancer of connective tissue.
**secondary structure** The local folding of regions of polypeptides/proteins or RNA.
**sedimentation coefficient** A measure of the rate at which a molecule or particle moves to the bottom of a centrifuge tube under the influence of a centrifugal force.
**segregation** The Mendelian principle that each pair of alleles of diploid parents separates into different gametes and then into different offspring.
**semiconservative replication** The form of replication of double-stranded DNA in which each parental strand becomes a part of the newly formed progeny

double helix; each newly synthesized duplex is composed of a parental strand and a newly synthesized strand; the Watson and Crick model of replication.

**semidiscontinuous replication** The mechanism by which double-stranded DNA is replicated; one new strand is synthesized continuously while the other is synthesized in fragments (Okazaki fragments), which are then joined.

**sequence tagged site** STS; a small segment of DNA identified using PCR with defined primers used to amplify the sequence.

**sequencing** The processes by which amino acid or nucleotide sequences are determined.

**sex chromosomes** The chromosomes involved in sex determination; in humans the X and Y chromosomes.

**sex-linked** Genes found on the sex chromosomes.

**Shine-Dalgarno (SD) sequence** The sequence at the 5′ end of *E. coli* mRNA that binds a complementary sequence at the 3′ end of the 16S rRNA of ribosomes.

**shuttle vector** A cloning vector capable of replicating in more than one host, allowing recombinant DNA to move from one host to another.

**sickle-cell anemia** A form of anemia in which the red blood cells are distorted because of a mutation in the beta globin gene; distortion is produced under low oxygen tension, which brings about aggregation of the mutated hemoglobin molecules.

**sigma factor** The polypeptide needed by *E. coli* RNA polymerase to recognize specific promoter sequences.

**signal peptide** A polypeptide segment of about 20 amino acids at the nitrogen terminus that helps to anchor the polypeptide and its polysome to the endoplasmic reticulum; these polypeptides are destined for export from the cell after passing through the Golgi apparatus, where they are modified by the addition of sugars.

**silent mutation** An undetectable mutation.

**single-copy DNA** Nucleotide sequences found only once or a few times in haploid DNA.

**single-strand binding proteins** SSB; proteins that bind DNA during replication, thereby preventing the reassociation of the separated DNA strands.

**site-directed mutagenesis** A procedure for the specific sequence alteration of cloned genes.

**small nuclear ribonucleoproteins** snRNPs; the structures within which primary RNA transcripts are modified; consist of polypeptides tightly bound to snRNAs.

**small nuclear RNAs** snRNAs; a class of RNAs that participate in the modification of primary RNA transcripts; they are a part of the snRNPs.

**snRNAs** *See* small nuclear RNAs.

**snRNPs** *See* small nuclear ribonucleoproteins.

**solenoid** The structure formed by the coiling of nucleosomes.

**somatic cells** The nonsex cells.

**somatic mutation** Mutation of somatic cells; these mutations are not inherited by progeny.

**SOS response** A response of *E. coli* to DNA damage caused by UV light or other mutagens.

**Southern blotting** A procedure (named after its inventor) for the transfer of DNA to a support medium for use in hybridizing the DNA to a labeled probe.

**spacer DNA** The DNA found between genes or less often within genes; it is nontranscribed DNA.

**spliceosome** The structure within which precursor mRNA is processed to remove introns.

**splicing** The process by which introns of mRNA are removed and the exons are joined.

***src*** The Rous sarcoma virus oncogene encoding a tyrosine protein kinase.

**SSB** *See* single-strand binding proteins.

**stem-loop** *See* hairpin.

**sticky ends** Complementary single-stranded ends produced by many restriction endonucleases; cos sites.

**stop codons** Triplets that do not code for amino acids; transcription termination sequences: UAA, UAG, UGA.

**streptomycin** An antibiotic produced by a bacterium that causes misreading of mRNA.

**structural genes** Genes that encode polypeptides or RNA needed for the normal metabolic activities of the cell. *Compare* regulatory genes.

**structural proteins** Molecules whose role is to maintain the structural integrity of other molecules or organelles, such as ribosomes.

**STS** *See* sequence tagged site.

**superhelix** The structure produced by double-stranded DNA when it coils back on itself.

**suppression** Compensation for one mutation by another mutation.

**suppressor mutation** A mutation that negates or lessens the effect of a previous mutation.

**SV40** Simian virus 40; a small double-stranded DNA virus able to cause tumors in rodents.

# T

**TATA box** *See* Hogness box.

**tautomeric shift** The reversible shift from one isomer to another of DNA bases.

**tautomerization** A tautomeric shift.

**tautomers** Isomers of a molecule differing in the position of a proton and double bond: e.g., keto and enol isomers or amino and imino isomers.

**T-DNA** The tumor-inducing segment of the Ti plasmid.

**telomerase** The enzyme that extends the ends of chromosomes, telomeres, after DNA synthesis.

**telomere** The sequence found at the end of eukaryotic chromosomes that folds back and base pairs with itself.

**temperate phage** A bacterial virus capable of a lysogenic stage in which a prophage is produced.

**temperate virus** *See* temperate phage.

**template** The sequence of nucleotides (DNA or RNA) used as the guide in synthesizing a complementary strand.

**template DNA strand** The anticoding strand of DNA; the complementary DNA strand guiding mRNA synthesis.

**terminal transferase** An enzyme that can add nucleotides one at a time to the 3′ ends of DNA; used in recombinant DNA work.

**termination sequences** *See* stop codons.

**tertiary structure** The three-dimensional structure of a protein or RNA molecule.

**tetramer (protein)** A protein composed of four polypeptide subunits.

**TFIIIA and C** Transcription factors that facilitate the binding of TFIIIB to 5S RNA genes to be transcribed by RNA polymerase III.

**TFIIIB** A transcription initiation factor required by RNA polymerase III.

**thalassemia** A genetic disorder in which the mRNA for one of the hemoglobin chains is nonfunctional.

**3′ end** The end of a DNA or RNA molecule having a free 3′ hydroxyl group (no phosphate group is present).

**thymidine** The nucleoside containing the base thymine.

**thymidine triphosphate** TTP; a nucleotide form of thymine containing three phosphate groups.

**thymine** A nitrogenous base found in DNA but not in RNA; a pyrimidine; base pairs with adenine.

**thymine dimer** Two adjacent thymines covalently linked; the most common form of UV damage.

**Ti (tumor-inducing) plasmid** The plasmid found in the bacterium *Agrobacterium tumefaciens* capable of inducing tumors in plants; often used as a vector to carry genes into plant cells.

**TMV** Tobacco mosaic virus.

***trans*-acting** A genetic element such as a repressor gene on one chromosome whose product is capable of influencing genes on other chromosomes.

**transcript** The RNA product of a gene.

**transcription** The process of decoding a gene resulting in the transfer of information from the DNA molecule via synthesis of an RNA molecule (mRNA).

**transcription factor** A protein needed for transcription activity of eukaryotic genes; they bind to promoters or enhancers.

**transcription terminator** One of the transcription terminating sequences: UAA, UAG, UGA.

**transfer RNA** A species of RNA that carries a specific amino acid and contains the anticodon necessary to read a specific mRNA codon; there is a tRNA for each amino acid and for N-formyl-methionine.

**transformation (genetic/malignant)** The alteration of a cell's genotype by the incorporation of exogenous DNA. The process whereby a normal cell is converted into a cancer cell.

**transgenic organism** An organism into which new genes have been inserted.

**transition** A mutation in which one purine or pyrimidine replaces another purine or pyrimidine: e.g., AT ↔ GC.

**translation** The process by which mRNA is read and a polypeptide formed.

**translocation** The stage in translation in which the growing polypeptide chain is moved from the acceptor site to the peptidyl site and a new codon is placed in the acceptor site.

**translocation (chromosomal)** The movement of a segment of chromosome to another nonhomologous chromosome.

**transposable element** A transposon; a piece of DNA capable of moving from one DNA site to another; movable genes.

**transposases** A class of enzymes encoded in transposon DNAs that bring about transfer of transposons.

**transposition** *See* conservative transposition *and* replicative transposition.

**transposon** *See* transposable element.

**transversion** A mutation in which a purine replaces a pyrimidine or vice versa.

**tRNA** Transfer RNA.

**Trp** The designation for the amino acid tryptophan.

***trp* operon** The set of *E. coli* encoding the enzymes needed for tryptophan biosynthesis; this operon contains an attenuation sequence.

**TTP** *See* thymine triphosphate.

**TΨC loop** The tRNA sequence that contains the modified base pseudouridine.

# U

**UCE** *See* upstream control element.

**ultraviolet radiation** Relatively low-energy radiation of wavelengths between 100 nm and 400 nm, just below violet; causes primarily thymine dimer DNA damage; DNA absorbs maximally at 260 nm and protein at 280 nm.

**unidirectional replication** DNA replication in which there is only one replication fork.

**unique DNA** Single-copy DNA; sequences found once or very infrequently at most in DNA.

**upstream** In the 5′ direction with regard to a reference sequence.

**upstream control element** UCE; sequences in eukaryotic promoters recognized by RNA polymerase I and located about 100–150 bp upstream of the transcription start site.

**uracil** A nitrogenous base found in RNA but not in DNA; a pyrimidine base pairs with adenine.

**uridine** The nucleoside form of uracil.

**uridine triphosphate** The nucleotide form of uracil containing three phosphate groups.

**UTP** *See* uridine triphosphate.

# V

**variable loop** The tRNA sequence lying between the anticodon sequence and the TΨC loop.

**vector** A segment or piece of DNA that is the carrier of genes in cloning procedures.

**virulent virus** A virus whose reproduction results in the lysis of the host cell.

# W

**western blotting** A procedure in which a protein is transferred to a support medium after electrophoresis and identified using labeled antibody.

**wild type** A nonmutated genotype of an organism.

**wobble hypothesis** Crick's hypothesis that explains how one tRNA can read two codons; it is based on the idea that the third base of codons has some play, allowing for unusual base pairing with the anticodon; e.g., an anticodon G may base pair with cytosine or uracil in the third position of a codon.

**wobble position** The third nucleotide or base of a codon.

# X

**xeroderma pigmentosum** A skin disorder resulting from a faulty DNA excision repair, making thymine dimer repair ineffective; these individuals have a higher than average susceptibility to skin cancer.

**x-rays** High-energy ionizing radiation that can produce chromosome breaks; wavelengths are between UV and gamma rays.

# Y

**YAC** *See* yeast artificial chromosome.

**yeast artificial chromosome** An artificial vector of high carrying capacity consisting of yeast telomeres and a centromere; foreign DNA placed between the centromere and one telomere behaves in yeast as a normal yeast chromosome.

# Z

**Z-DNA** A left-handed helical form of double-stranded DNA; the backbone has a zigzag appearance.

**zinc fingers** DNA-binding proteins (transcription factors) containing a zinc ion bound by cysteines and histidines in fingerlike projections of the protein.

**zygote** A fertilized egg containing the diploid number of chromosomes.

# INDEX

Page numbers followed by *f*, *g*, and *t*, indicate figures, glossary, and table, respectively.

## A



## B

# C

# D

# N



# O

# U



# V